P9-ELG-125

MORE SHOW

DEREK FAGERSTROM, LAUREN SMITH & THE SHOW ME TEAM

HOW

EVERYTHING WE COULDN'T FIT IN THE FIRST BOOK

INSTRUCTIONS FOR LIFE FROM THE EVERYDAY TO THE EXOTIC

COLLINS DESIGN
An Imprint of HarperCollins*Publishers*

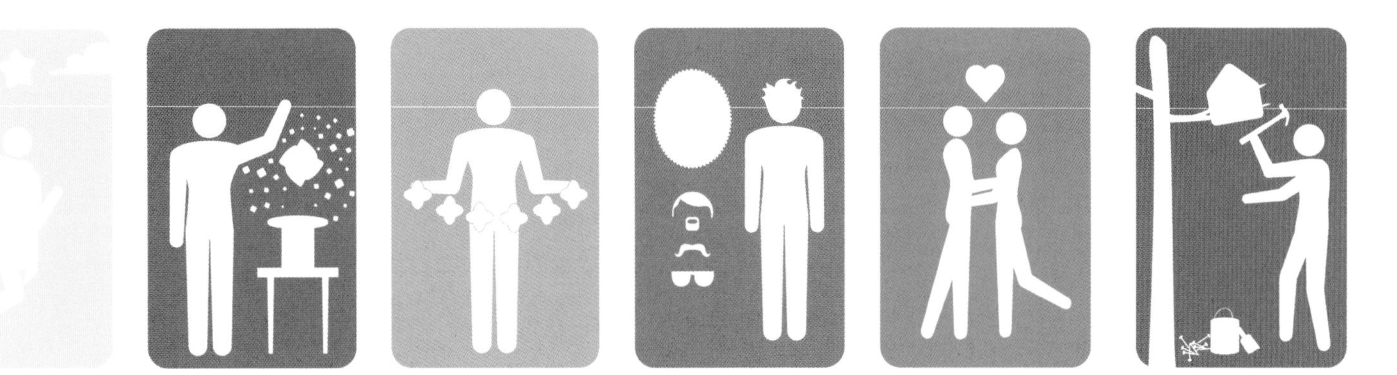

show me how to...

slick moves

wild stunts

cool tricks

magical illusions

sneaky cheats

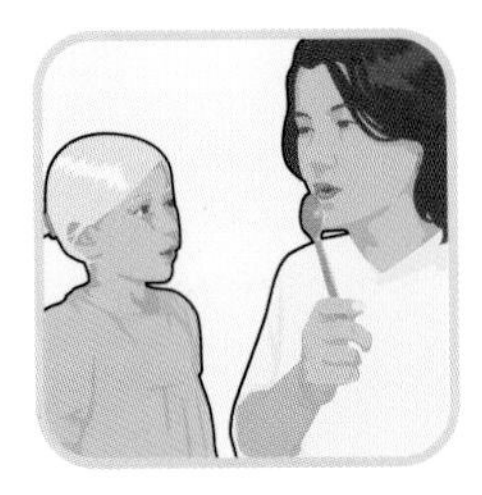

silly pranks

amaze

trick

create

style

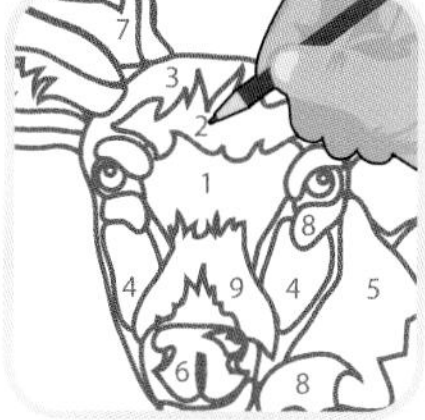

quirky feats

street music

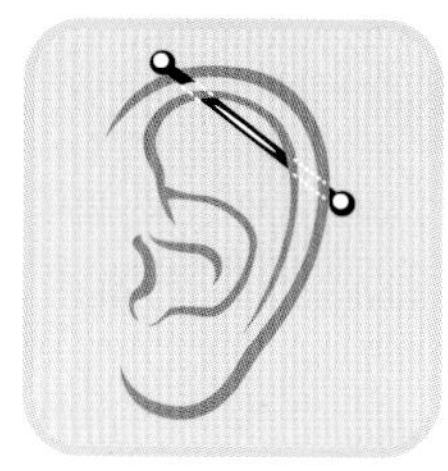

fierce trends

makeup advice

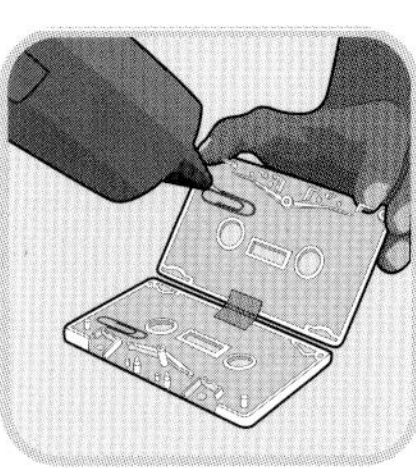

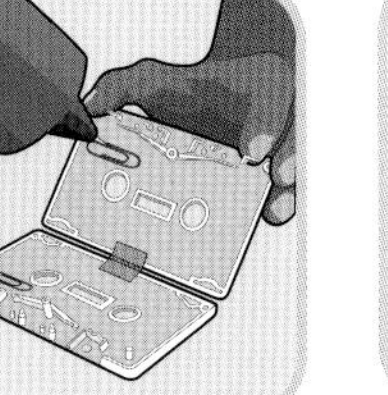

crafty accents

romance starters

trysting tips

big-day advice

furniture fixes

love

nest

grow

pet basics

wilderness survival

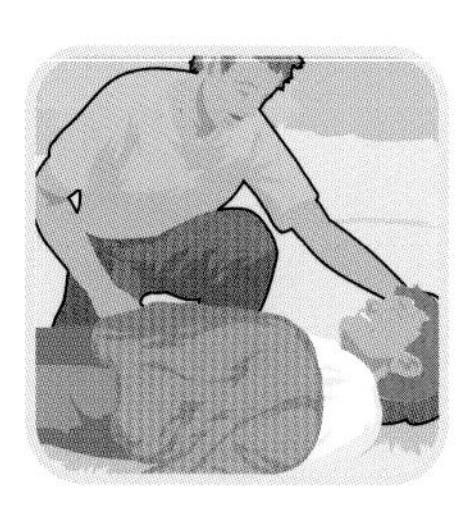

crucial first aid

disaster prep

world culture

offbeat travel

creative transport

help

go

celebrate

indulge

family feasts

holiday crafts

festive dances

simple snacks

decadent desserts

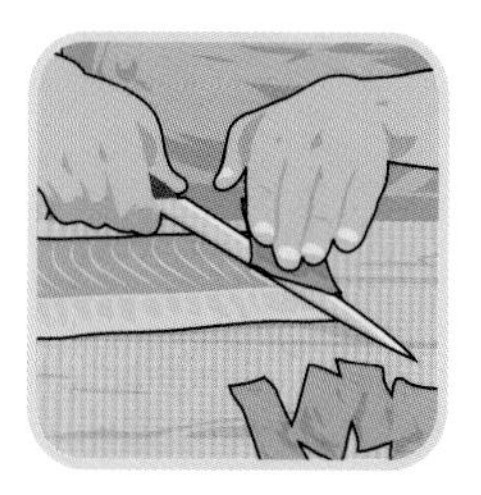

kitchen tips

fresh recipes

amazing drinks

easy exercises

extreme sports

impressive maneuvers

classic moves

move

a note from derek and lauren

189 cut out wallpaper silhouettes

89 play the glass harp

Welcome to **More Show Me How**—the second volume of **Show Me How**! We're so excited that you've joined us on our continuing adventure. You're one of the hundreds of thousands of people around the world whose curiosity and enthusiasm for learning has made this series an ever-growing international phenomenon. In these pages you will find that great **Show Me How** balance of the practical and the magical. Where do all these wild ideas come from? Glad you asked! We get plenty of our inspiration close to home (#189 cut out wallpaper silhouettes), but we've also scoured the globe to bring you fascinating crafts and activities from different cultures (#324 cut festive papel picado, and #294 perform an attaya ceremony). Along the way we've discovered more things everyone should know how to do (#205 streamline a home office), plus an amazing collection of ideas you've never heard of (#89 play the glass harp).

In these pages you will find things to improve your own life (#337 pick seasonal produce) and things that will impress those around you (#6 pour three drinks at once). We've discovered exciting activities for kids, grown-ups, food lovers, artists, and adventure seekers—and of course we've also included a few things we hope you'll never need to know (#270 free myself from an anaconda), just in case. In addition to good old-fashioned know-how, we hope you'll find both inspiration and amusement in the pages that follow. Most of all, we hope this book encourages you to keep exploring all the wild and wonderful things life has to teach us.

270 free myself from an anaconda

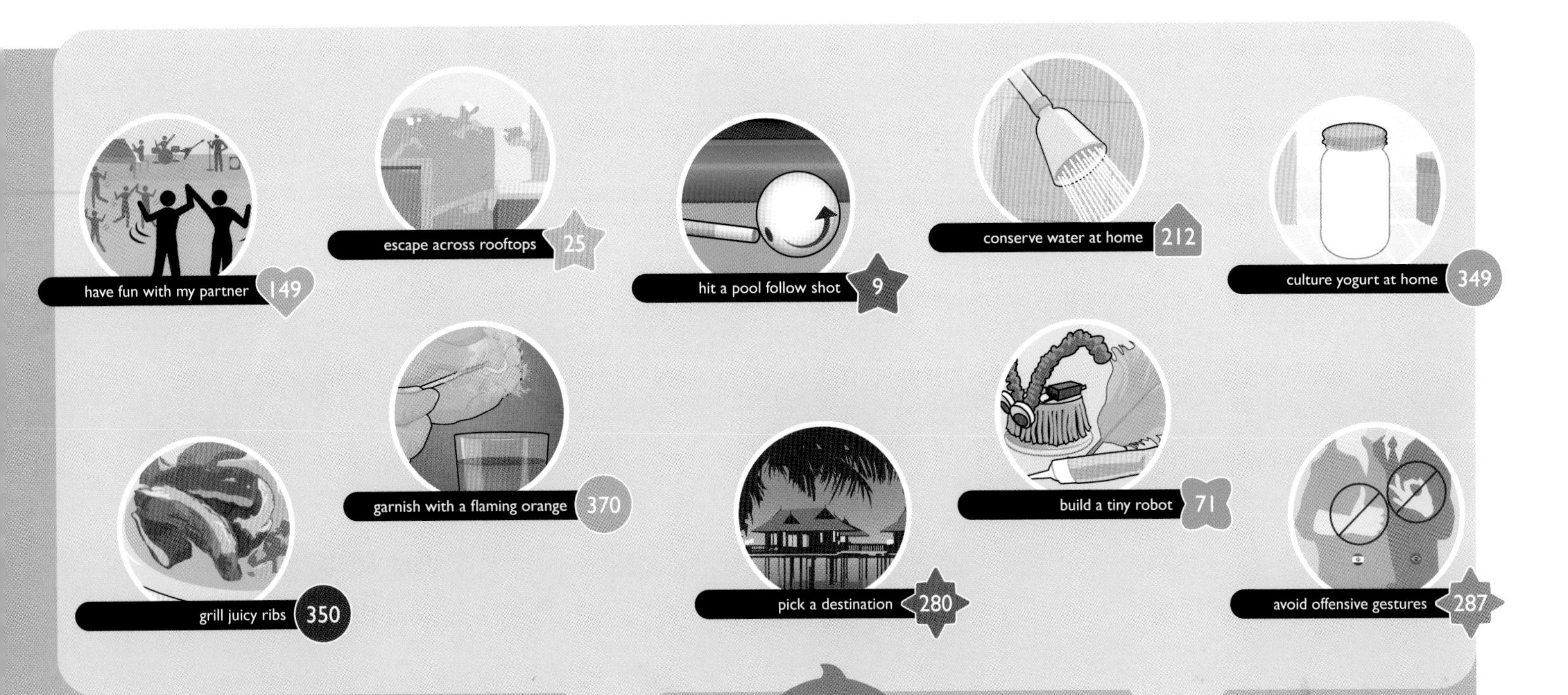

Since meeting in college fifteen years ago (#149), **Derek** and **Lauren** have shared an insatiable curiosity about . . . well . . . just about everything! Working on this book series has reinforced their belief that the passion for learning is always worthwhile, regardless of how practical (#212) or ridiculous (#25) the topic may be.

In this book, they share their interests—from Lauren's mastery of homemade yogurt (#349) to Derek's skills as a pool shark (#9)—but it's also been an excuse to discover other things. Their newfound loves of grilling (#350) and mixology (#370) have made them popular at parties, while their skill at crafting mini robots (#71) has earned them serious cred with the under-ten crowd. This year they hope to fulfill their lifelong dream of traveling the globe (#280) and keeping out of trouble once they get there (#287).

More Show Me How is a different breed of reference book, in which virtually every piece of essential information is presented graphically. We let the illustrations speak for themselves, with a little help from the icons and treatments explained here.

MATHEMATICS When exact measurements matter, find them called out right in the box. Required measurements appear alongside an ingredient or next to a line. Handy "angle" icons help you do it right, or at least from the right direction!

1 head garlic   6 in (15 cm)   135°

make a bouquet garni 338

Cut cheesecloth. Add herbs. Fold up and tie. Add to soup at start of cooking.

peel garlic with hot oil 339

Separate cloves; put in jar. Heat olive oil. Pour hot oil over cloves. Let cool; discard skins. Strain; reserve infused oil.

prepare compound butter 340

Dice garlic and herbs. Mix herbs and butter. Spread onto parchment paper. Roll into a cylinder; freeze. Serve slices on meat or bread.

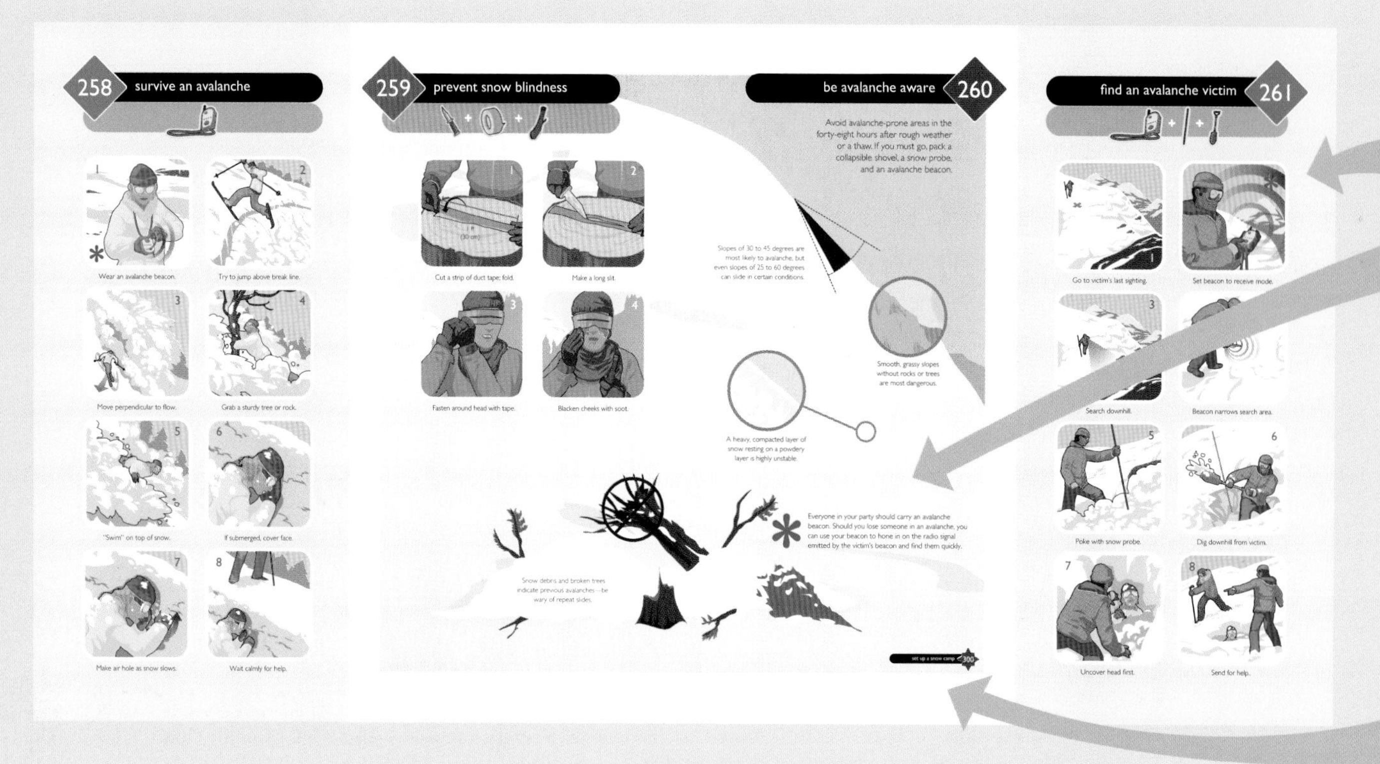

TOOLS The toolbar shows every object used during the course of an activity. Having a hard time deciphering an item? Turn to the tools glossary at the back of the book to find out what it is.

ZOOMS Count on these little circles within or near an activity's frames to show you up-close views, added information, or crucial "don'ts."

MORE INFORMATION Expect the * symbol to point the way to extra information that's vital, helpful, or just plain interesting.

*

CROSS REFERENCES Sometimes one thing just leads to another. Follow the cross references for related, helpful, or otherwise interesting information.

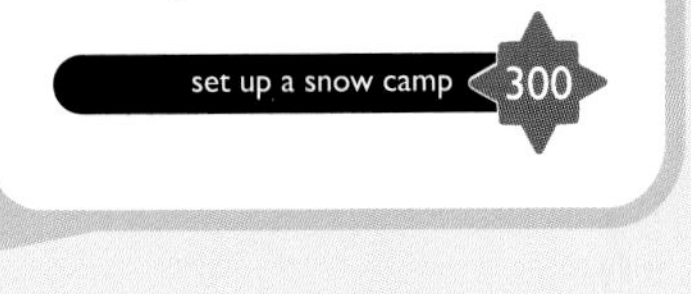

ICON GUIDE Whenever necessary, simple icons alert you to aspects of time, degree, safety, and more. Here are the useful symbols you'll encounter as you journey through the book.

Danger! Avoid this if you're not trained. (Or if you don't want to get into trouble!)

Check out the timer to learn how much time a relatively short task takes.

Call 9-1-1 to seek professional help if you find yourself in this situation.

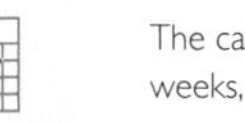

The calendar shows how many days, weeks, or months an activity requires.

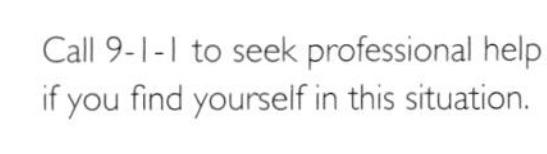

×2

Repeat the depicted action the designated number of times.

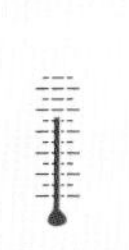

Look to the thermometer to learn the proper temperature for a given action.

3:1

When you're combining ingredients in proportion, the ratio symbol will indicate how many parts of each to add.

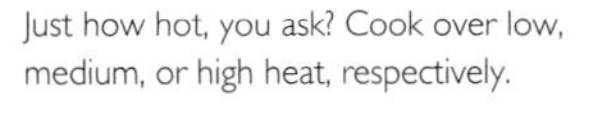

Just how hot, you ask? Cook over low, medium, or high heat, respectively.

copy me!

This page serves as a pattern. Photocopy or scan it, blowing it up as necessary, then follow the instructions.

Phew—fumes! Open a window before performing this activity.

A NOTE TO READERS The depictions in **More Show Me How** are presented purely for entertainment value. Please keep the following in mind if attempting any of these activities:

- **RISKY ACTIVITIES** Certain activities in this book are not just risky but downright weird. Before attempting any new activity, make sure you are aware of your own limitations and have adequately researched all applicable risks. (And just don't do #21. Really.)
- **PROFESSIONAL ADVICE** While every item has been carefully researched, **More Show Me How** is not intended to replace professional advice or training of a medical, culinary, sartorial, veterinary, mixological, athletic, automotive, or romantic nature—or any other professional advice, for that matter.
- **PHYSICAL AND HEALTH-RELATED ACTIVITIES** Be sure to consult a physician before attempting any health- or diet-related activity, or any activity involving physical exertion, particularly if you have a condition that could impair or limit your ability to engage in such an activity. Or if you don't want to look silly (see #380).
- **ADULT SUPERVISION** The activities in this book are intended for adults only, and they should not be performed by children without responsible adult supervision. Many of them shouldn't really even be performed by adults if they can possibly help it (see #268).
- **BREAKING THE LAW** The information provided in this book should not be used to break any applicable law or regulation. In other words, don't try #393 in New York City.

amaze

2 demonstrate awesome strength

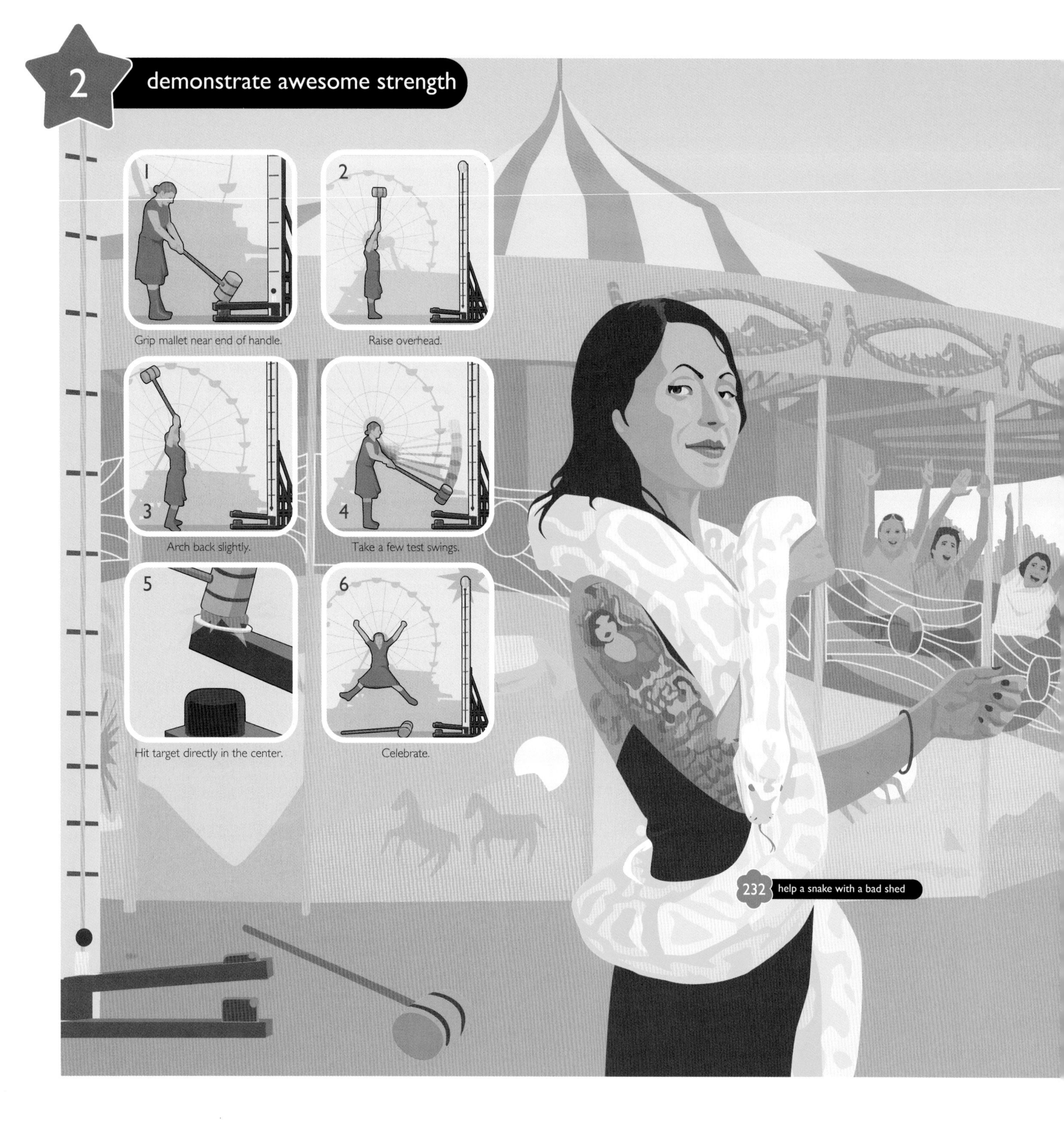

232 help a snake with a bad shed

3 win at the coin toss

1 Secretly lick coin.

2 Lean in as far as possible.

3 Throw in a high arc.

* Carnival workers will put a slick coating on the plates. Foil them by wetting your coin to make it stickier.

4 ace the balloon darts

1 Aim at a balloon near the edge.

2 Throw the dart in a high curve.

3 These balloons are thick; hit one repeatedly to weaken and puncture.

5 ride a mechanical bull

Swing right leg over bull.

Start on lowest setting.

Squeeze thighs; relax torso.

Wave nondominant arm.

Swing leg over; slide down.

6 pour three drinks at once

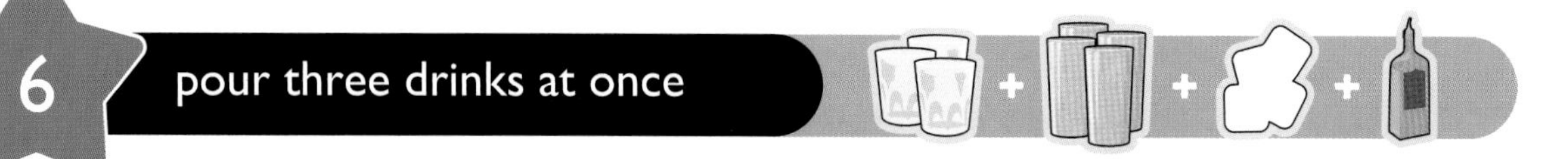

Set down three glasses.

Fill three shakers halfway with ice.

Pour a shot into each.

Stack the shakers.

Place an empty shaker on top.

Hold both ends; form an arc.

Tilt to begin pouring.

Lower arc to pour all three.

7 shotgun a beer

Tilt can to form air pocket.

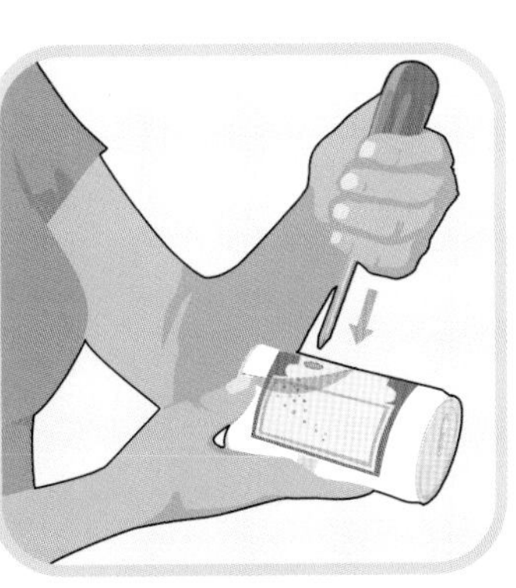

Jab hole in air pocket.

Lean; put mouth over hole.

Tilt up, opening tab. Chug.

8 dominate a dart match

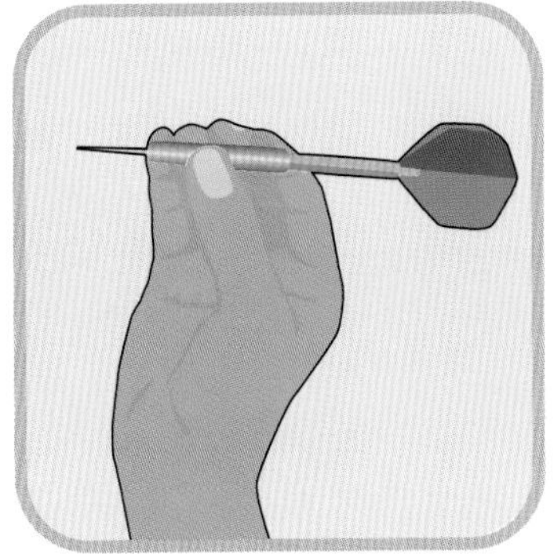

Grip dart with at least two fingers.

Toe the line; stand up straight.

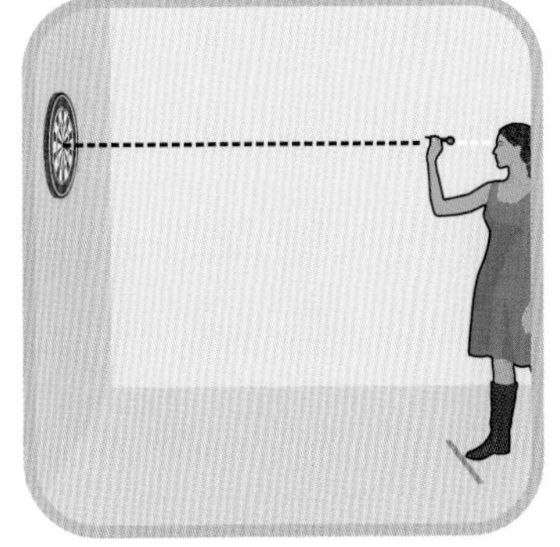

Hold dart in line with target.

Cock your arm.

Thrust, holding shoulder still.

Release with tip pointed slightly up.

Follow through with arm.

Bull's-eye!

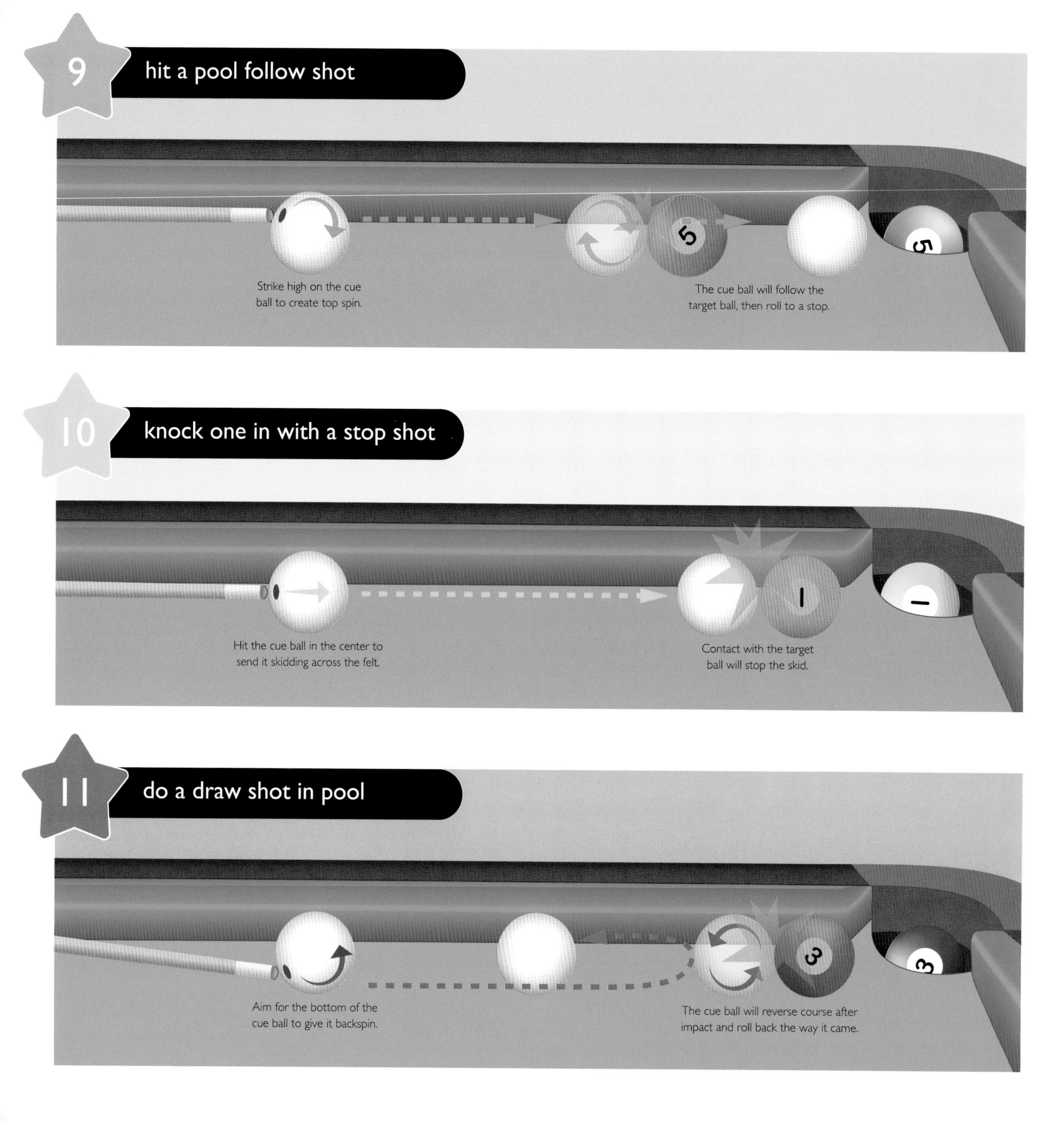
9
hit a pool follow shot
5
5
Strike high on the cue ball to create top spin.
The cue ball will follow the target ball, then roll to a stop.
10
knock one in with a stop shot
1
1
Hit the cue ball in the center to send it skidding across the felt.
Contact with the target ball will stop the skid.
11
do a draw shot in pool
3
3
Aim for the bottom of the cue ball to give it backspin.
The cue ball will reverse course after impact and roll back the way it came.

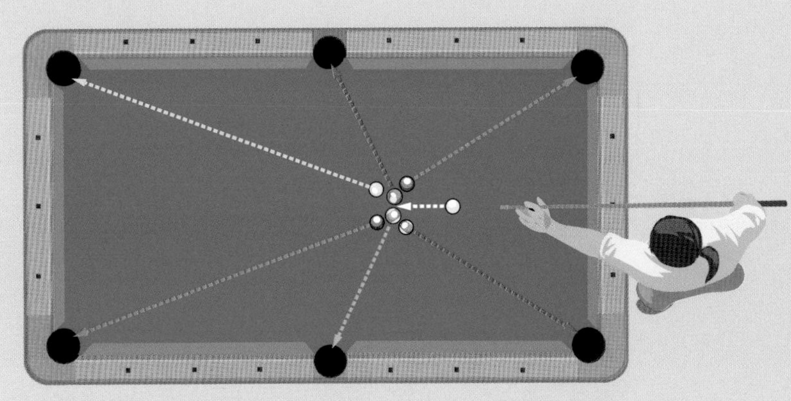

the sneaky pete

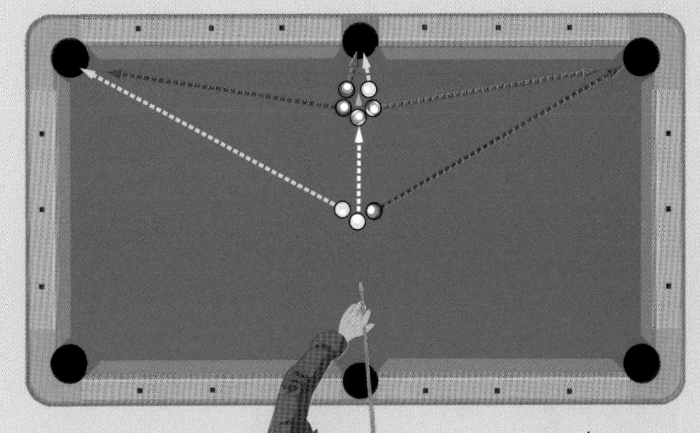

the top gun

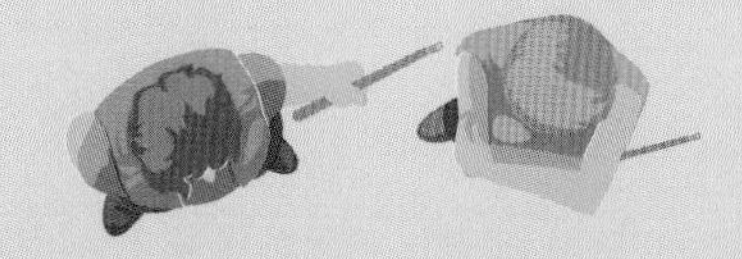

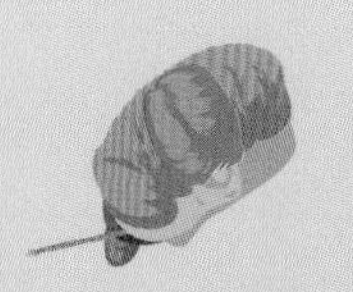

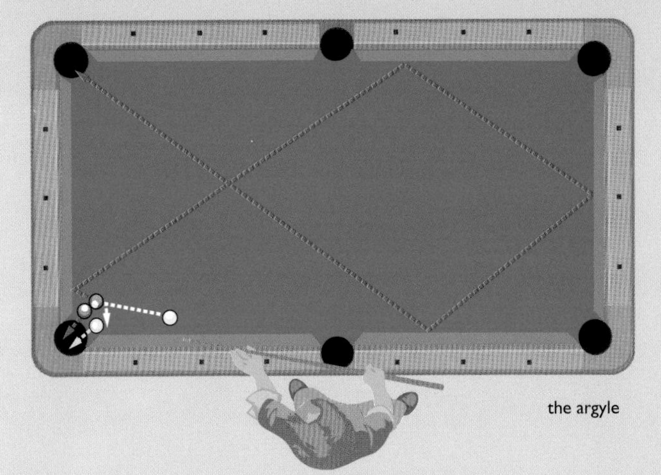

the argyle

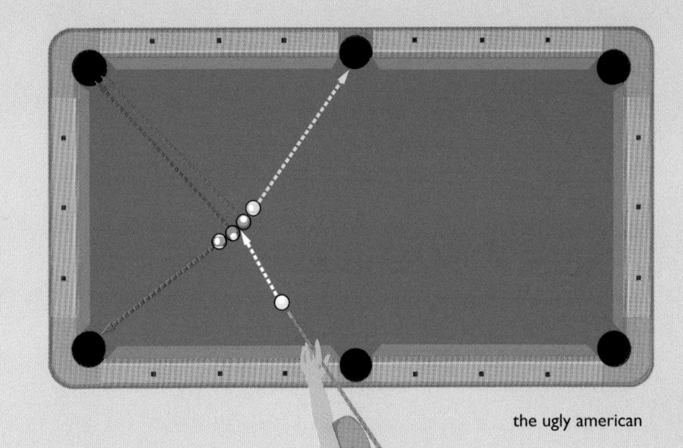

the ugly american

13 spin a guitar

Lock strap; hold at waist.

Step behind the cord.

Push base up and over.

Twist hips for momentum.

Catch neck; rock on.

14 give myself a metal makeover

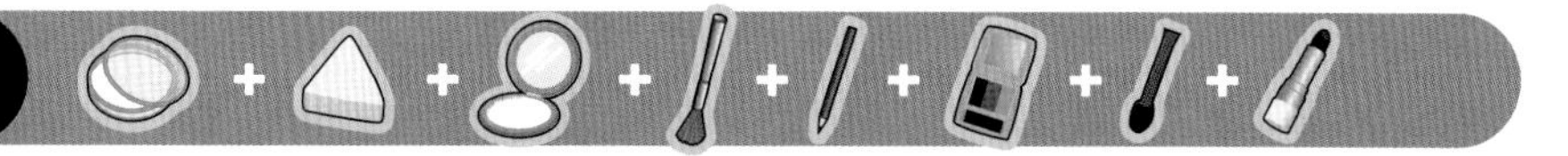

Apply base to face and neck.

Set with powder.

Draw bat shapes with liner.

97 draw on dramatic eyes

Line eyelids.

Fill the shapes with dark shadow.

Line lips in black.

Fill with dark lipstick.

Brutal.

"get on up" from a split 15

Leap with hips facing side.

Slide down into a split.

Land it.

Push up; bend leg.

Drag front heel to stand.

rock a dynamite stage spin 16

Put foot behind knee.

Put ball of back foot on ground.

Open back thigh; spin.

twirl a pirouette 413

Raise up as you turn.

End facing front.

Fall to knees in celebration.

17 shoot an arrow

Place hood on eagle.

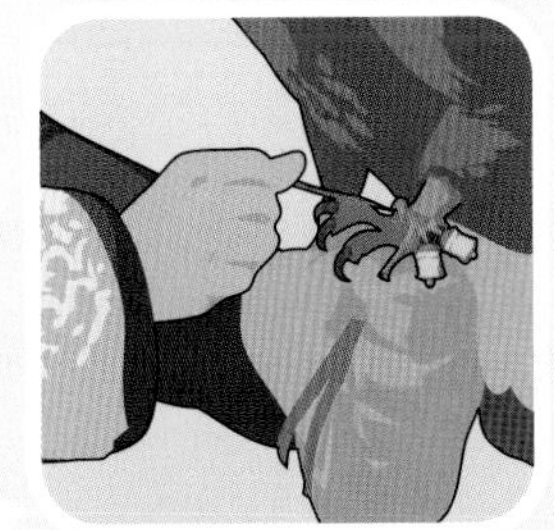
Affix flying jesses to talons.

Travel to hunting grounds.

De-hood bird; release jesses.

Ride to scare up game.

Screeeee!

Lure bird away with a treat.

Grab jesses; re-hood bird.

kazakh-style hood
keeps the bird calm

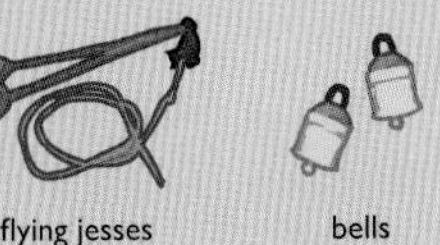
flying jesses
leash the bird

bells
help locate the bird

hunter's pouch
stores eagle treats

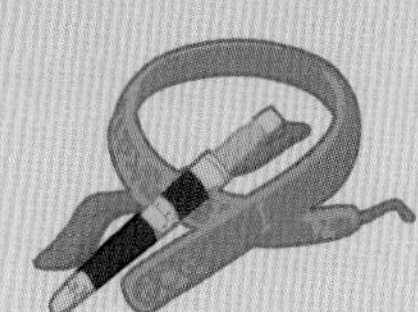
hunting knife
cuts meat

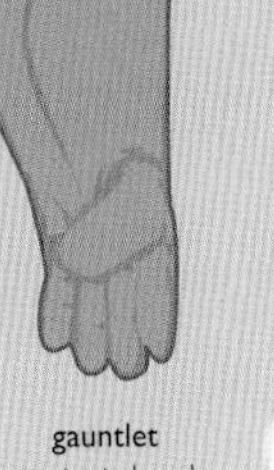
gauntlet
protects hand

19 build a hoverboard

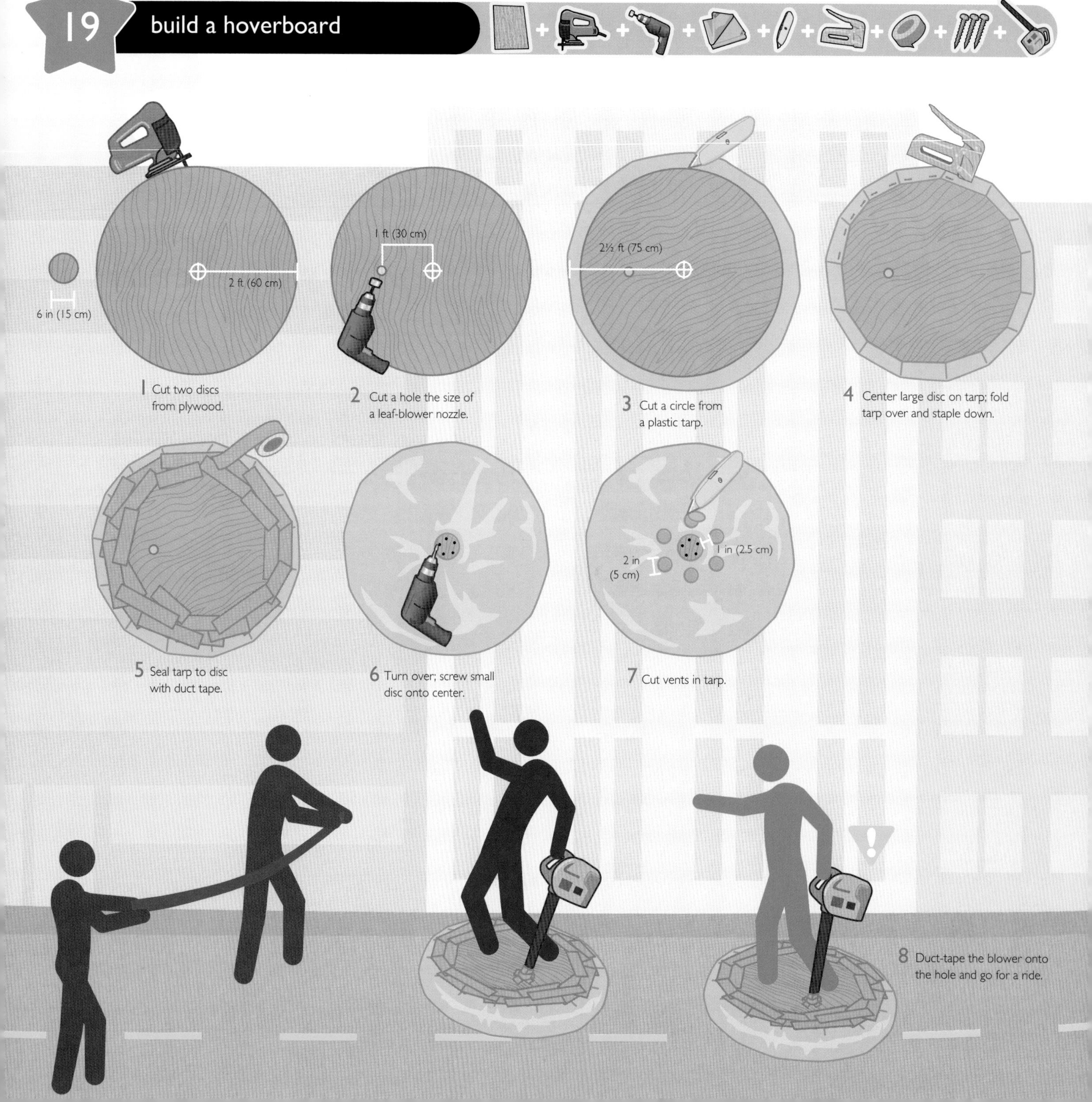

20 travel by jet pack

Strap in securely.

Pull throttle to take off.

Push bars to go forward.

Steer with left hand.

Push throttle to land.

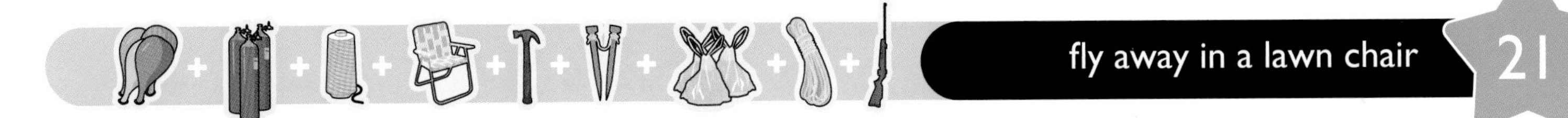

21 fly away in a lawn chair

Inflate balloons with helium.

Stake down lawn chair; add ballast.

Tie balloons to chair arms and back.

Sit with drop line and pellet gun.

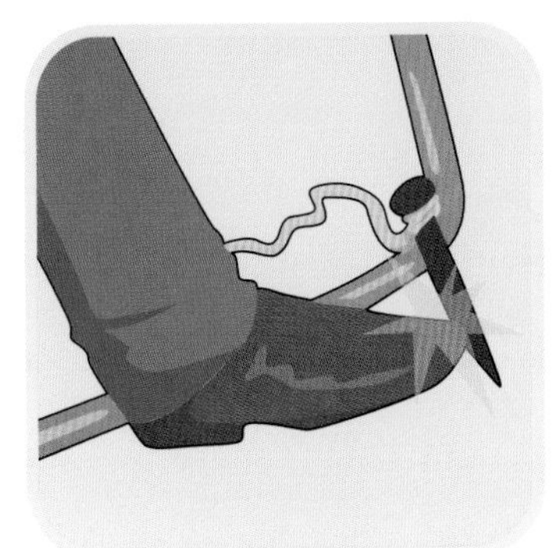

Release chair.

Puncture ballast bags to gain altitude.

join the mile-high club 156

Shoot balloons to descend.

Lower drop line for landing help.

22 cook up a sugar smoke screen

Grind into a fine powder.

Mix powder with sugar in a pan.

Cook into a brown paste, stirring.

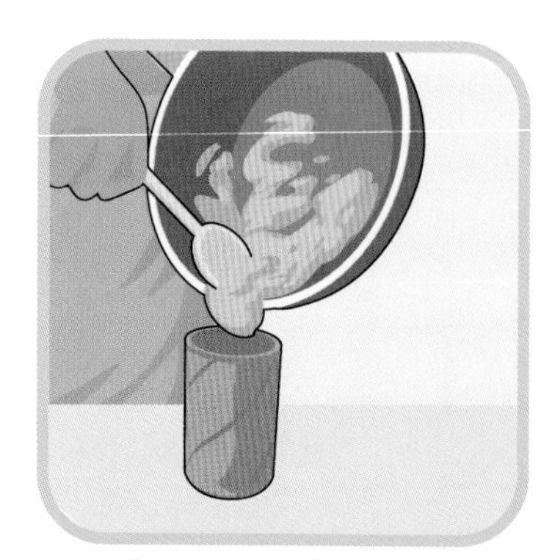

Spoon into a cardboard tube.

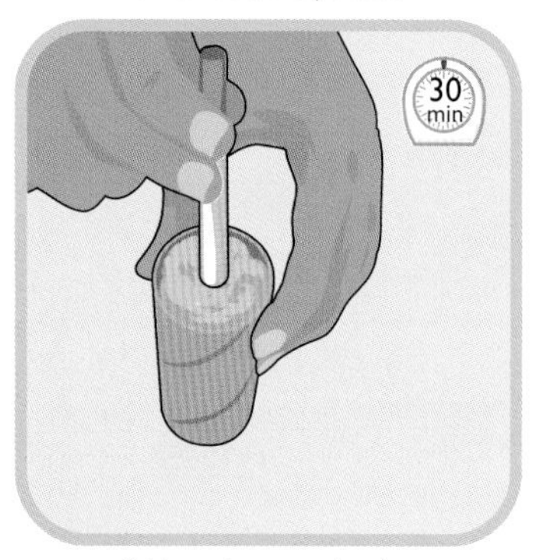

Stick pen in center; let dry.

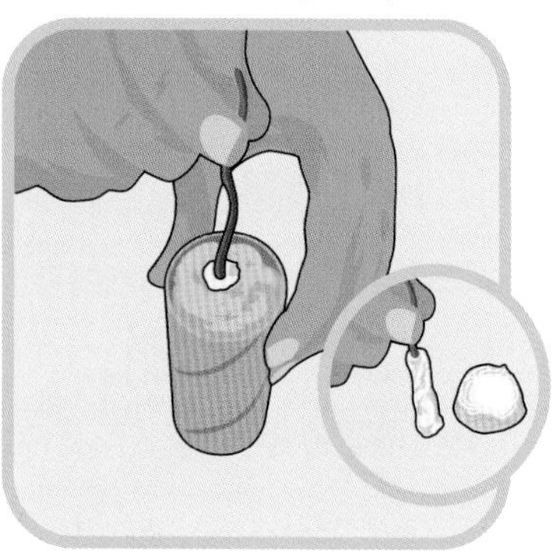

Remove pen; add cotton-wrapped fuse.

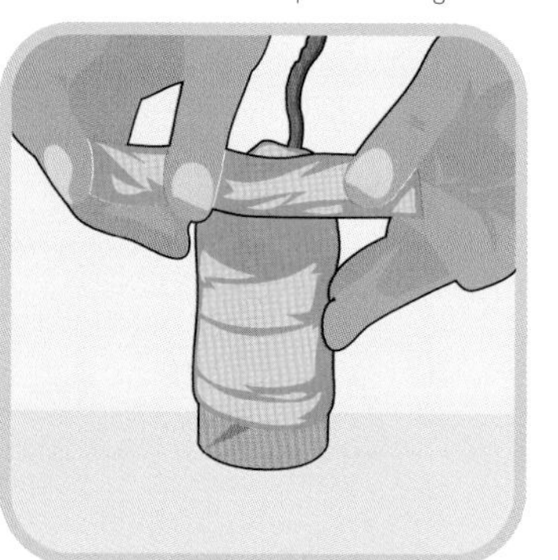

Cover with duct tape.

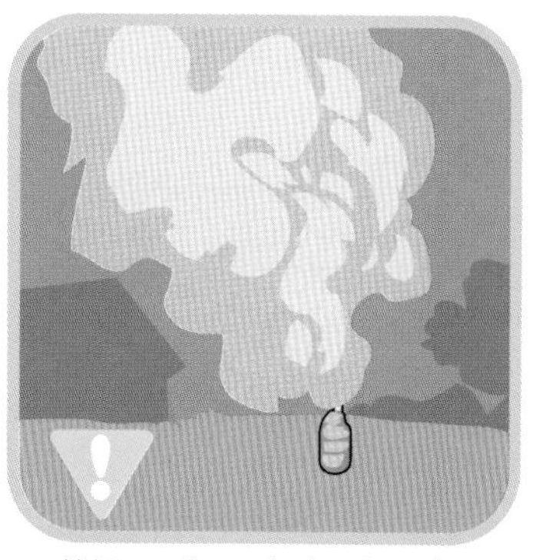

Light away from animals and people.

The active ingredient in this smoke bomb is potassium nitrate, which is found in stump remover (a substance used to dissolve unwanted tree stumps). It's also sold as saltpeter, niter, and E252.

23 create a smoky special effect

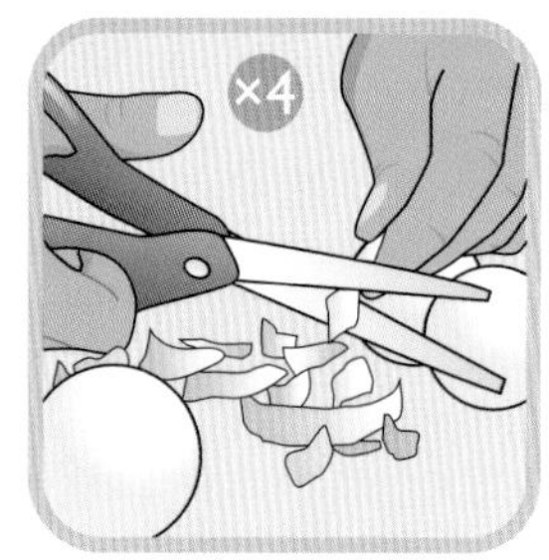

Cut up four ping-pong balls.

Cut hole in top of fifth ball.

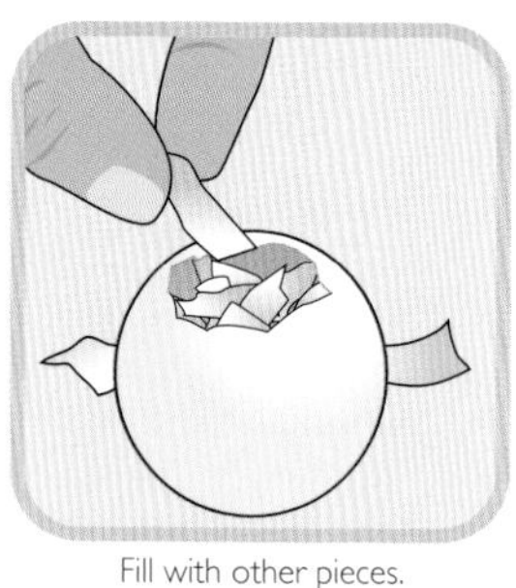

Fill with other pieces.

Add fuse; wrap in foil.

Light away from creatures.

Warm the diet cola in the sun. Keep the cap on.

2
Remove the cap. Drill a hole in it.

3
Widen the hole so that the candy will fit.

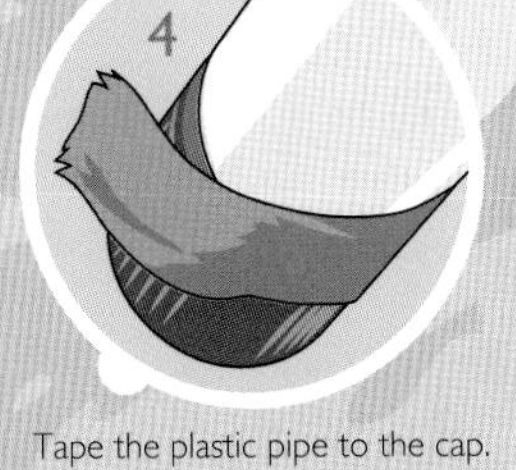

Tape the plastic pipe to the cap.

5
Drill a hole in the endcap.

Drill a hole through the pipe near the tape.

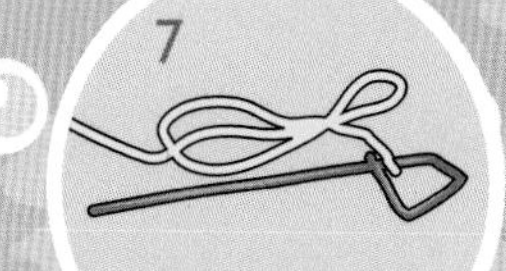

Bend the paper clip into a hook. Tie on a long string.

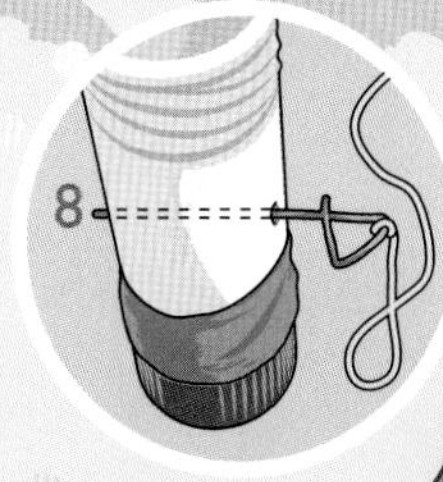

Stick the paper clip through the pipe.

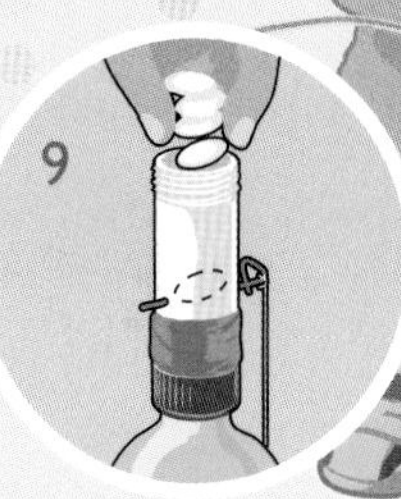

Put the cap on the bottle. Fill with Mentos™.

Add the endcap.

11 Stand back and yank the string.

make a fruity soda-pop float 368

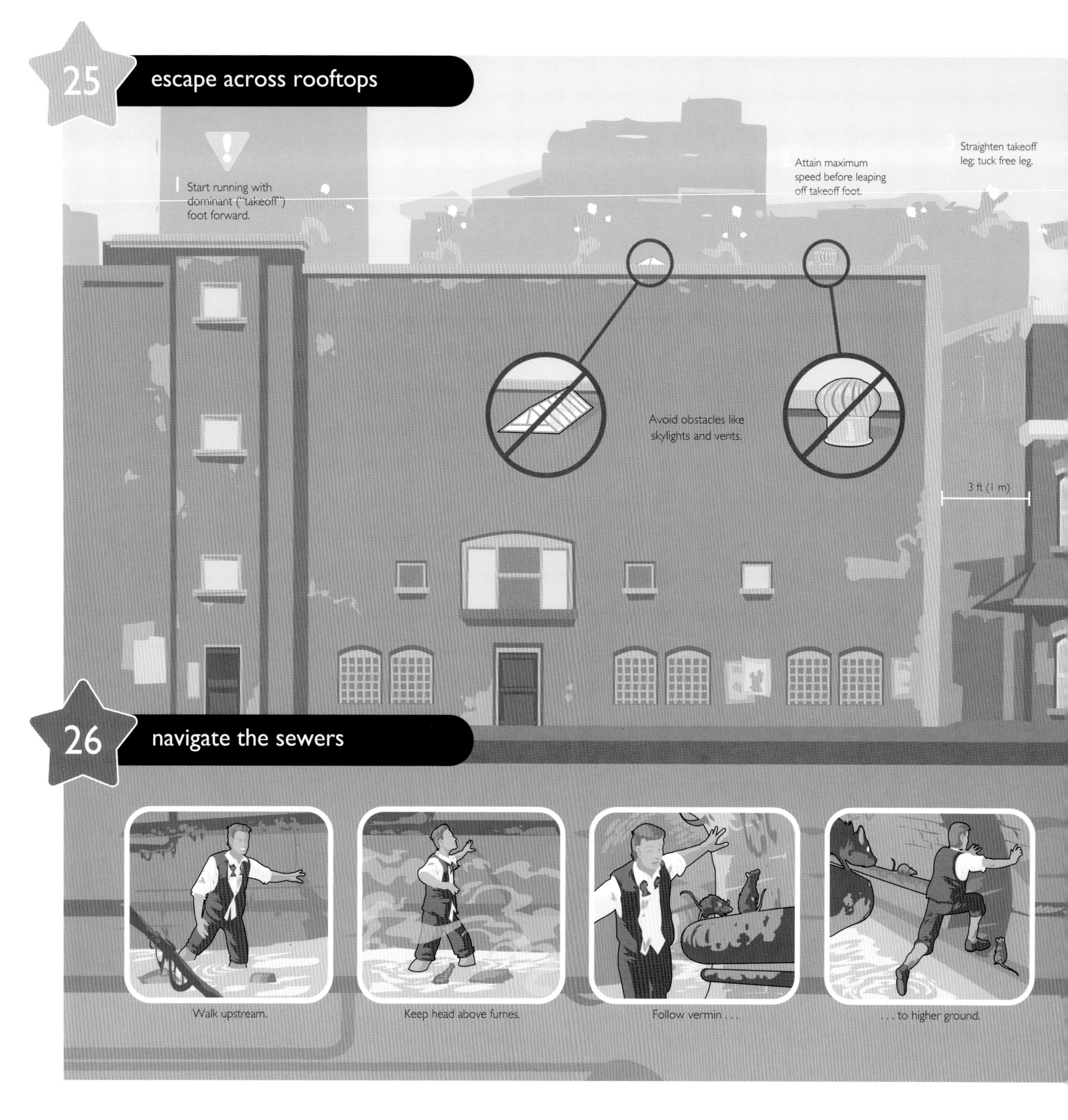
25
escape across rooftops
1 Start running with dominant ("takeoff") foot forward.
2 Attain maximum speed before leaping off takeoff foot.
3 Straighten takeoff leg; tuck free leg.
Avoid obstacles like skylights and vents.
3 ft (1 m)
26
navigate the sewers
Walk upstream.
Keep head above fumes.
Follow vermin . . .
. . . to higher ground.

Bring free leg forward.
5 Land heels first.
6 Spring up and continue running.
7 Locate a drainpipe to slide down.
8 Wrap your heels for extra stability.
9 Use awnings to slow your fall and disperse impact force.
10 Just before impact, twist your shoulder to roll onto your back.
Listen for traffic.
Look for ladders and manholes.
Brace; open with shoulder.
Carefully look for traffic. Climb out.

trick

27 perform a pig prank

Grease three pigs.

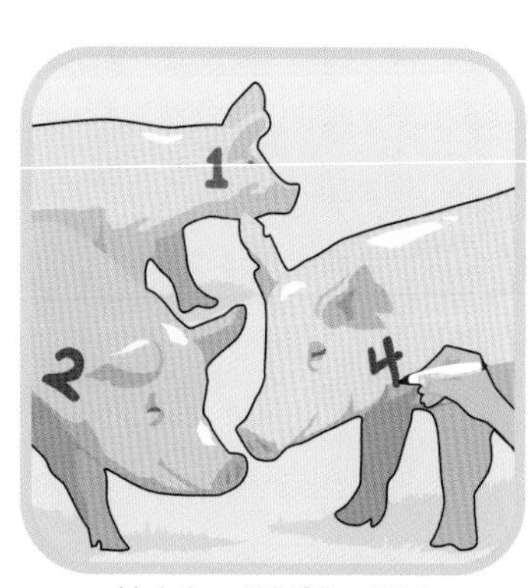
Mark them "1," "2," and "4."

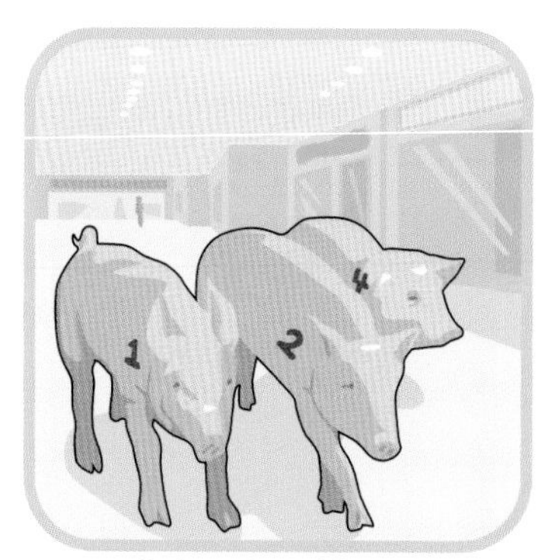
Set loose in a public area.

Watch the search for "3."

28 pull a rhino hoax

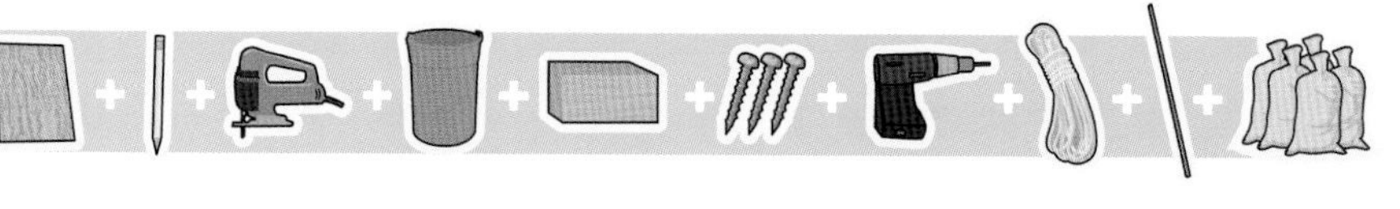

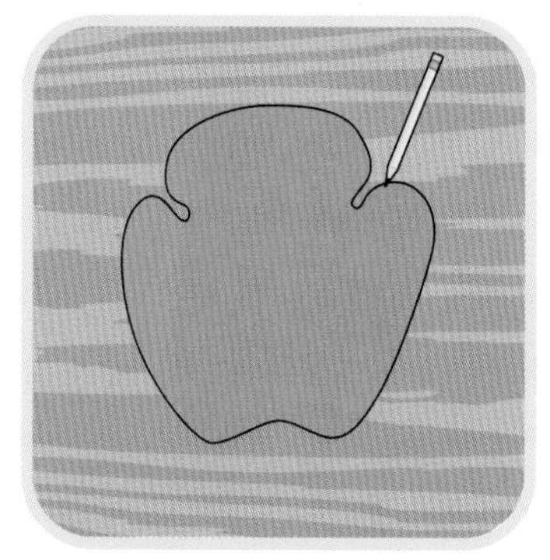
Trace rhino footprint on plywood.

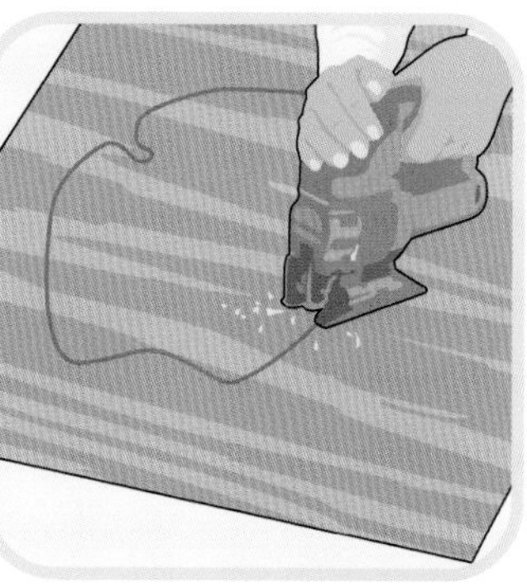
Cut.

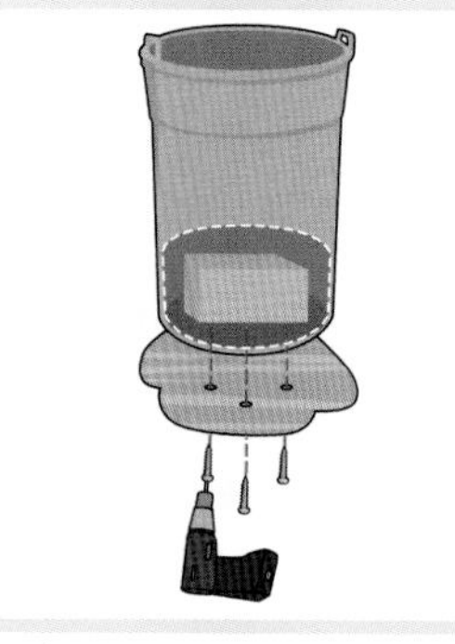
Attach to bottom of trash can.

Hang can from a very long pole.

Weight can with sandbags.

Make tracks through the snow . . .

. . . into a frozen pond.

Let the kids piece the story together.

29 make a straw wrapper slither

Remove wrapper so it bunches up.

Fill straw with water.

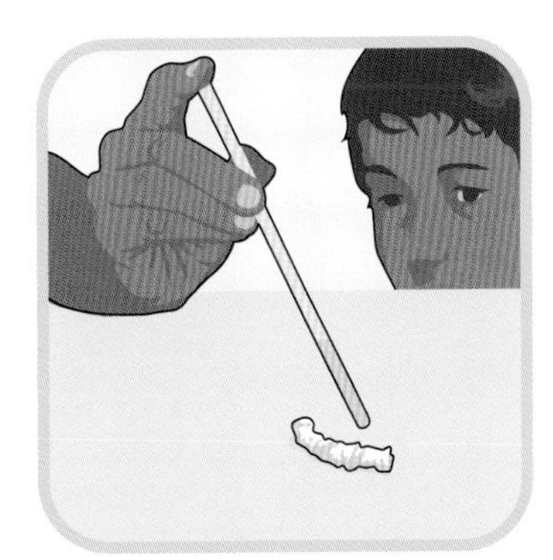

Put a few drops on the "snake."

free myself from an anaconda 270

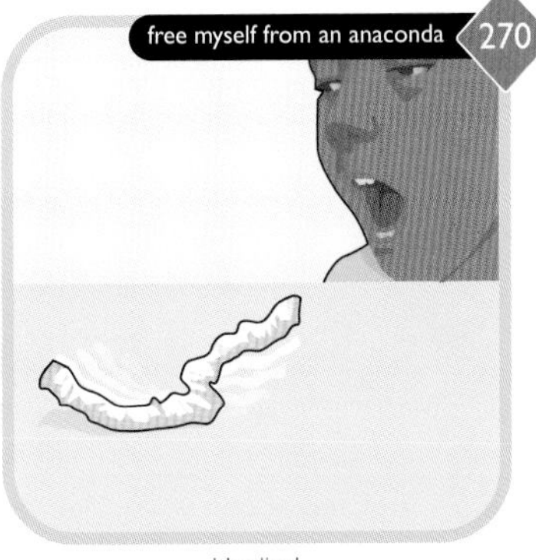

It's alive!

30 put a pen through my brain

Hold pen with fingers flat.

Set in ear; cover with hand.

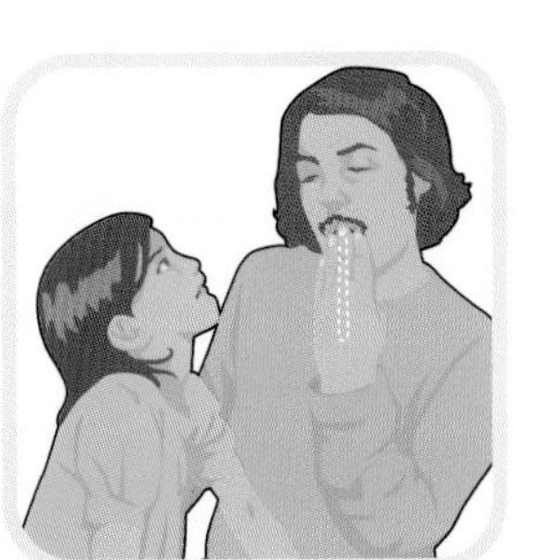

Bring hidden pen to mouth.

Bite pen; pull hand away.

Offer pen.

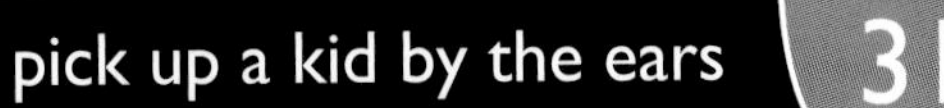

31 pick up a kid by the ears

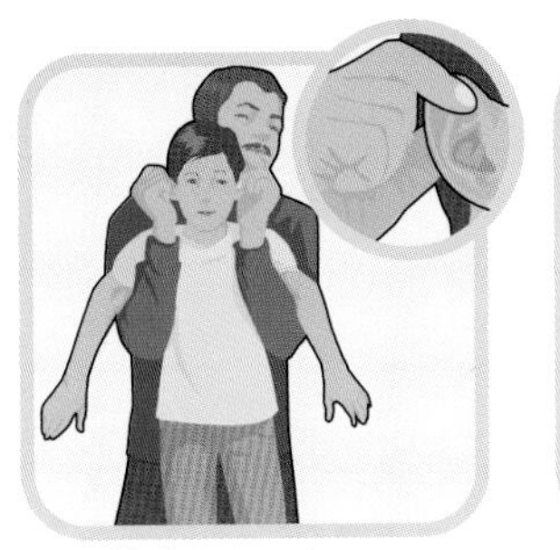

Hook arms; grab ears.

Lift under his arms.

32 hang a spoon from my nose

Fog spoon with breath.

Hang from tip of nose.

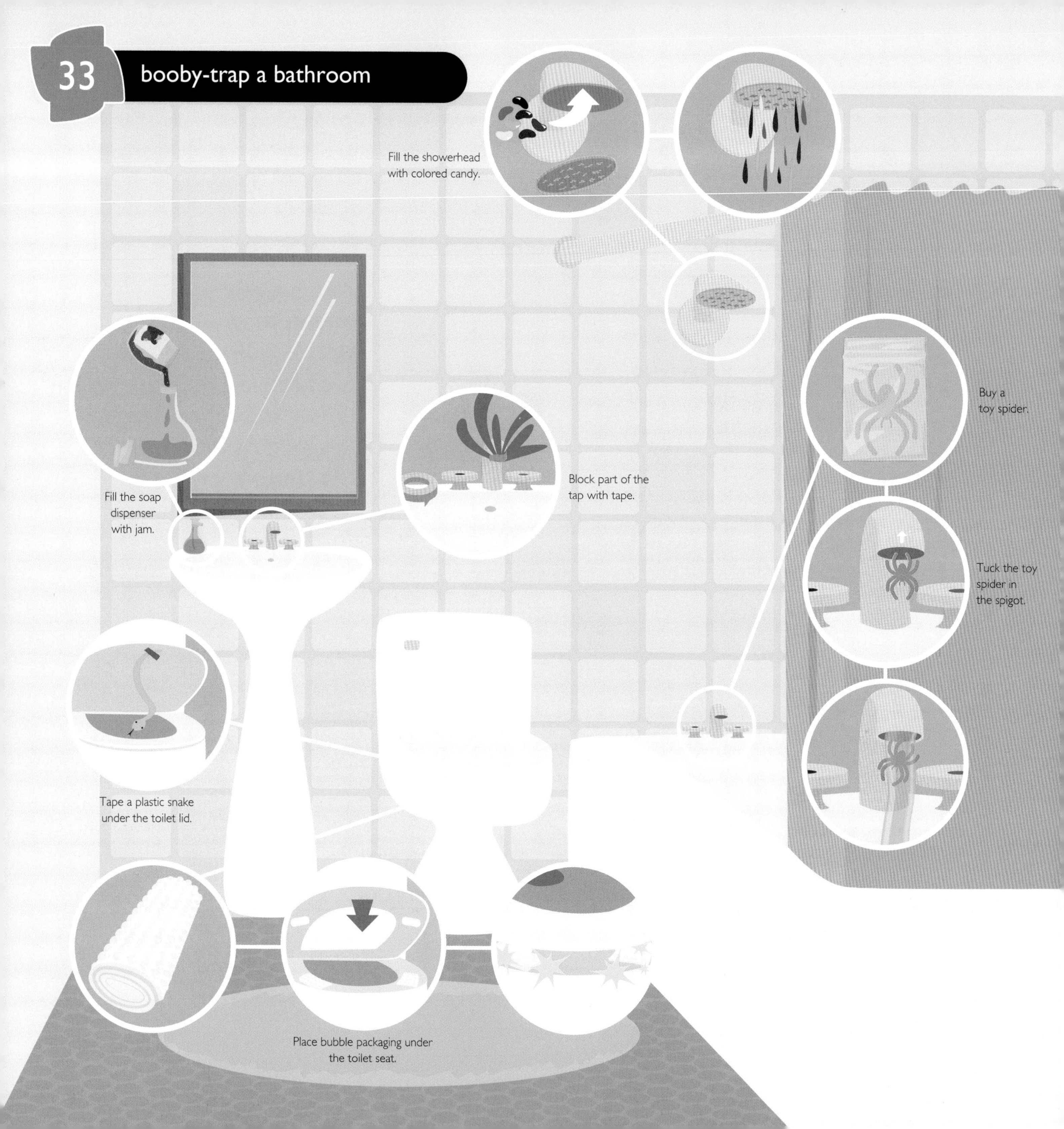
33
booby-trap a bathroom
Fill the showerhead
with colored candy.
Buy a
toy spider.
Fill the soap
dispenser
with jam.
Block part of the
tap with tape.
Tuck the toy
spider in
the spigot.
Tape a plastic snake
under the toilet lid.
Place bubble packaging under
the toilet seat.

punk a car 34

Fill the vents with confetti, then turn the heat up high—the confetti will erupt on ignition.

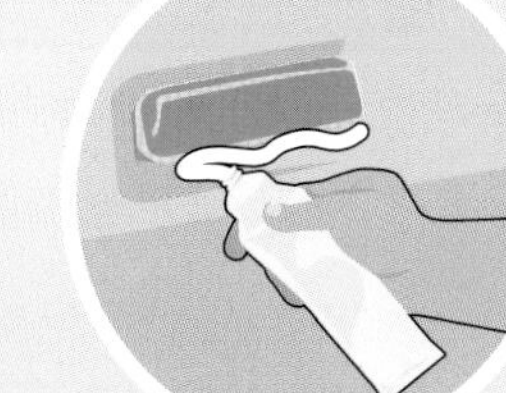

Squirt toothpaste under the door handles for a minty surprise.

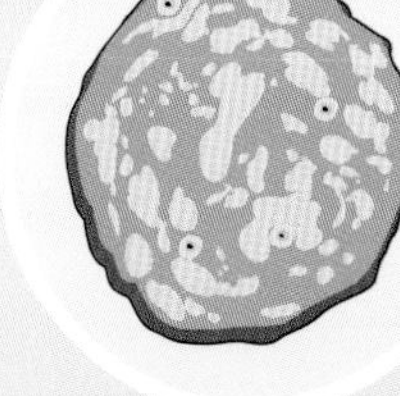

Put salami slices on the windshield for enduring grease spots.

Put a condom on the tailpipe. It'll inflate impressively.

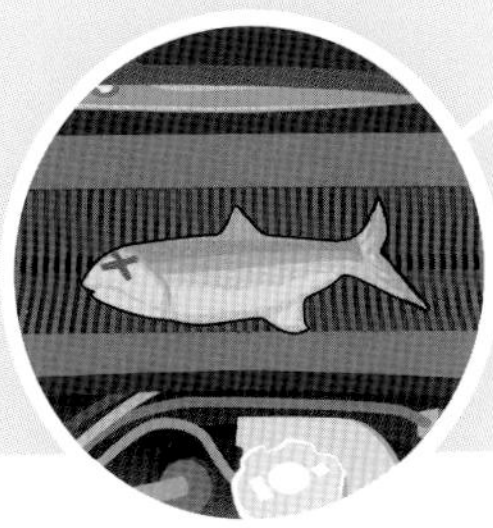

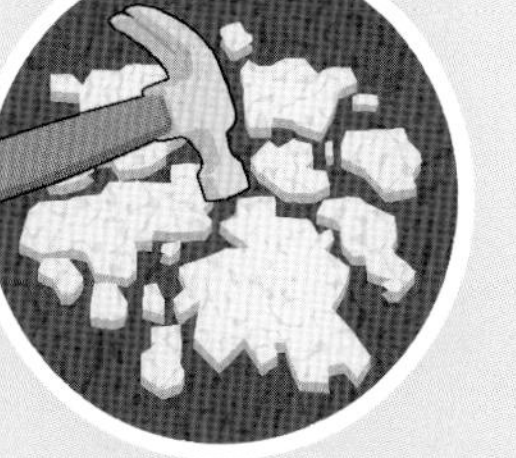

For a lingering stench, place a dead fish on the air intake.

paint an art car 66

Roll down the window, then sprinkle broken safety glass on the ground.

rearrange a morning commute 35

Collect traffic cones.

At night, reroute roads.

Direct traffic into blind alley.

escape across rooftops 25

Relish the havoc from above.

36 make a cigarette disappear

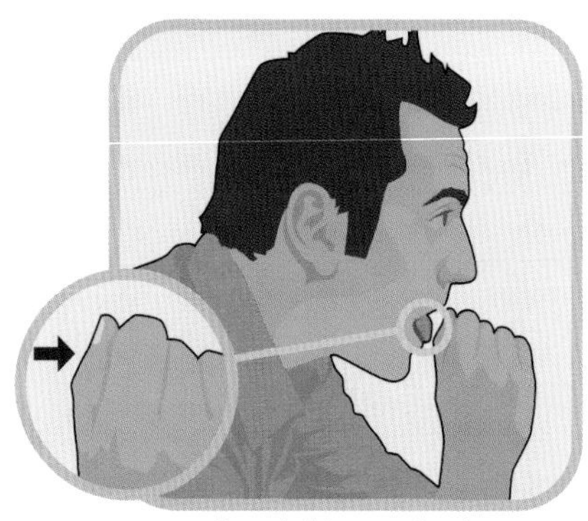

Covertly lick upper thumb.

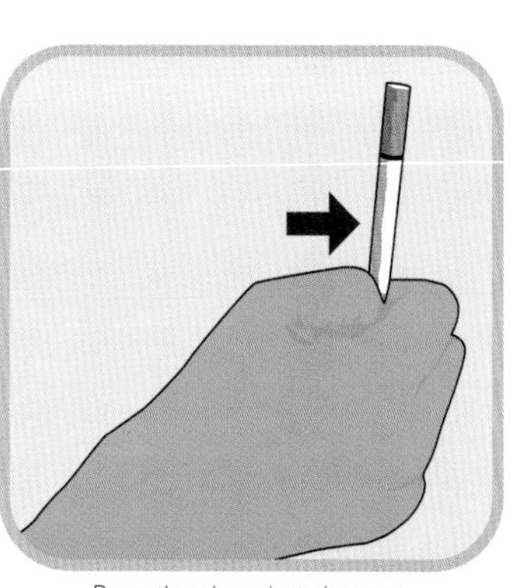

Press thumb against cigarette.

Offer smoke.

Open hand with palm forward.

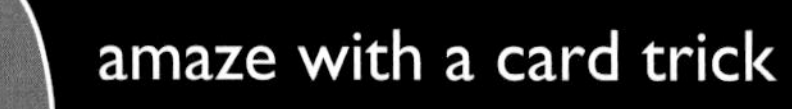

37 amaze with a card trick

* If you find your card on the bottom of the deck, the mark's card will be at the top.

Note card at bottom of deck.

Ask mark to pick a card. Avert eyes.

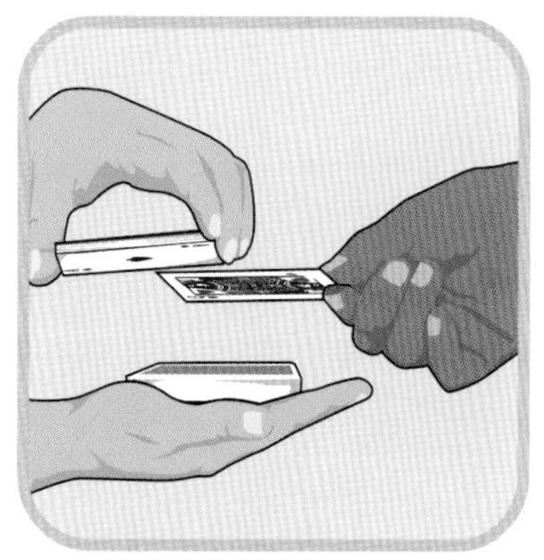

Cut deck; have mark put card back.

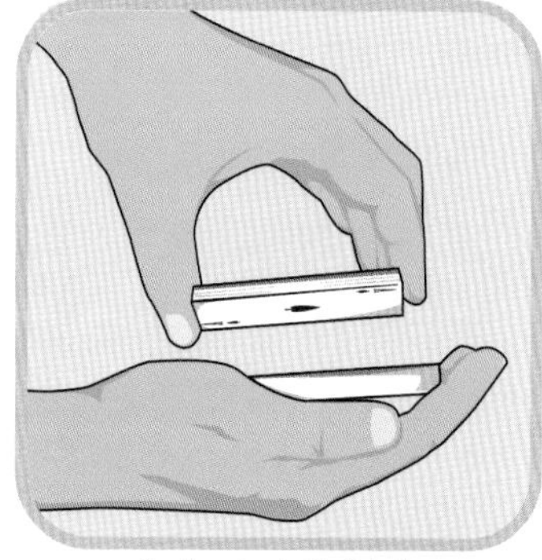

Replace top half of deck.

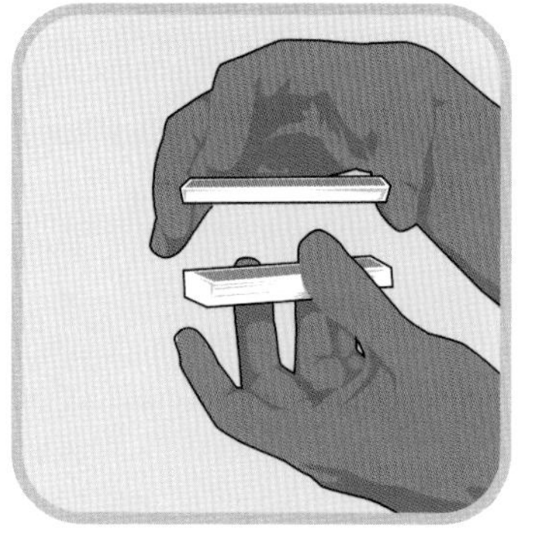

Ask mark to cut deck.

Find your card; his is beneath it.

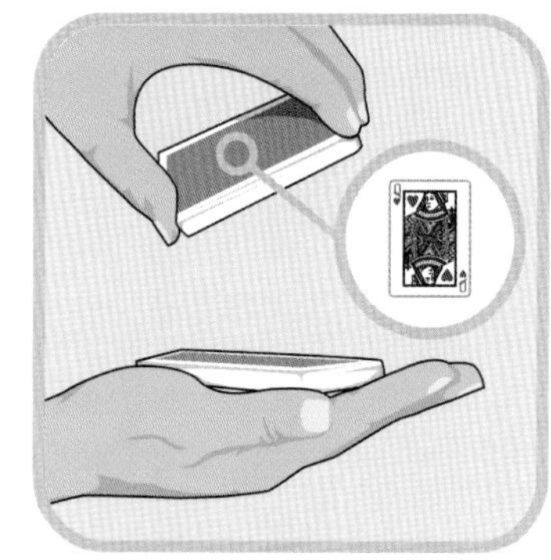

Cut mark's card to top of deck.

Have mark draw top card.

bend a spoon with ease 38

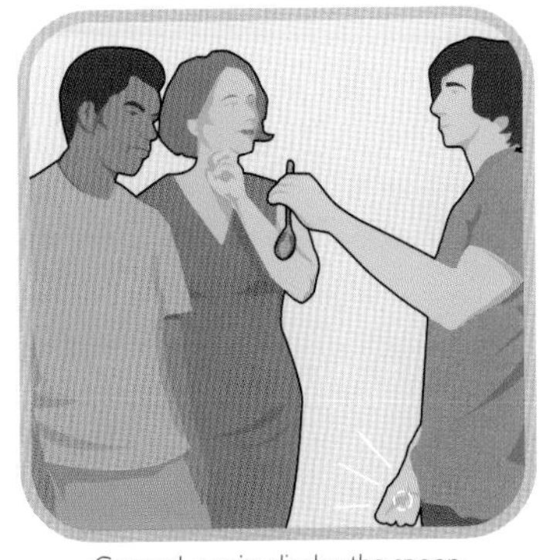
Conceal a coin; display the spoon.

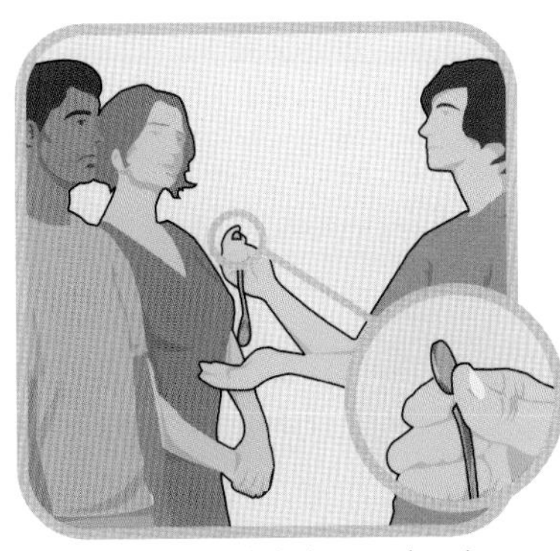
Pinch a coin to mimic the spoon's end.

Hold spoon in fist; "bend" on palm.

Hide the coin before opening hand.

bring a match to life 39

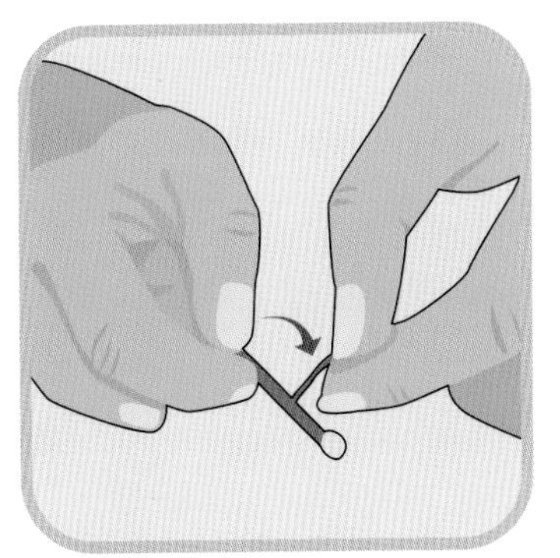
Peel layer of cardboard forward.

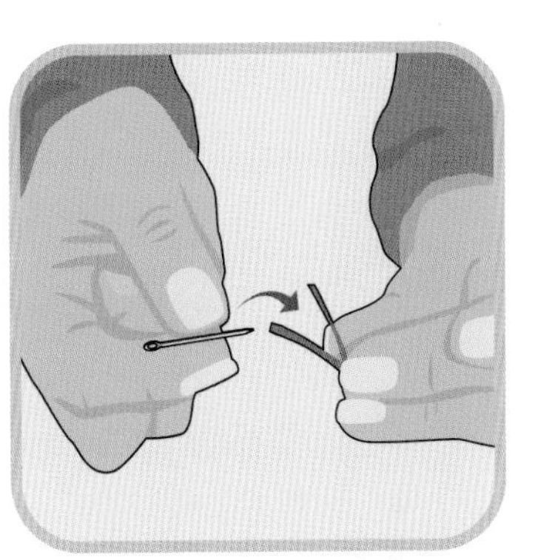
Set a magnetized needle inside.

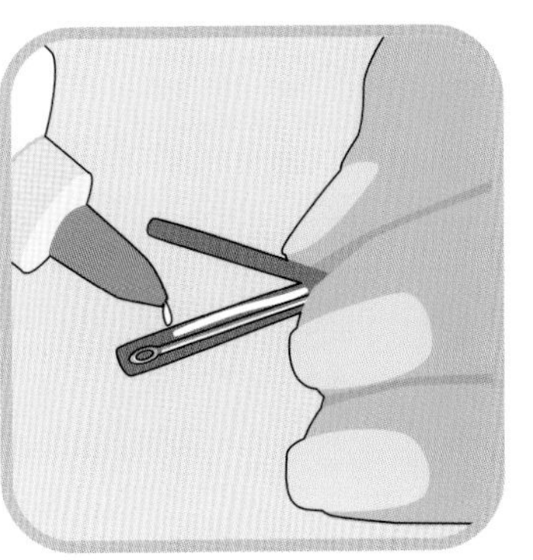
Glue needle inside match.

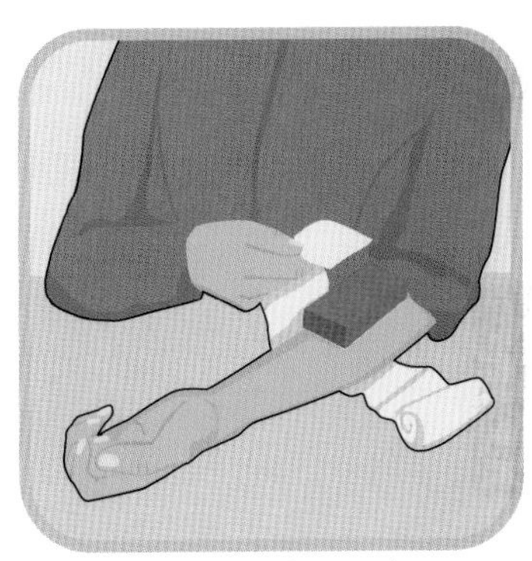
Hide a magnet under your sleeve.

Palm rigged match; approach mark.

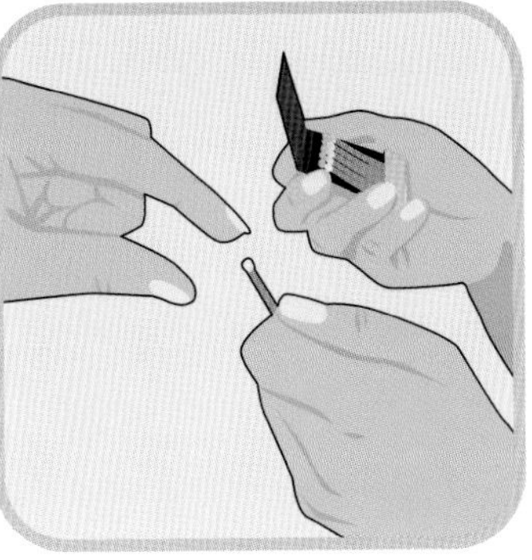
Accept her match.

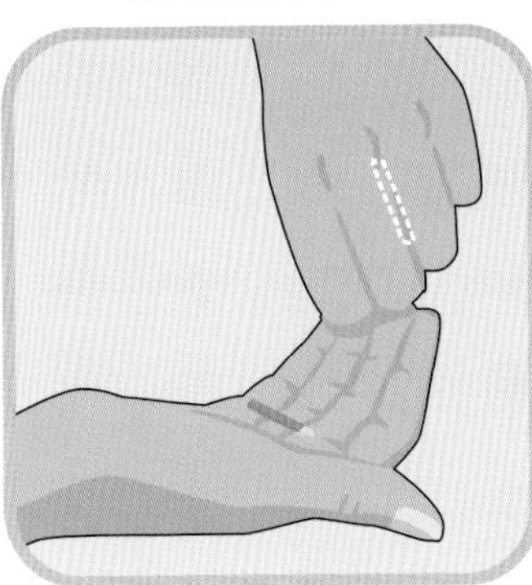
Switch dummy for rigged match.

Wave hand over hidden magnet.

40 spin a plate

41 saw a lady in half

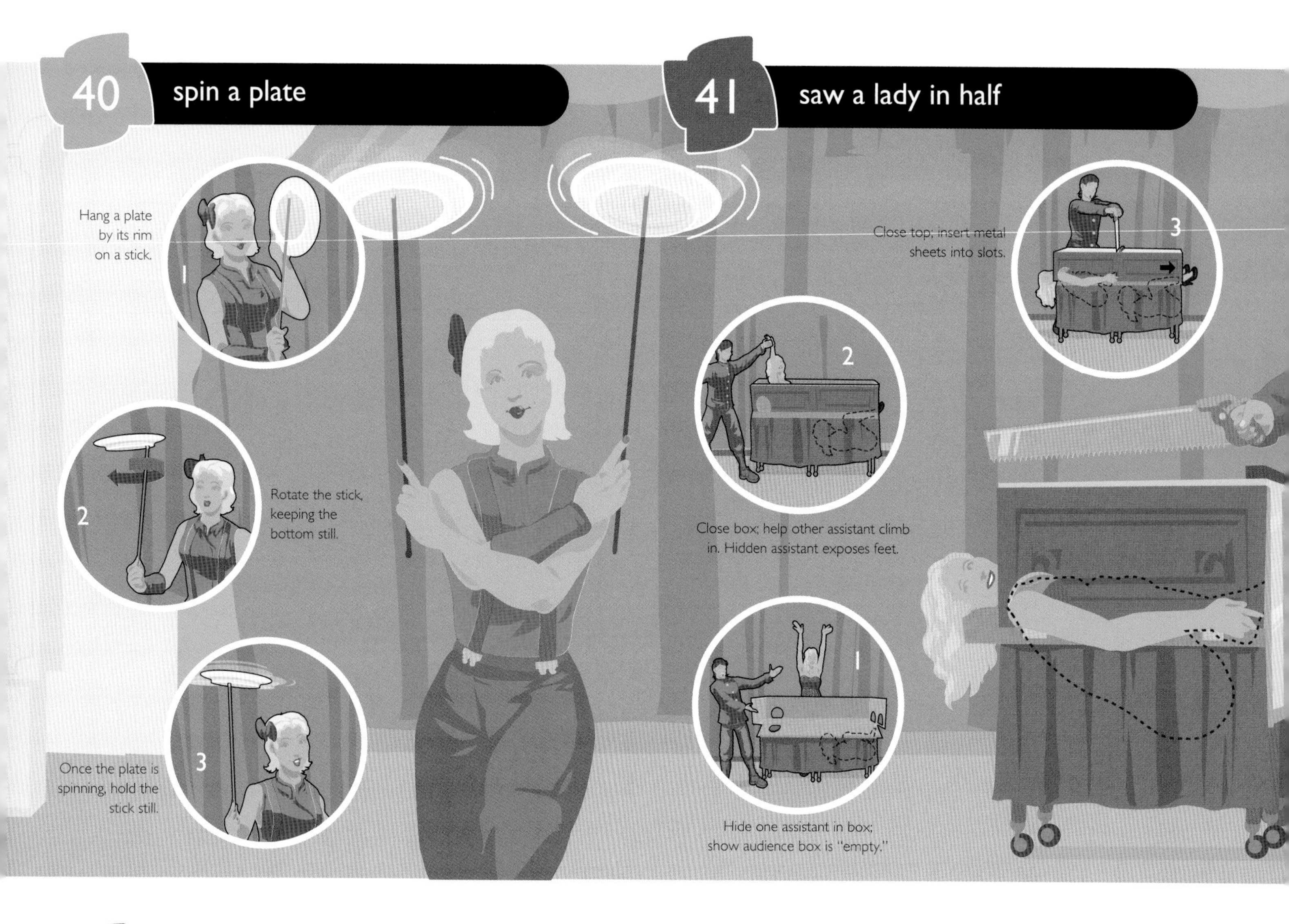

42 rig a knife throw

Toss knife to set the scene.

Pick audience volunteer.

Show her the knife.

Place assistant on target.

341 master knife techniques

Hand knife to volunteer.

"Saw" the box in half.

Pry boxes apart; display halves.

Push halves together; remove sheets.

Open top; help assistant climb out.

Blindfold her.

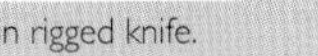
Switch in rigged knife.

Reel it in after she throws.

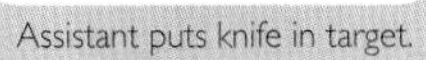
Assistant puts knife in target.

Tell her she's got great aim!

43 apply lipstick with my cleavage

Place lipstick in cleavage.

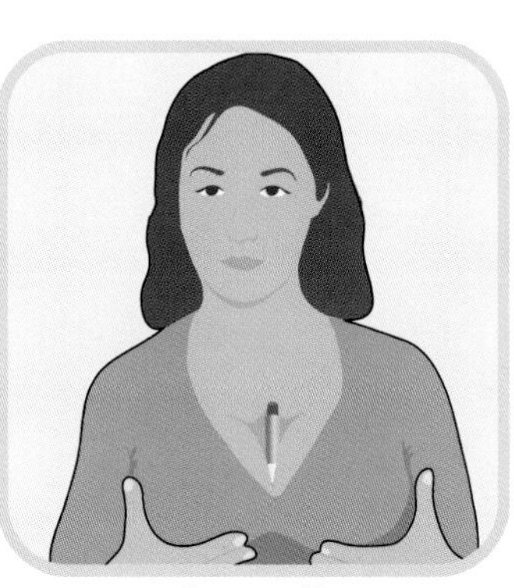
Compress; lift.

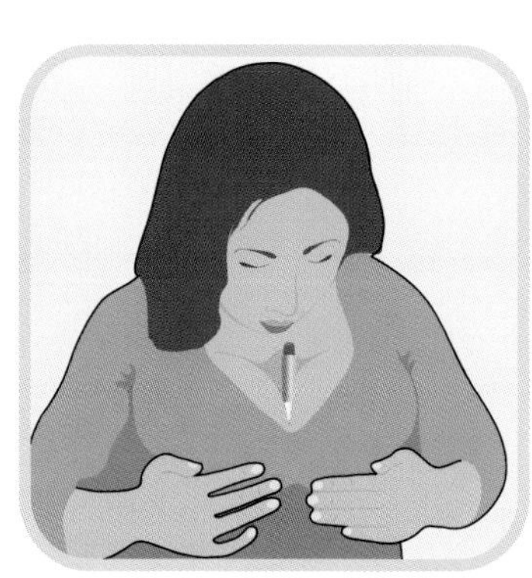
Apply to lower lip.

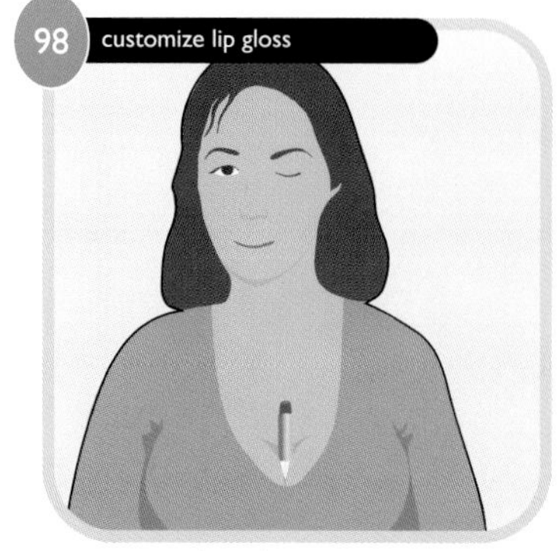
98 customize lip gloss
Press lips together.

44 ditch a bad date

Excuse yourself from the table.

Add layers.

Buy clothes from a bystander.

Alter your hairdo.

Change your makeup.

Don hat; lose glasses.

Don't make eye contact.

Give the waiter your share of the bill.

propose from jail 45

Stop by local jail; call her for help.

Say you're a prisoner . . .

. . . of love.

Give her the ring.

crash a wedding

Look your best.

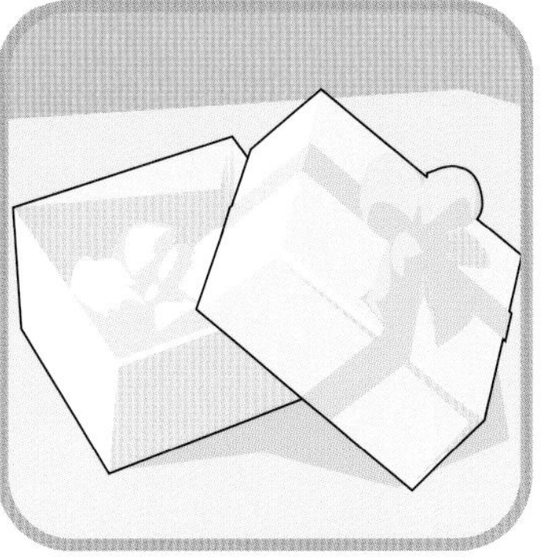
Carry an empty gift box.

Head to a fancy hotel.

Check lobby for wedding signs.

Stay at the bar; don't mingle.

Step out during seated meal.

Avoid the bride and groom.

Dance with the elderly for goodwill.

47 fake an antiquity

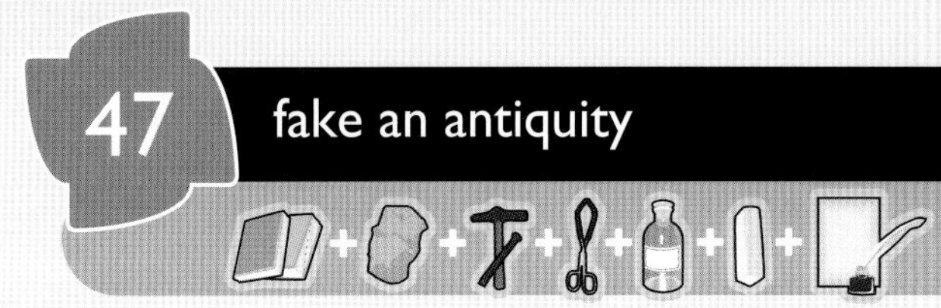

Study art history.

Get stone at ancient quarry.

Sculpt with period tools.

Dip in hydrochloric acid.

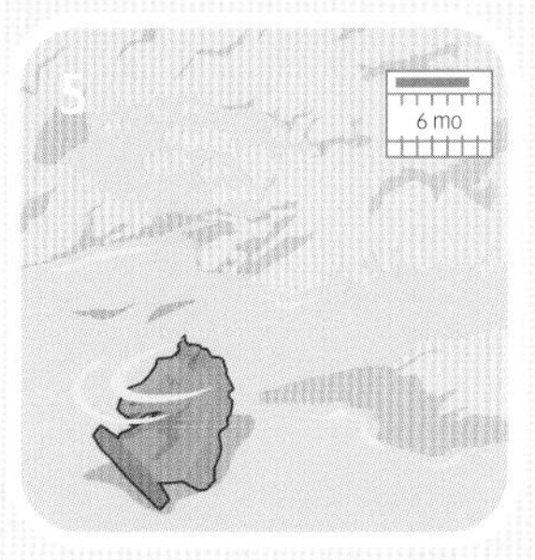

Leave in river to add limescale.

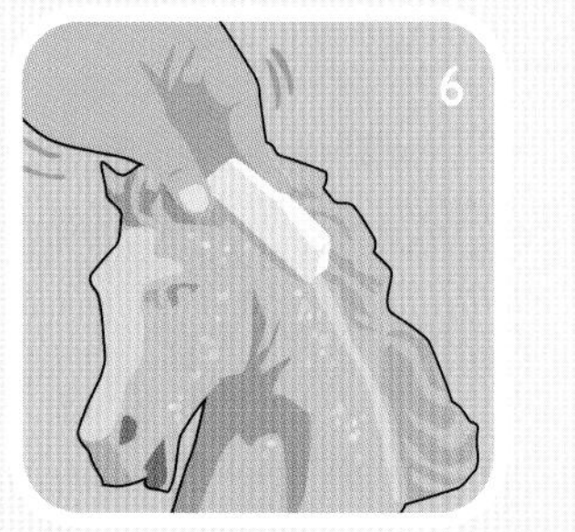

Coat in wax to foil UV test.

Document object's history.

Add to museum archives.

48 improvise an art show

Art is everywhere . . . so why not celebrate the beauty of your surroundings with a cheeky DIY art show? No gallery needed!

Hang labels on posters, pay phones, manholes—whatever piece of "art" catches your eye.

coatrack

hang my art in a museum

Pick a spot for your painting.

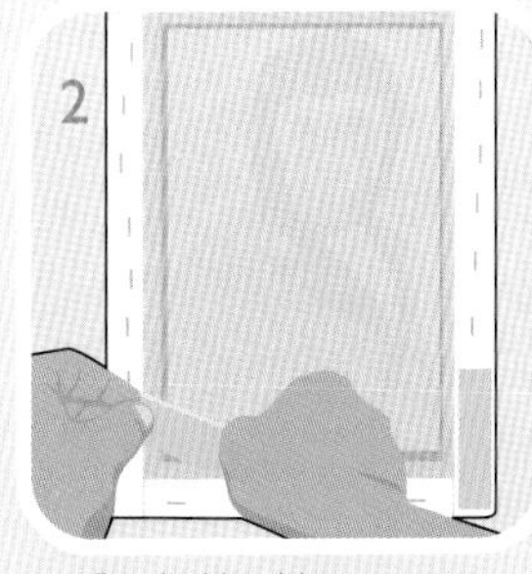

Put double-stick tape on art.

Pack art; dress like a tourist.

Cover your art with a map.

Peel off tape backing.

Holding map, press art to wall.

make a paint-by-number 83

Document your achievement.

Exit casually.

50 catch a table-talker

When two card players are in cahoots, they'll sometimes signal each other with subtle hand gestures. Learn these positions and catch them red-handed.

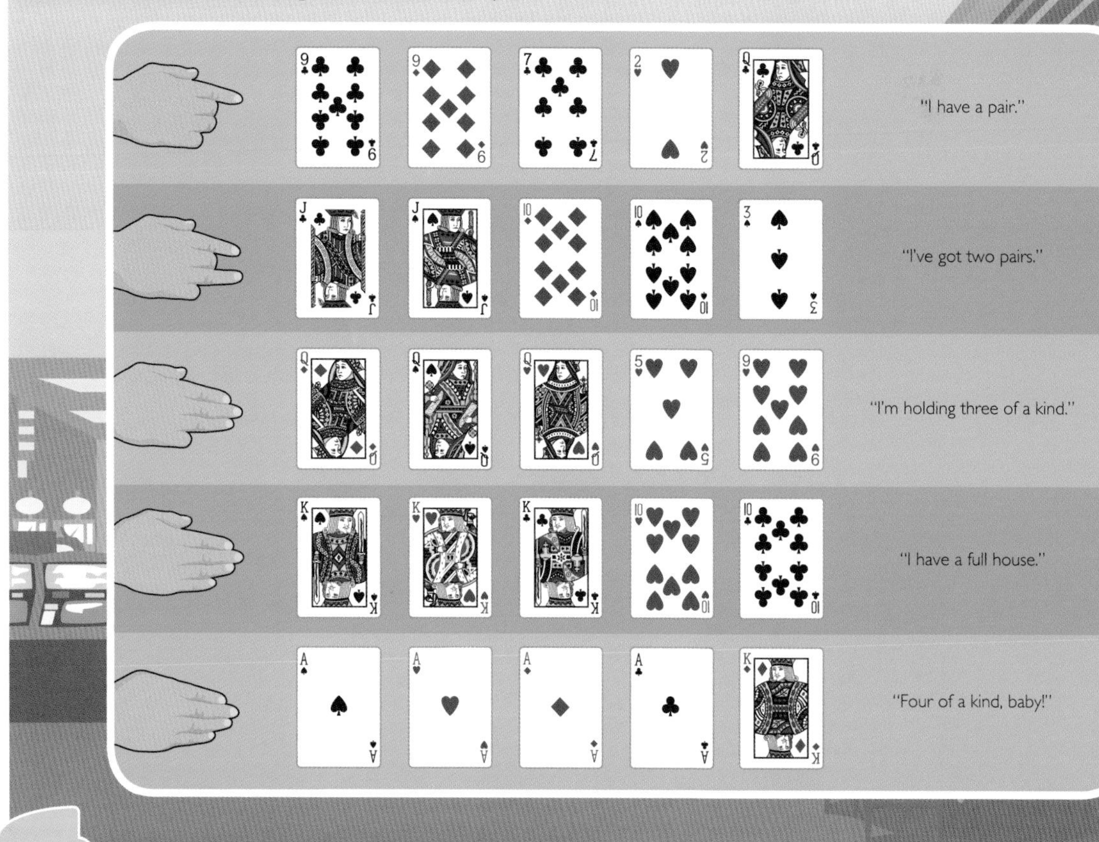

51 make loaded dice

Choose your lucky number.

Heat with number facing up.

Dice will melt slightly.

Bet on your number.

excessive chit-chat
good hand
throbbing forehead vein
good hand
slouching posture
weak hand
avoiding eye contact
weak hand

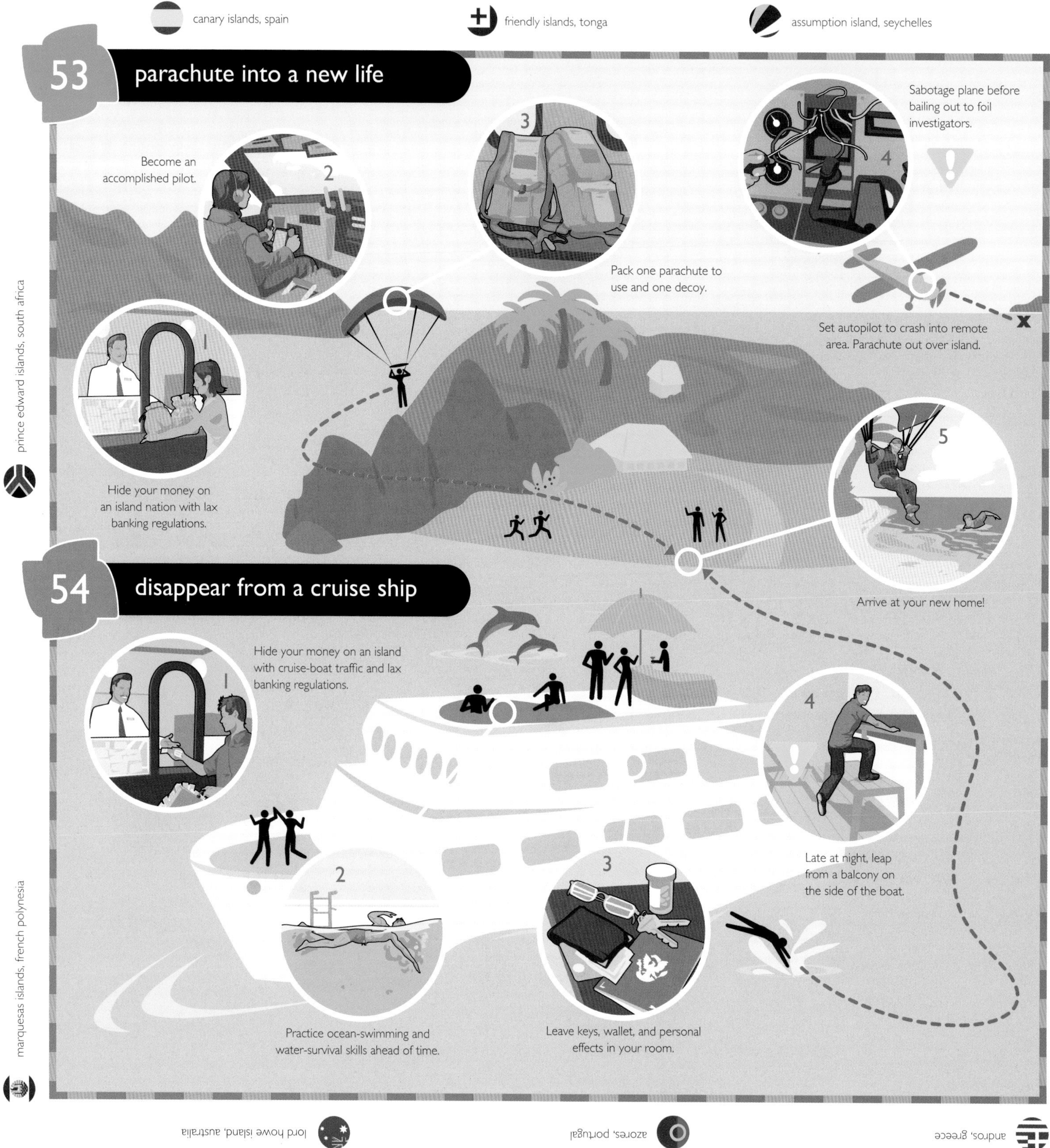
canary islands, spain
friendly islands, tonga
assumption island, seychelles
53
parachute into a new life
Become an accomplished pilot.
2
3
Sabotage plane before bailing out to foil investigators.
4
Pack one parachute to use and one decoy.
1
Set autopilot to crash into remote area. Parachute out over island.
Hide your money on an island nation with lax banking regulations.
5
Arrive at your new home!
prince edward islands, south africa
54
disappear from a cruise ship
Hide your money on an island with cruise-boat traffic and lax banking regulations.
1
4
Late at night, leap from a balcony on the side of the boat.
2
3
Practice ocean-swimming and water-survival skills ahead of time.
Leave keys, wallet, and personal effects in your room.
marquesas islands, french polynesia
lord howe island, australia
azores, portugal
andros, greece

abscond in a suitcase

Cut large notches in ends of suitcases.

Cover the holes with stickers.

hop a train 303

Arrange a distraction; sneak on the train.

Set up with the holes facing each other.

Climb up and close yourself in.

Stay quiet during the ride.

fool a polygraph

* During control questions, use these tricks to raise your vital signs. Your lies will be less obvious.

Mentally do complex math.

Hide tack in shoe; poke toe.

Clench your bottom tightly.

Give one-word answers.

Have your lawyer present.

create

57 twist a paper boutonniere

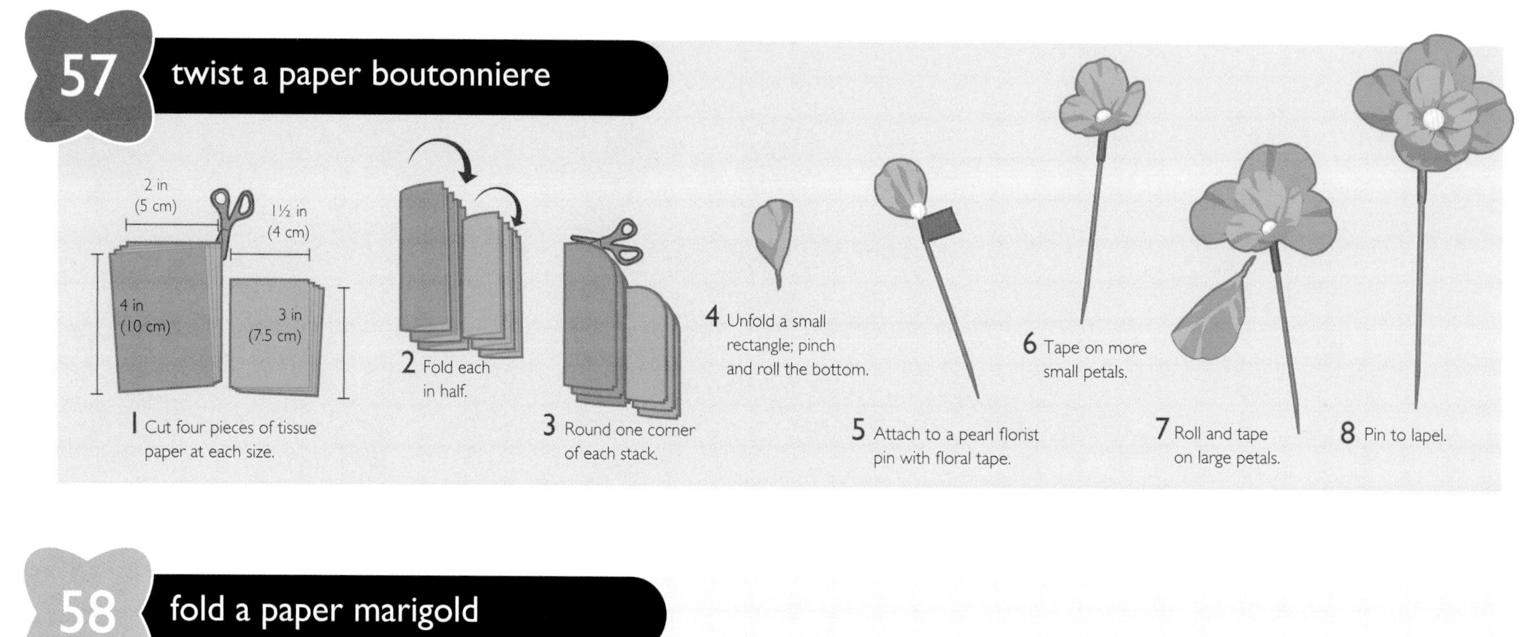

58 fold a paper marigold

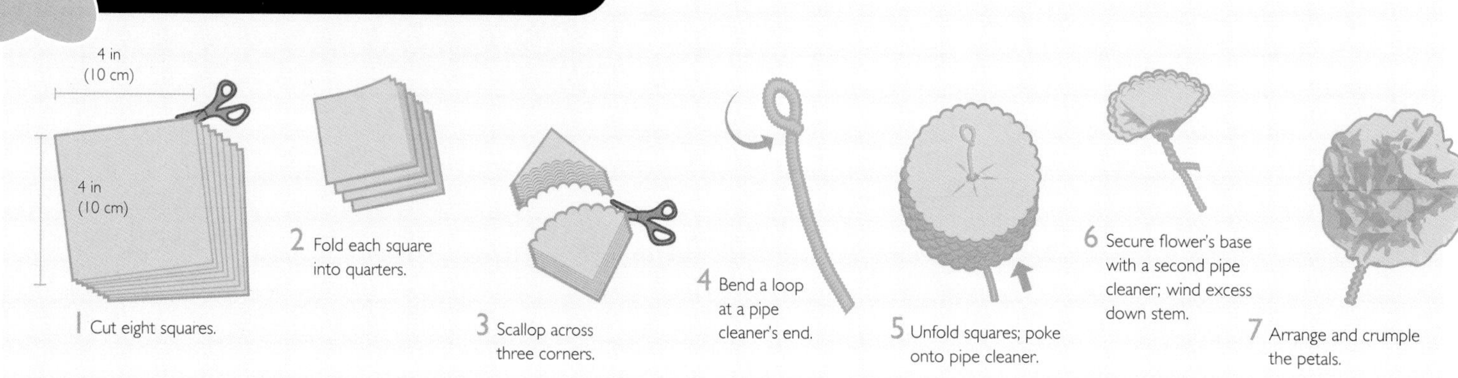

59 wrap up a duct-tape rose

345 create a sashimi rose

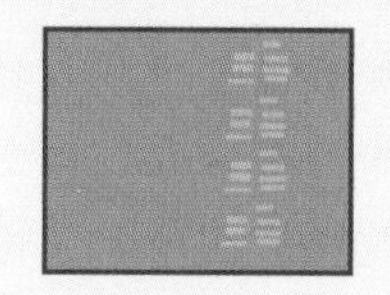

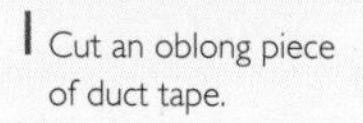

1 Cut an oblong piece of duct tape.

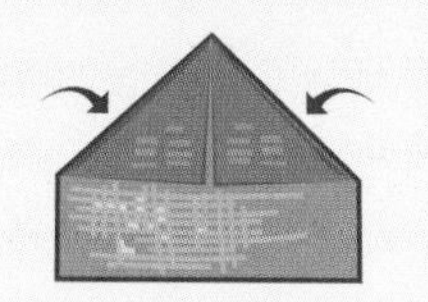

2 Fold two corners down.

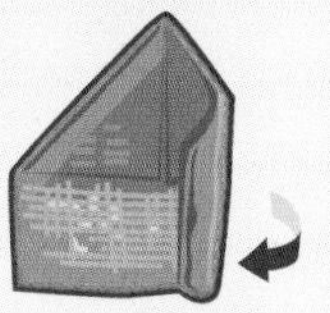

3 Loosely roll.

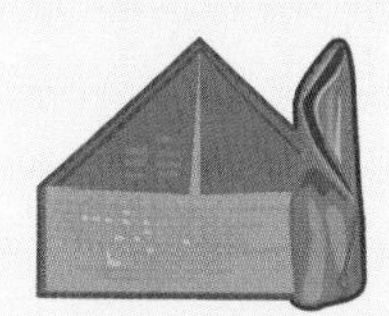

4 Cut and fold a new petal; wrap it loosely around first roll.

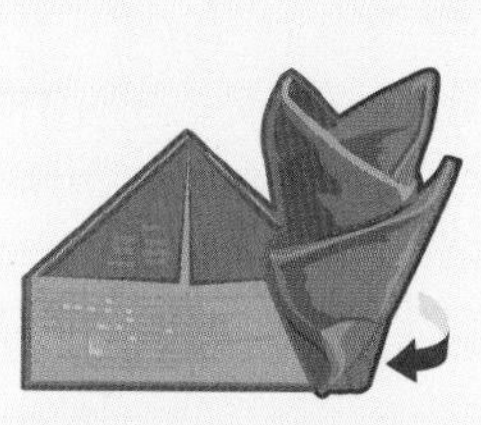

5 Fold and wrap new petals, pulling the points out as you go.

6 Continue to desired size.

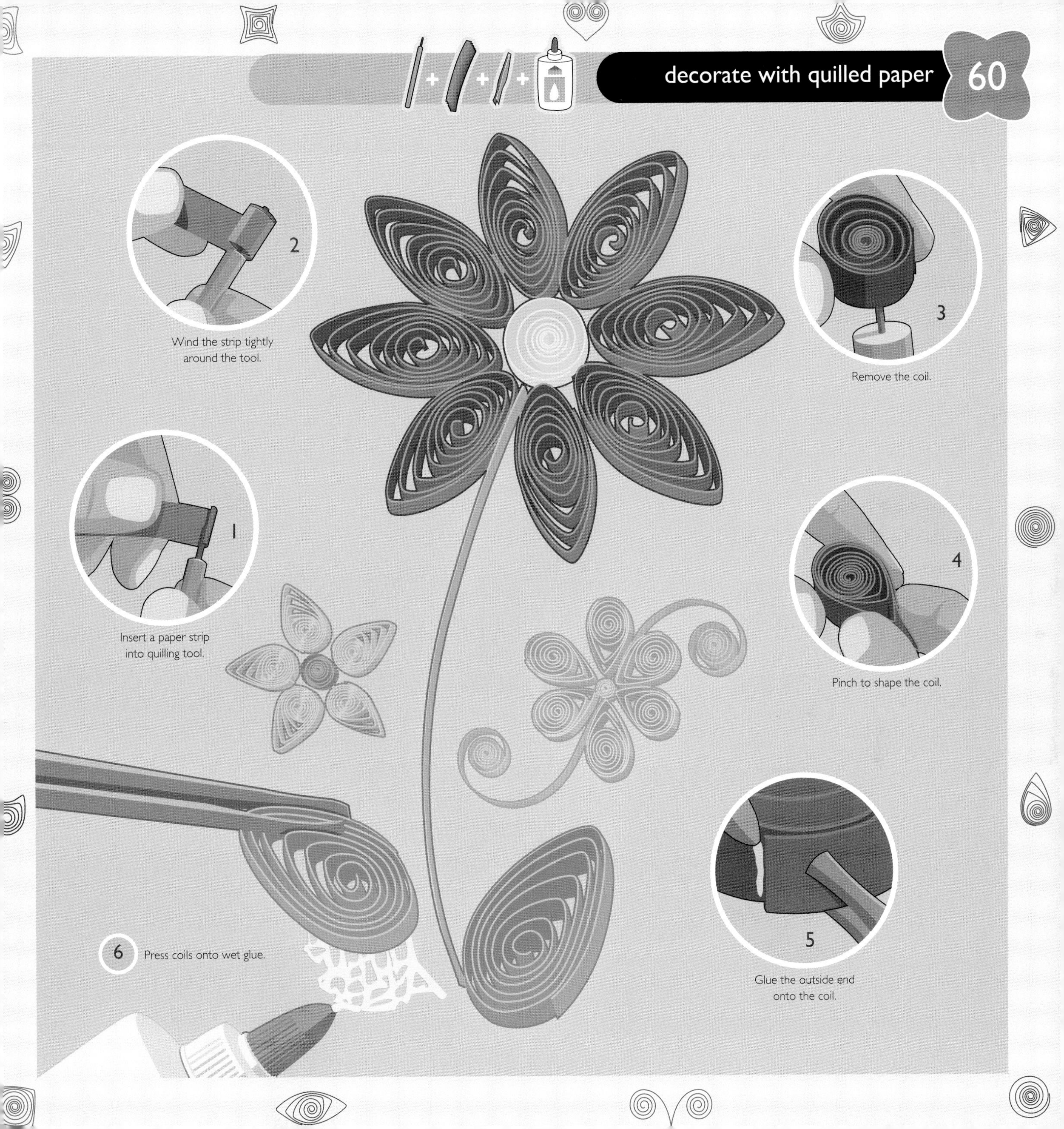

1
Insert a paper strip into quilling tool.
2
Wind the strip tightly around the tool.
3
Remove the coil.
4
Pinch to shape the coil.
5
Glue the outside end onto the coil.
6
Press coils onto wet glue.

61 make a linoleum-block print

Sketch onto a linoleum block.

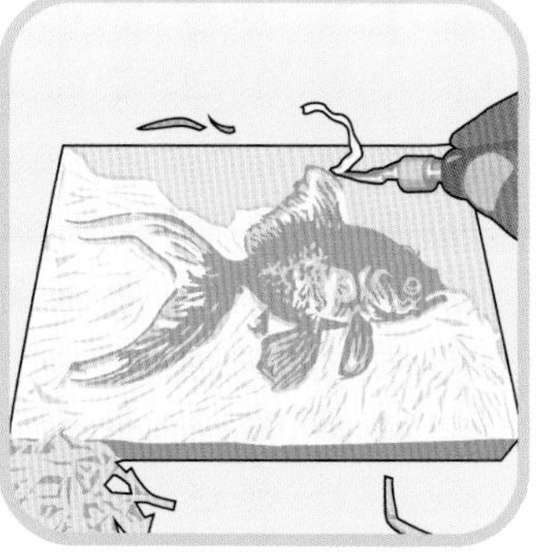

Carve out around the drawing.

Roll ink to coat roller evenly.

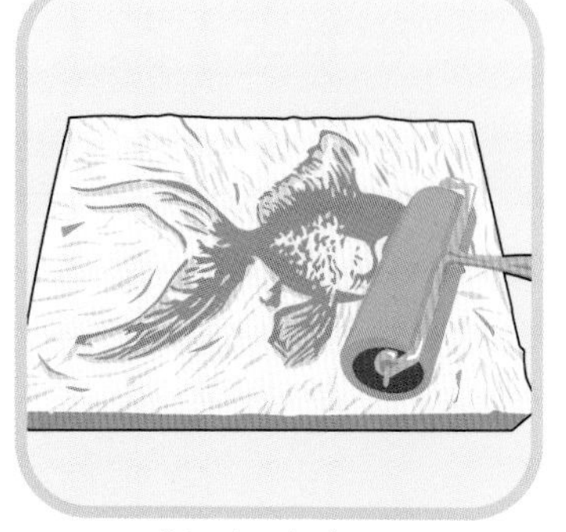

Paint the raised area.

Set paper on the block.

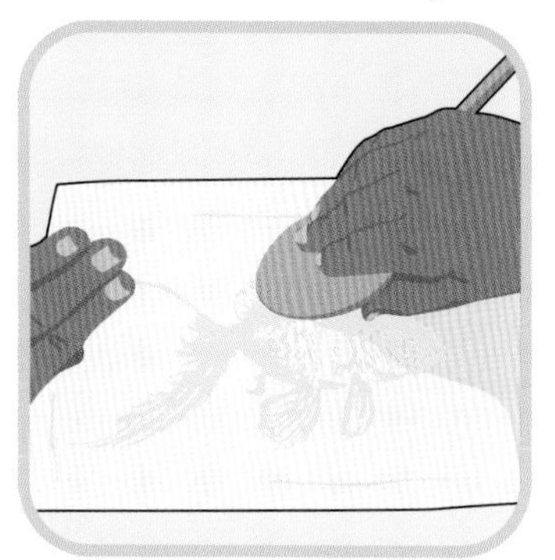

Rub with a spoon to transfer image.

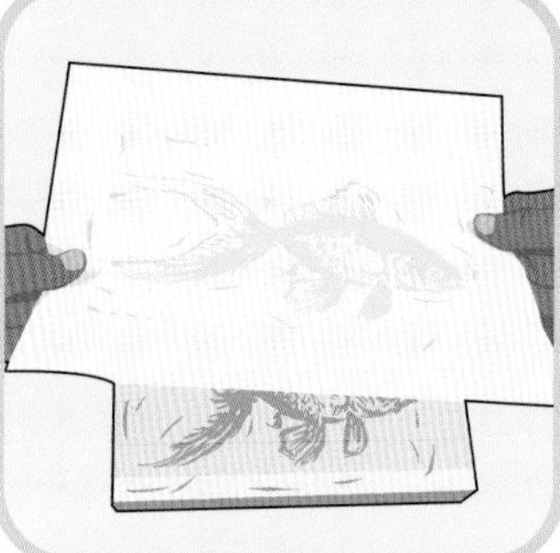

Gently lift paper straight up.

Hang to dry.

62 blow bubble prints

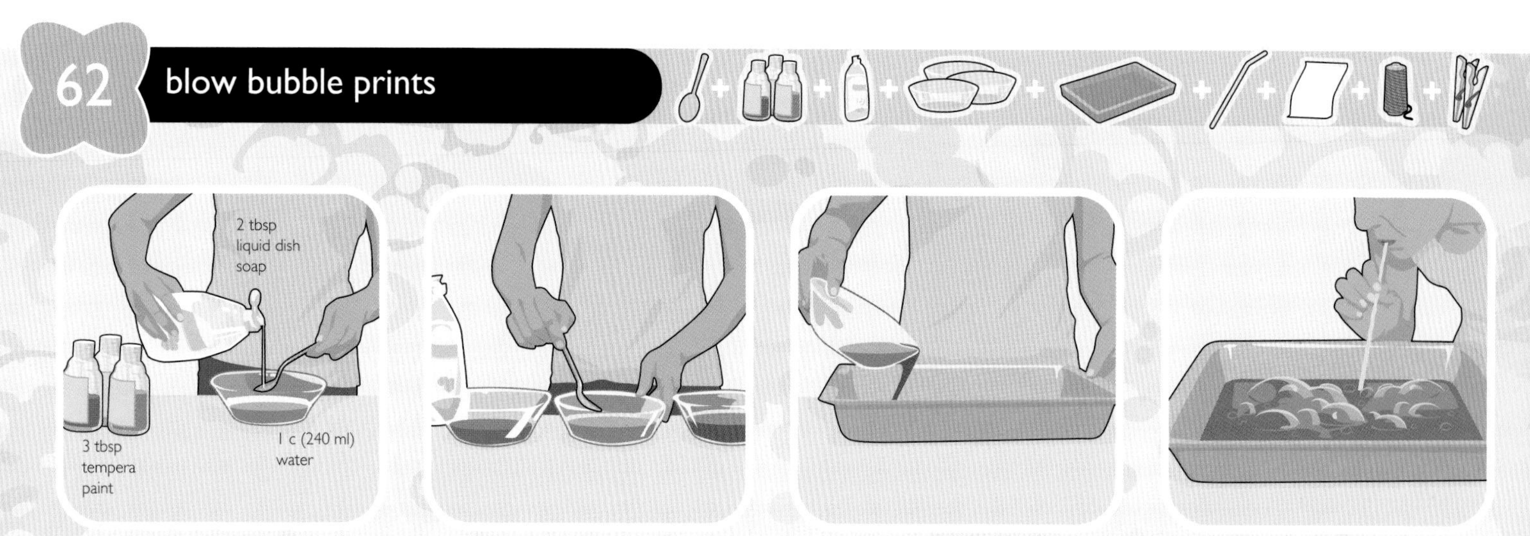

Mix one color with water and soap.

Repeat with additional colors.

Pour one color mix into a pan.

Blow a thick layer of bubbles.

Cut squares of black and white paper.

Tape one piece of each together.

Remove stems from mushrooms.

Center caps on papers.

Cover with bowls; let sit.

Remove bowls and mushroom caps.

Spray with hair spray to set.

gild an old picure frame 193

Display your spore decor.

Every species of mushroom creates a signature spore print, with a unique shape, pattern, and color. Some will show better on dark paper, and some on light.

Pop bubbles onto paper.

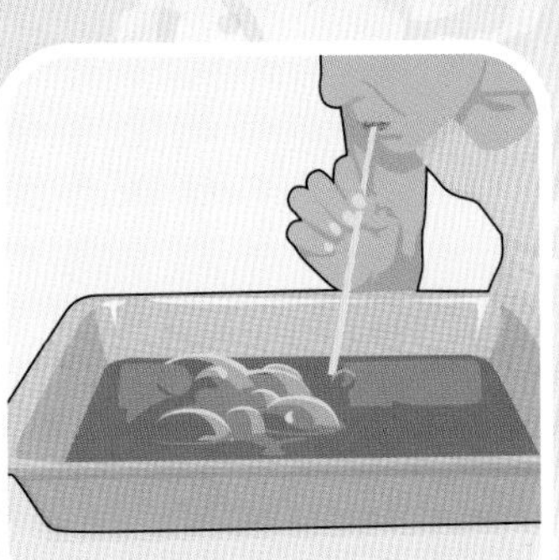

Agitate next color.

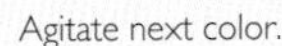

Drop bubbles on paper for new effect.

Repeat with third color; hang to dry.

64 turn t-shirts into quilt squares

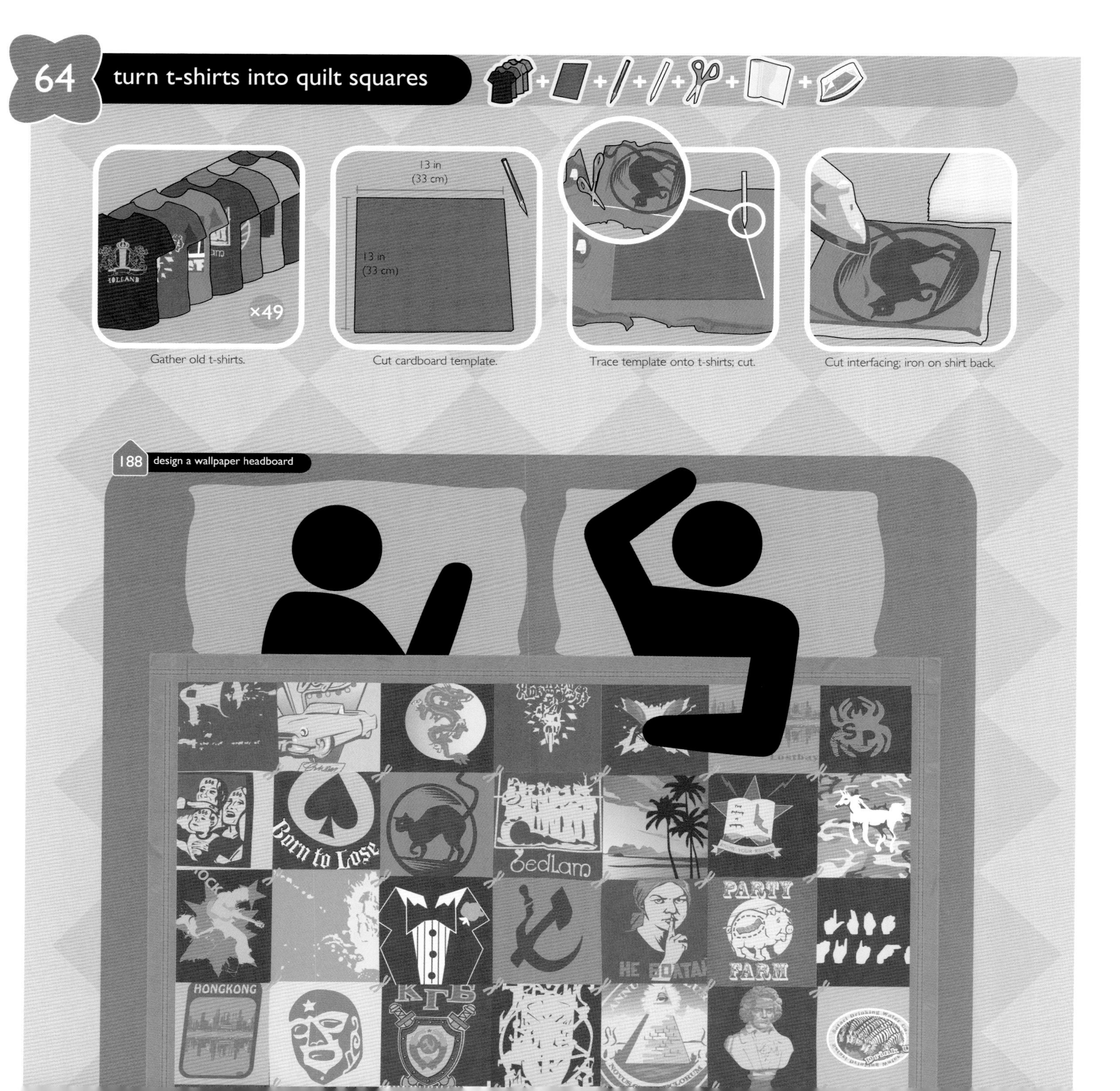

Gather old t-shirts.

Cut cardboard template.

Trace template onto t-shirts; cut.

Cut interfacing; iron on shirt back.

188 design a wallpaper headboard

65 sew a rockin' quilt

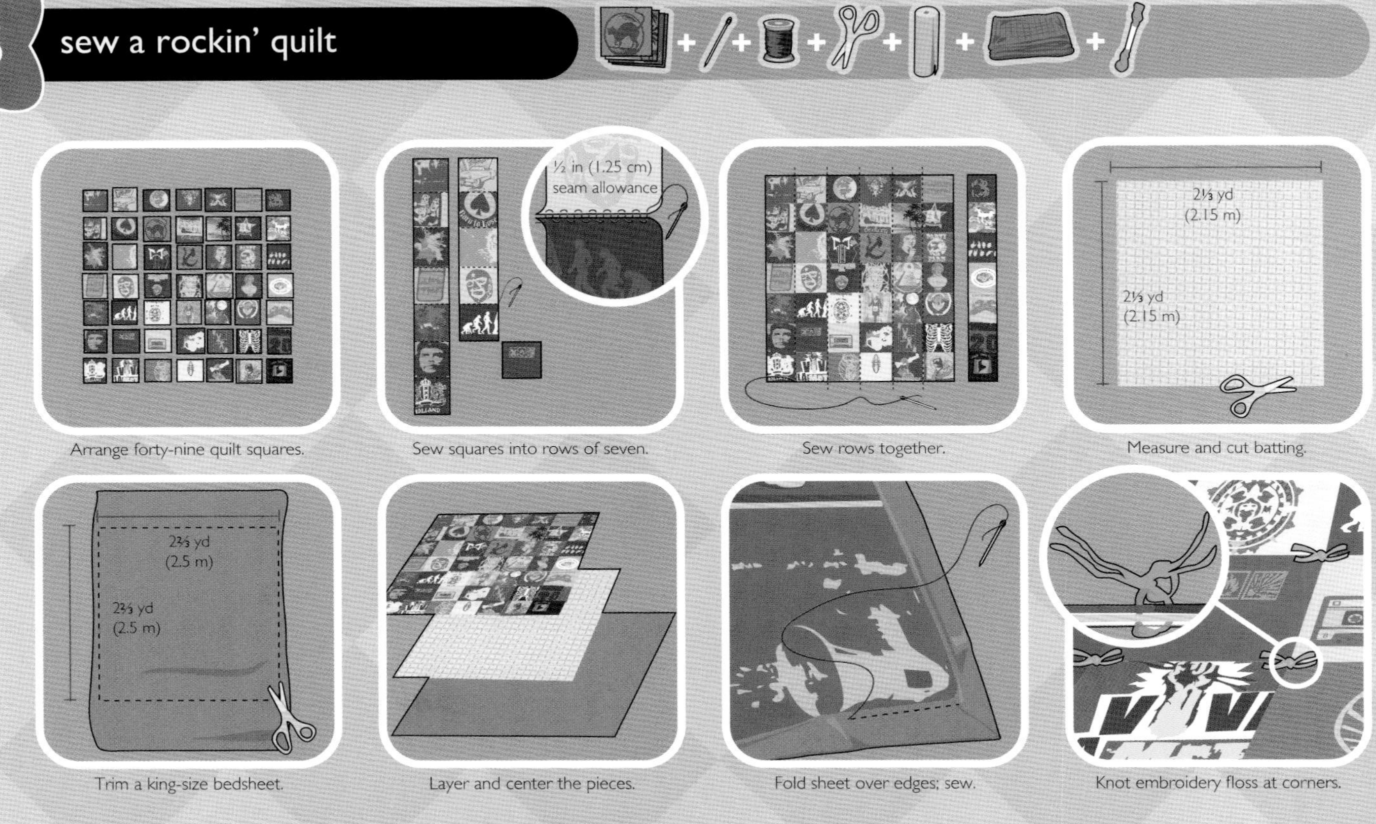

Arrange forty-nine quilt squares.

Sew squares into rows of seven.

Sew rows together.

Measure and cut batting.

Trim a king-size bedsheet.

Layer and center the pieces.

Fold sheet over edges; sew.

Knot embroidery floss at corners.

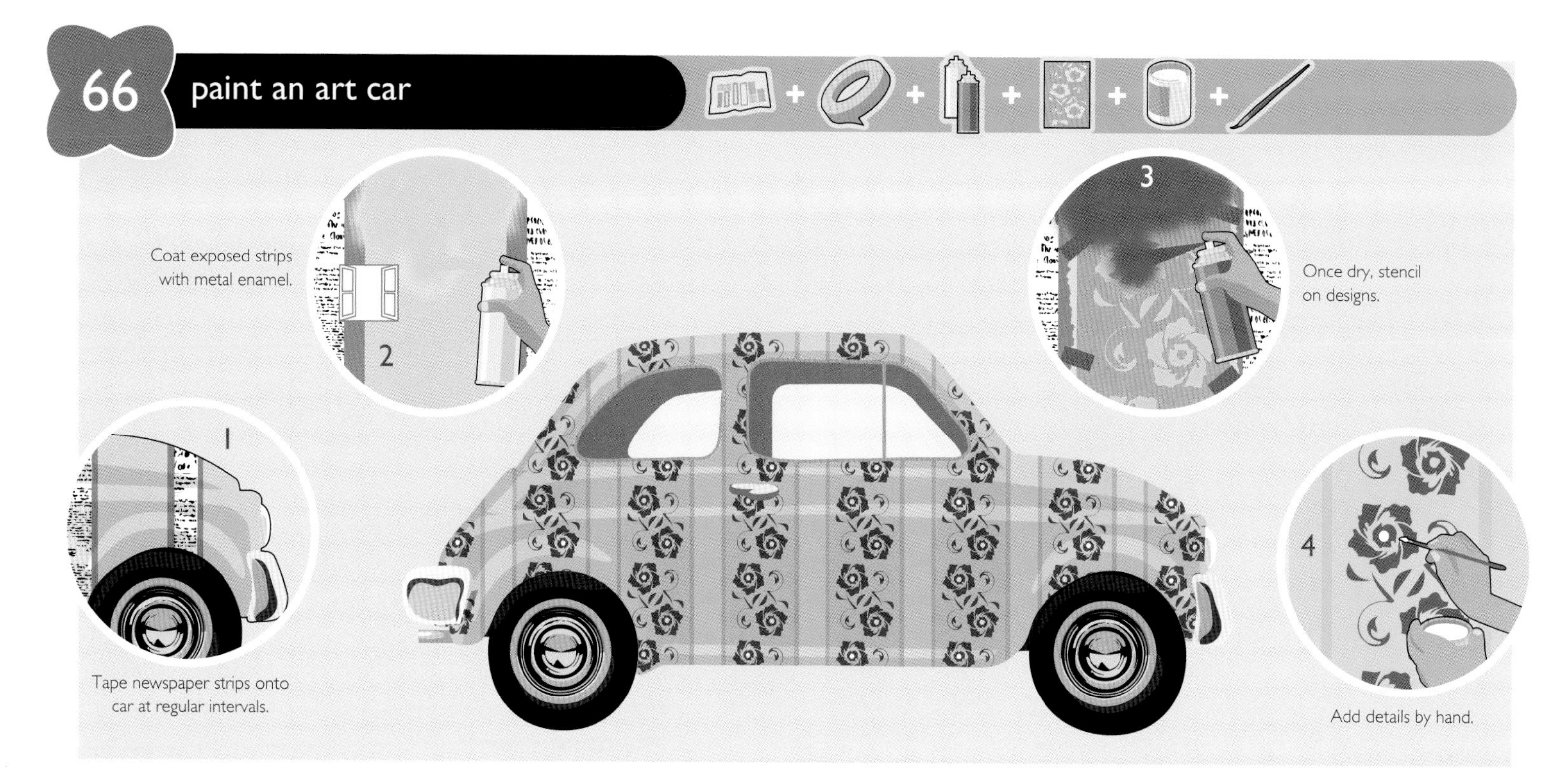
66
paint an art car
1
2
3
4
Coat exposed strips with metal enamel.
Once dry, stencil on designs.
Tape newspaper strips onto car at regular intervals.
Add details by hand.

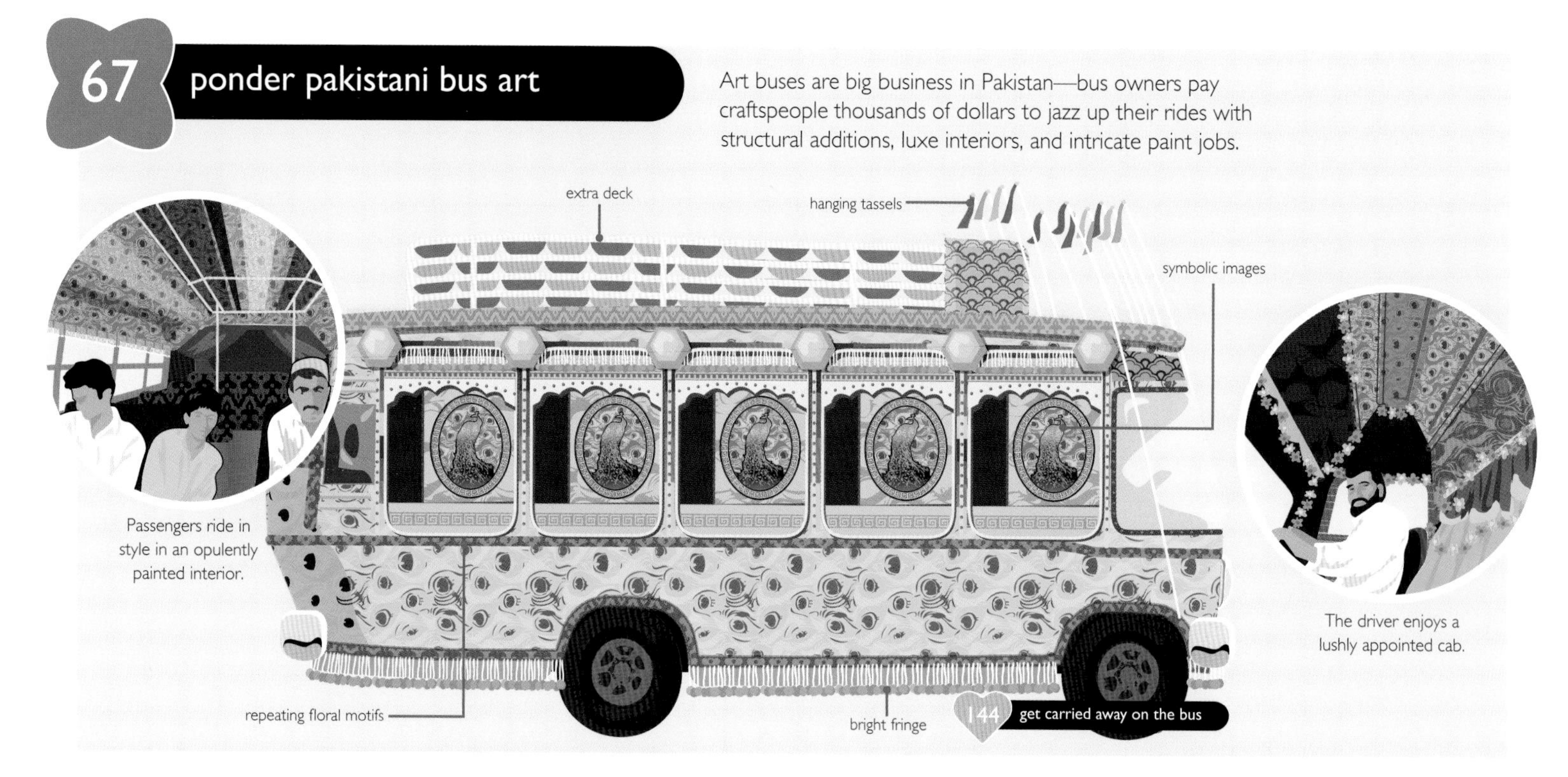
67
ponder pakistani bus art
Art buses are big business in Pakistan—bus owners pay craftspeople thousands of dollars to jazz up their rides with structural additions, luxe interiors, and intricate paint jobs.
extra deck
hanging tassels
symbolic images
Passengers ride in style in an opulently painted interior.
The driver enjoys a lushly appointed cab.
repeating floral motifs
bright fringe
144 get carried away on the bus

68 construct a gluey

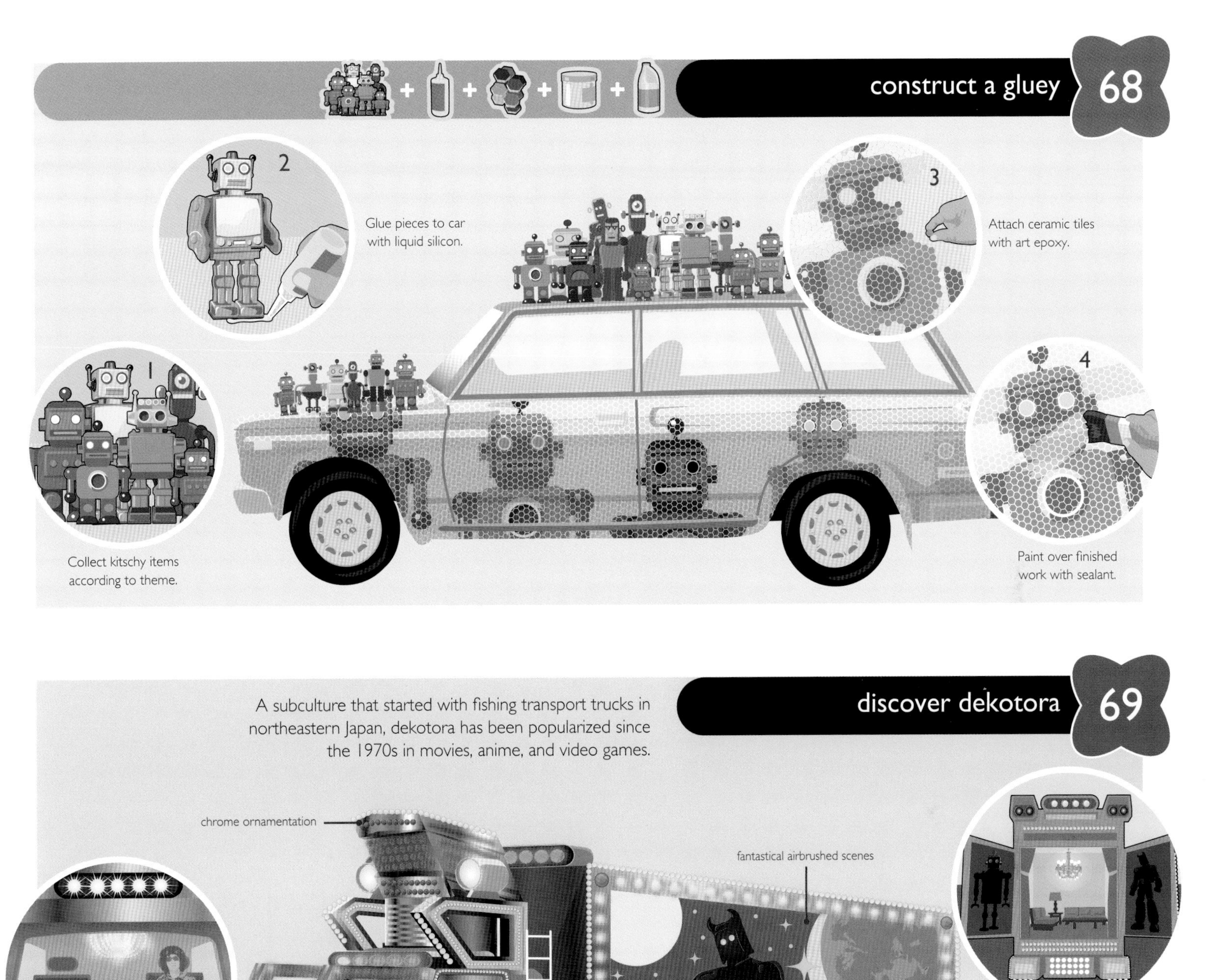

69 discover dekotora

A subculture that started with fishing transport trucks in northeastern Japan, dekotora has been popularized since the 1970s in movies, anime, and video games.

chrome ornamentation

fantastical airbrushed scenes

hundreds of lights flashing in sequence

A chandelier lights the driver's way.

The living quarters inside are as blinging as the exterior.

70 make a matchbook flashlight

39 bring a match to life

Remove the matches.

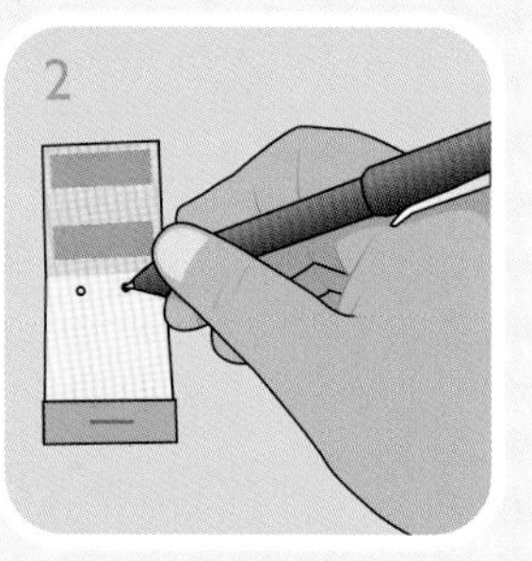

Mark two dots in the fold.

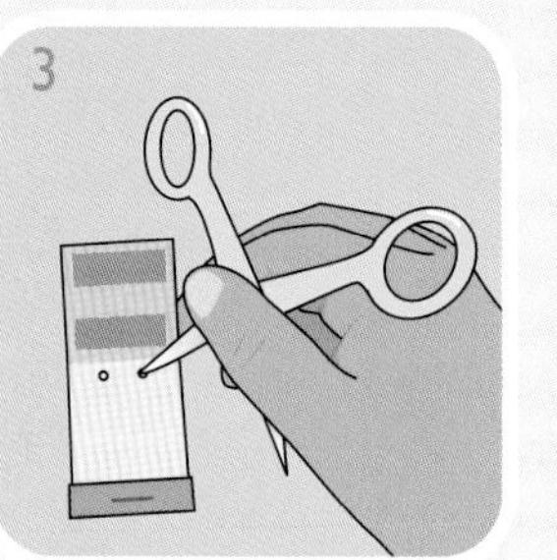

Poke out the dots.

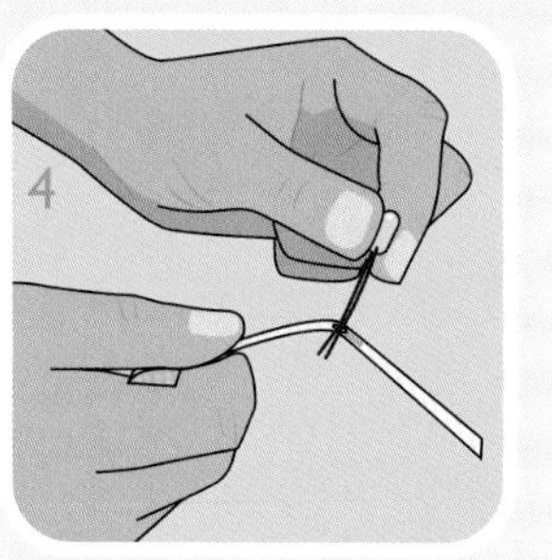

Insert two LED wires.

Tape the left wires down.

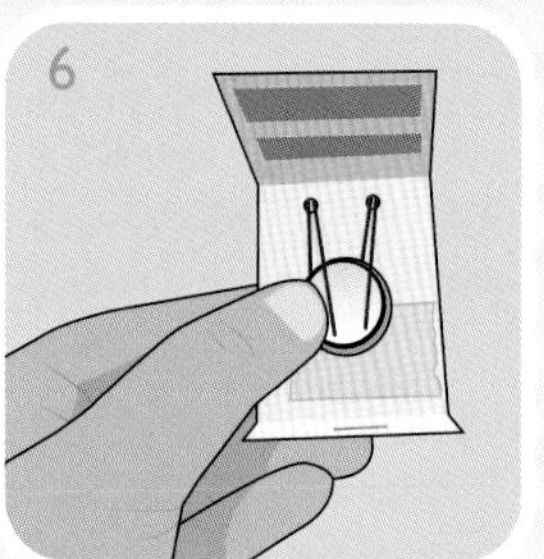

Set battery under loose wires.

Pinch to test the light.

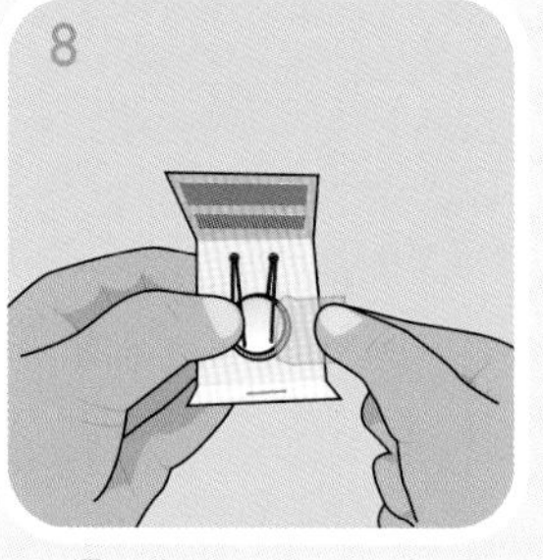

Tape battery edges down.

71 build a tiny robot

Glue fringe to salsa cup.

Put double-stick tape on top.

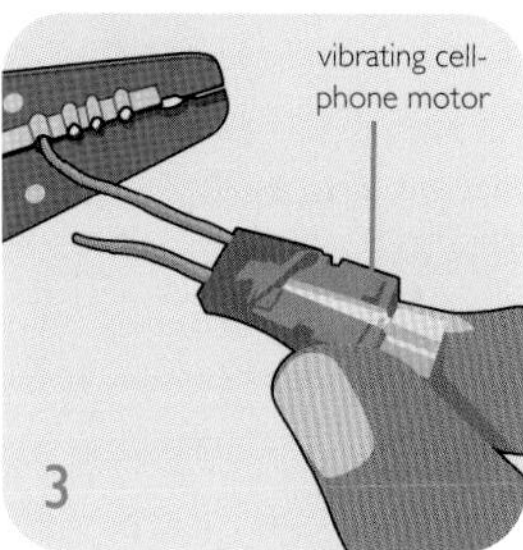
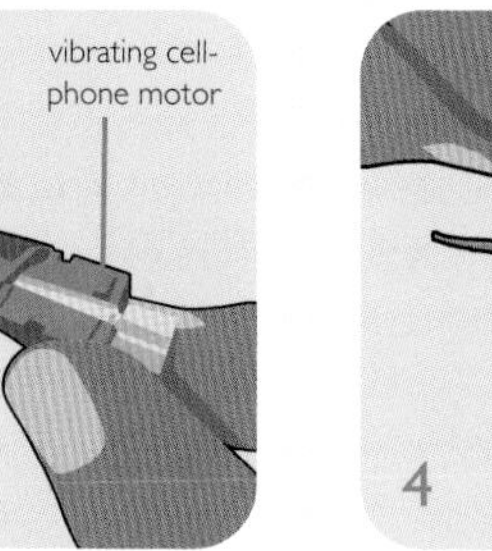

Strip wire tips.

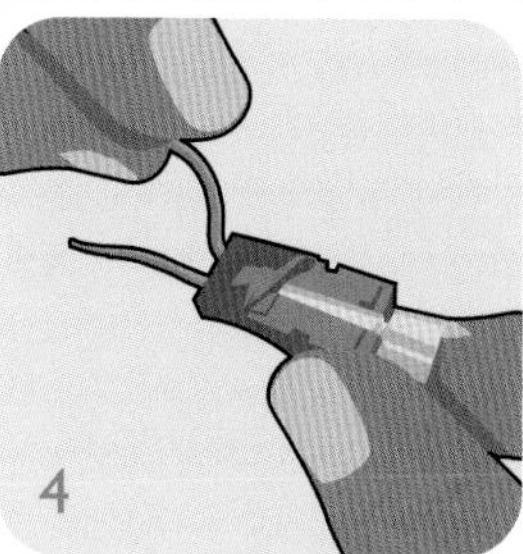

Bend one wire up.

Attach motor to cup.

Decorate your 'bot.

Slip battery between wires.

Watch him go!

hide a flash drive

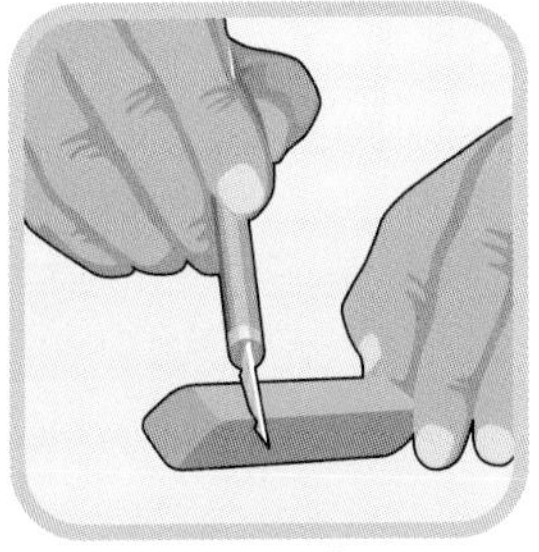

Cut one-third off eraser.

Hollow out both pieces.

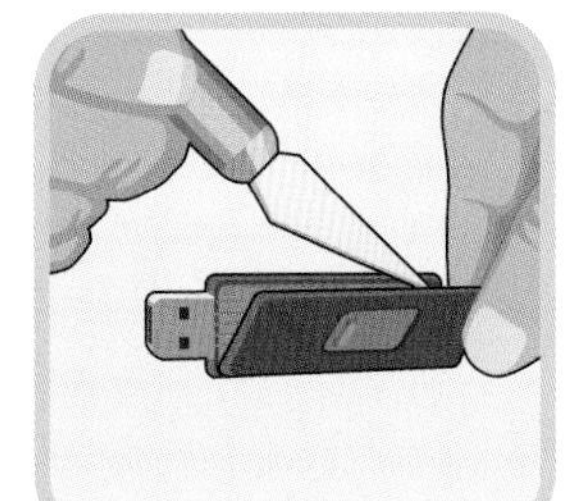

Slice casing off flash drive.

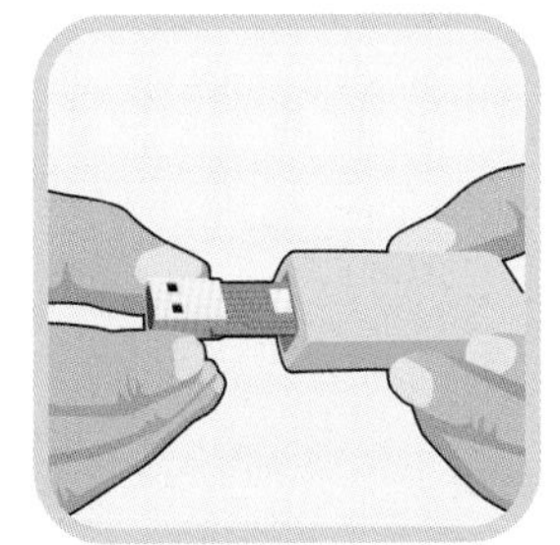

Insert flash drive board.

Conceal when not in use.

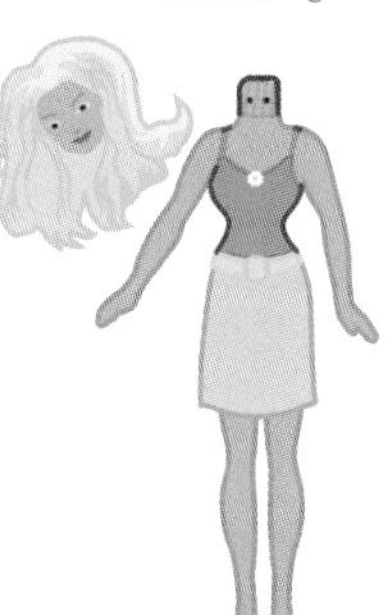

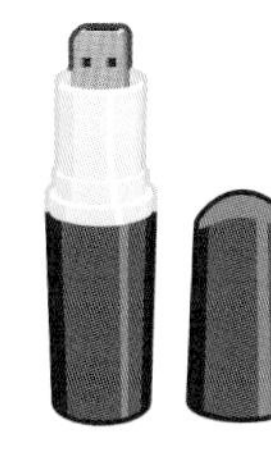

craft a metal detector 73

Apply double-stick tape.

Tape both in CD case.

Turn radio to AM band.

Tune to static at top of dial.

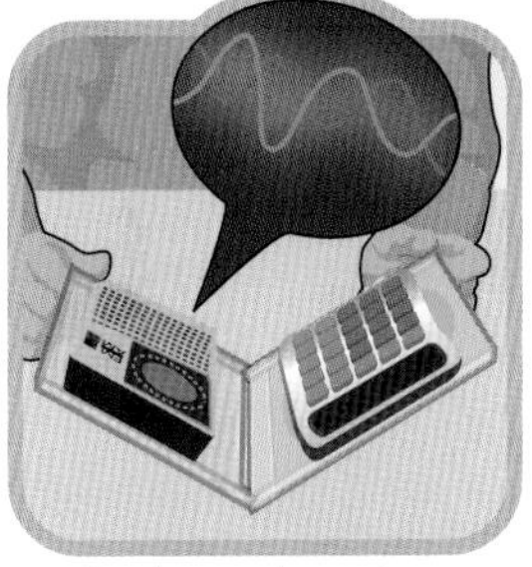

Bend case to hear a tone.

Unbend until tone stops.

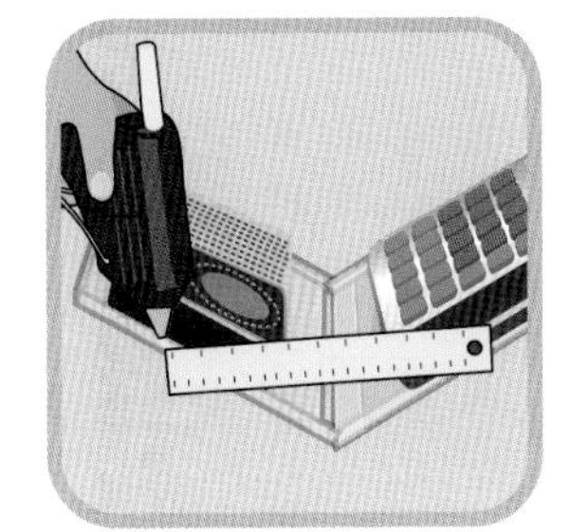

Brace case at that angle.

Hunt for valuable metal.

Tone resumes near metal.

Search for treasure!

74 rig a cat-feeding device

Why feed kitty the boring old way? A few short hours and some household odds and ends is all it takes to make a fantastical Rube Goldberg–style cat-feeding machine!

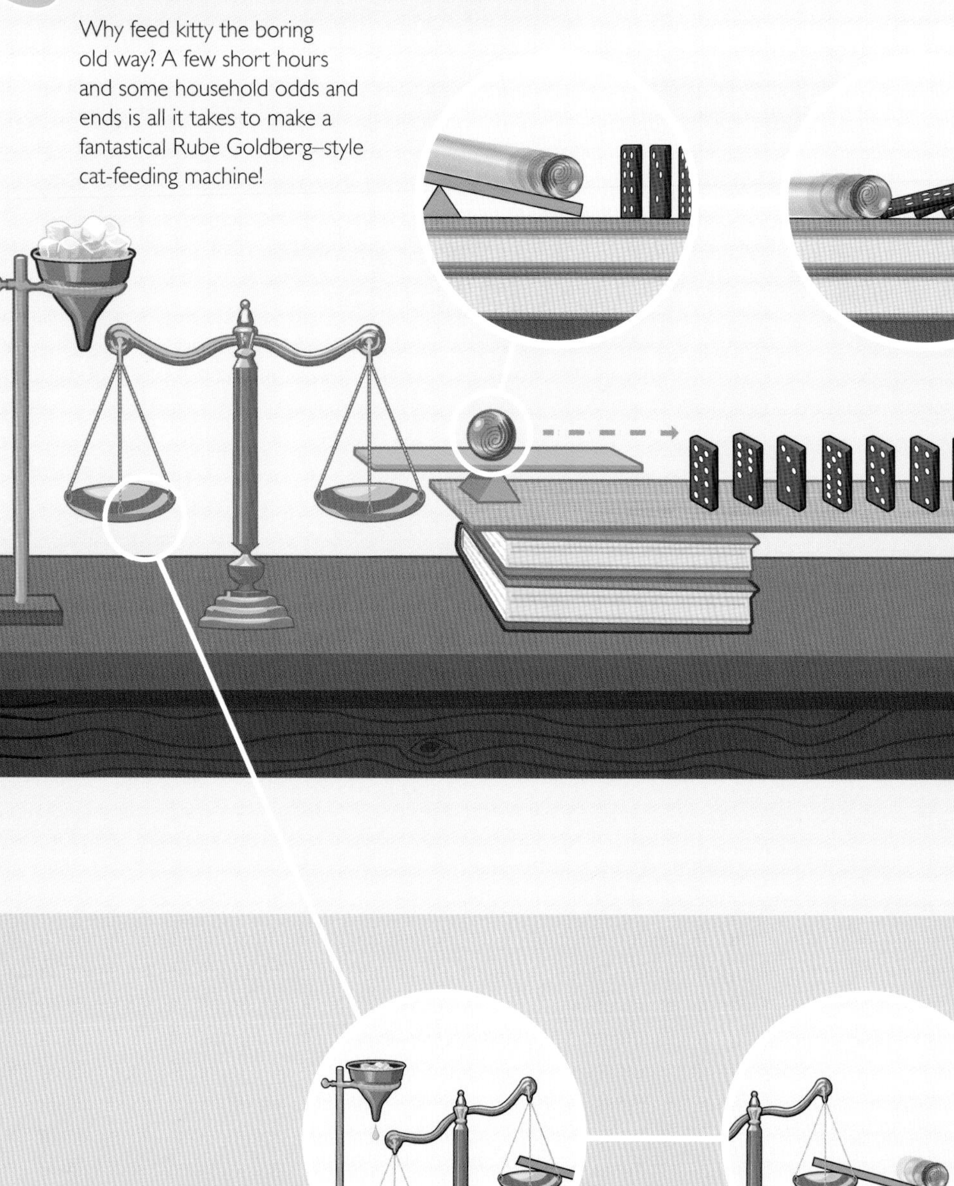

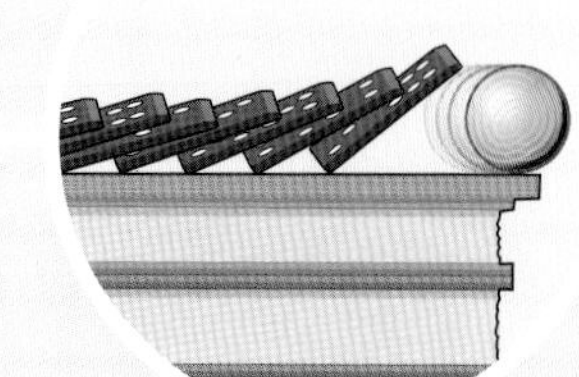

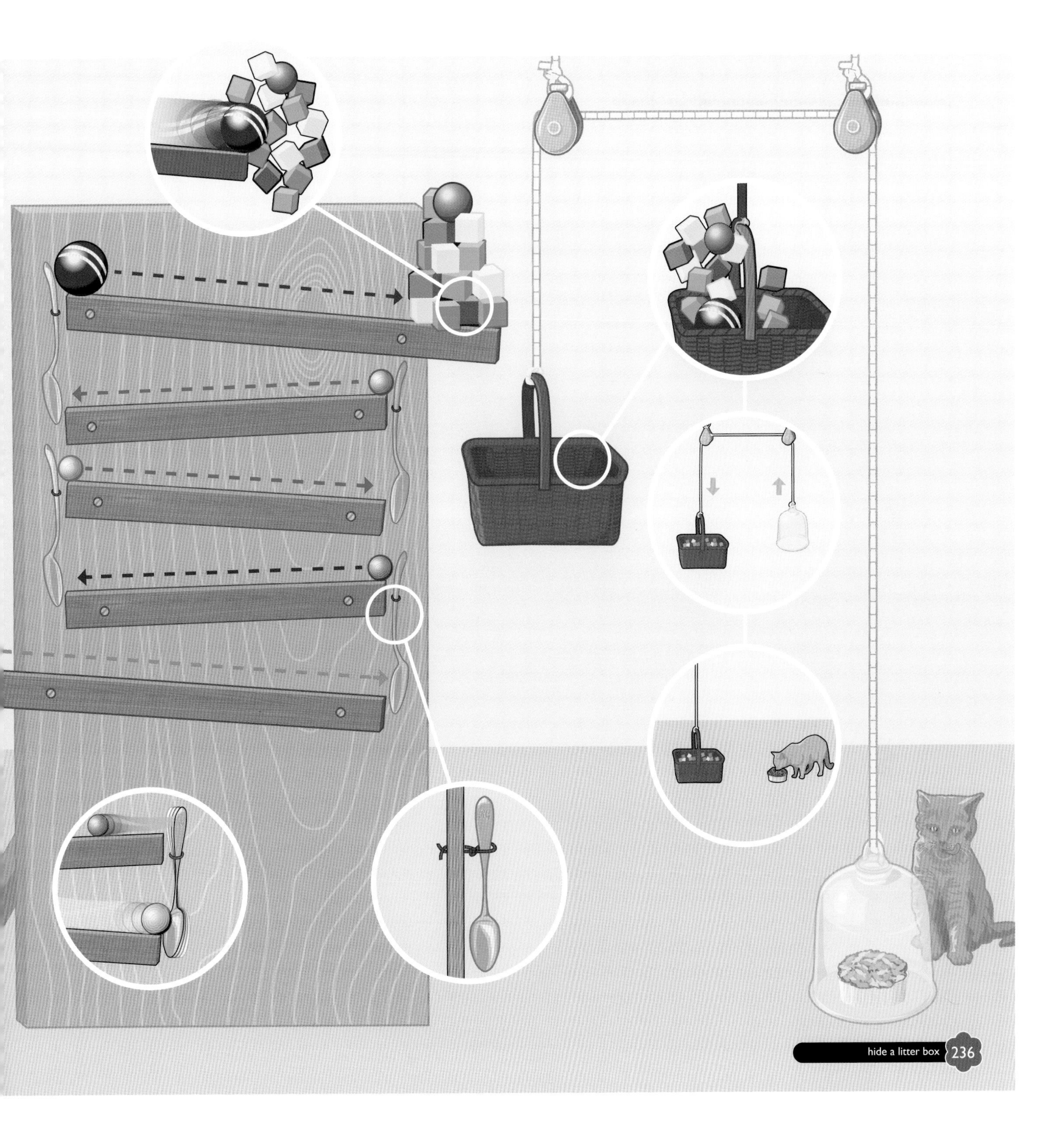

75 carve chainsaw art

Secure a log with sandbags.

Make a rough sketch in chalk.

Don safety equipment.

Roughly cut the basic shape.

Whittle off thin strips.

Burn details with a torch.

Refine features with saw tip.

Treat with teak oil.

76 craft dice coasters

Cut squares from plywood.

Sand to smooth edges.

Mark and drill divots.

Paint divots and trim.

Spray on waterproof finish.

77 put a ship in a bottle

1

Carve boat from a cork.

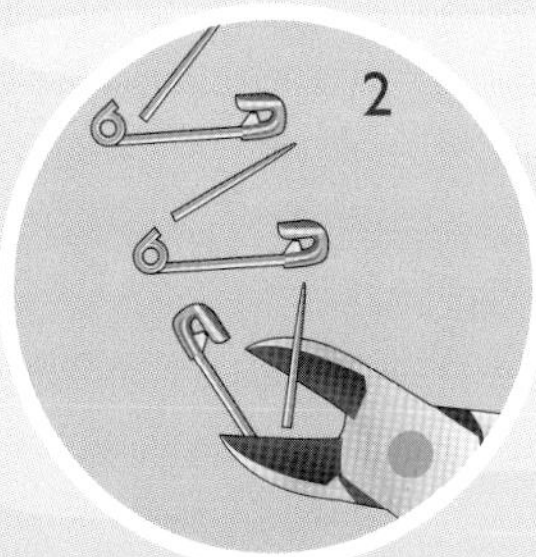

Cut off three safety-pin arms.

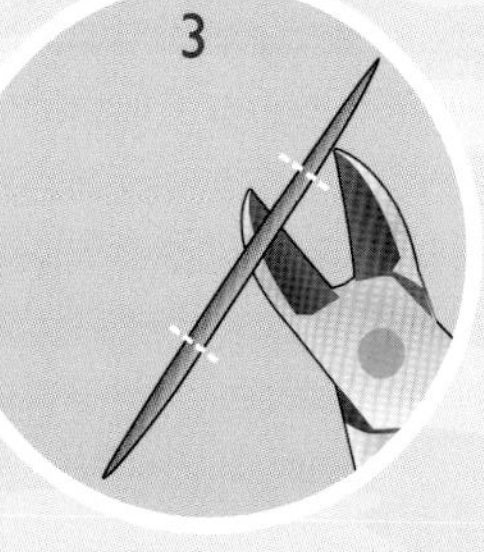

Cut bamboo skewer into thirds.

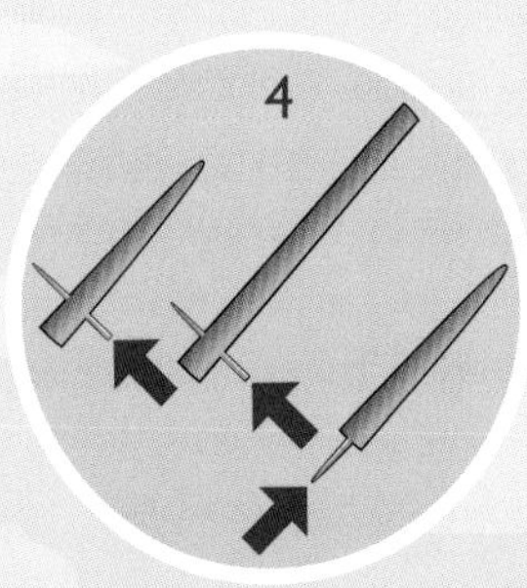

Insert arms into a skewer.

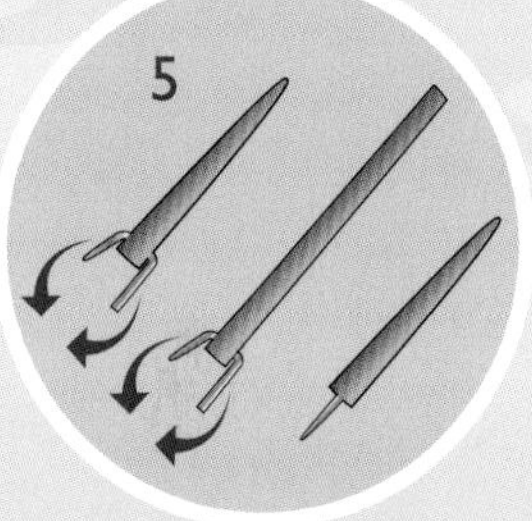

Bend two into hooks.

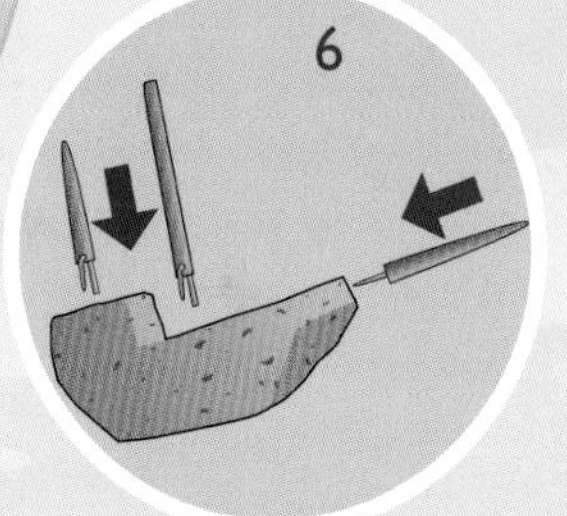

Attach to boat.

Design sails; cut them from paper.

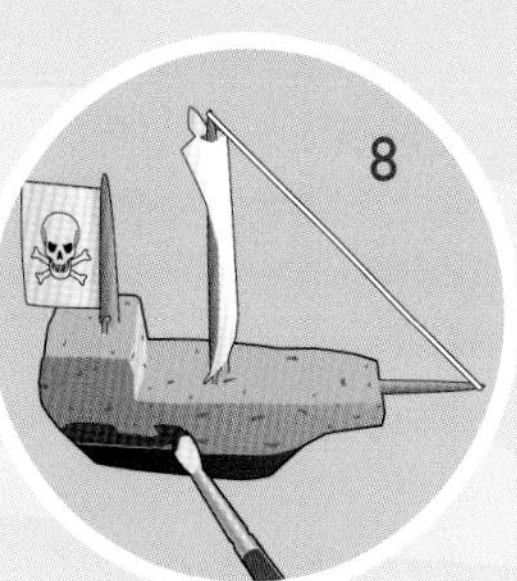

Attach sails and decorate ship.

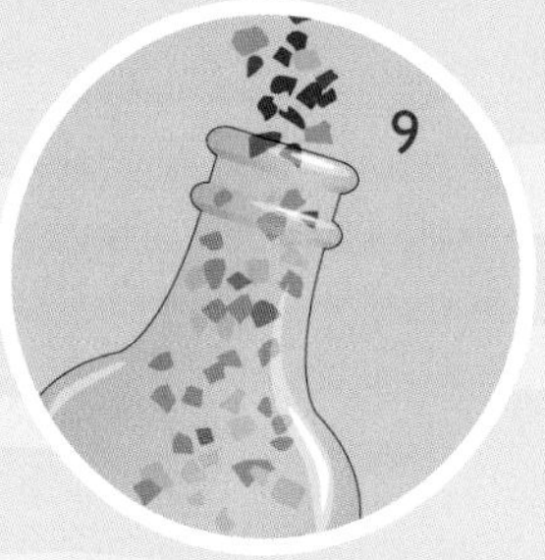

Add colored gravel to bottle.

10

Bend down masts; insert the ship into the bottle.

11 Pinch masts and push upright. Bon voyage!

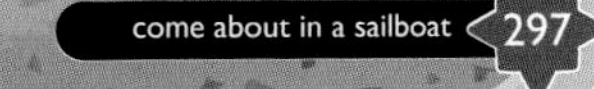
come about in a sailboat 297

78 build a bead loom

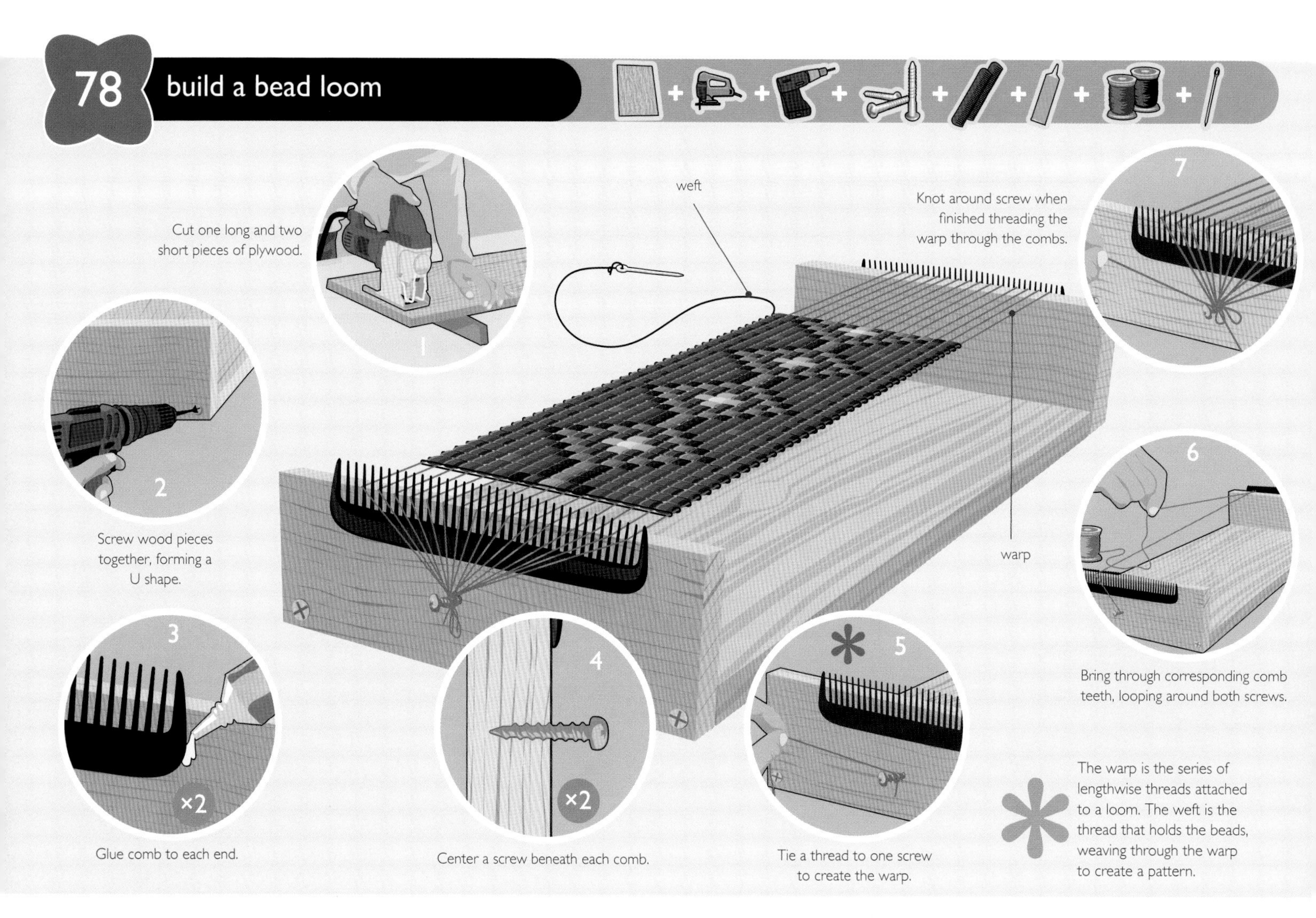

80 string a safety-pin bracelet

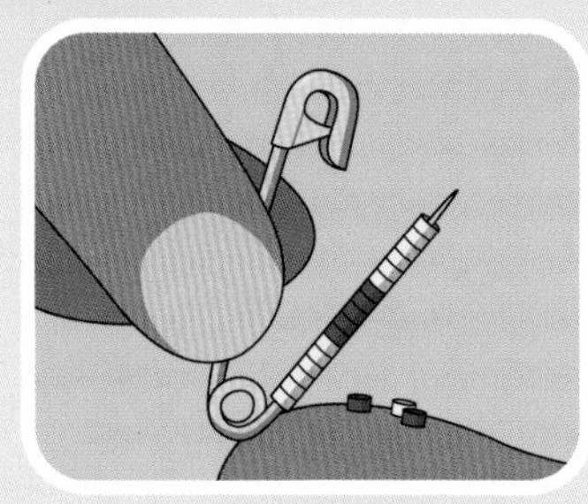
Fill a pin arm with beads; close pin.

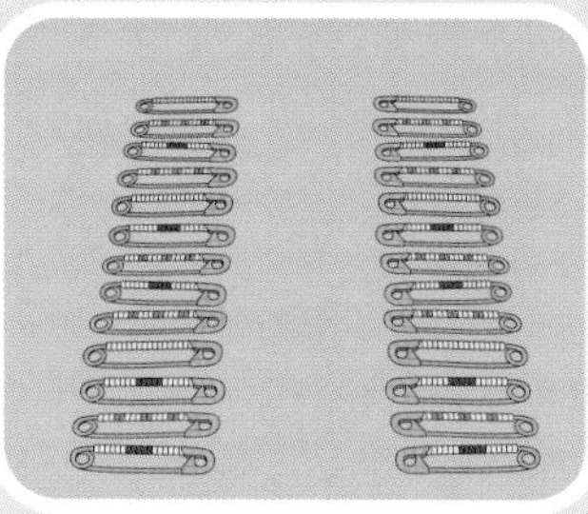
Repeat, beading many pins.

Cut wrist-size lengths of elastic cord.

String beads and pins onto cord.

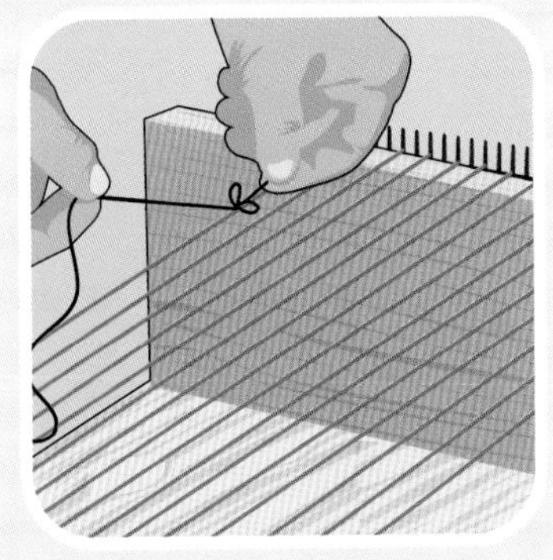
Tie a weft thread to edge warp string.

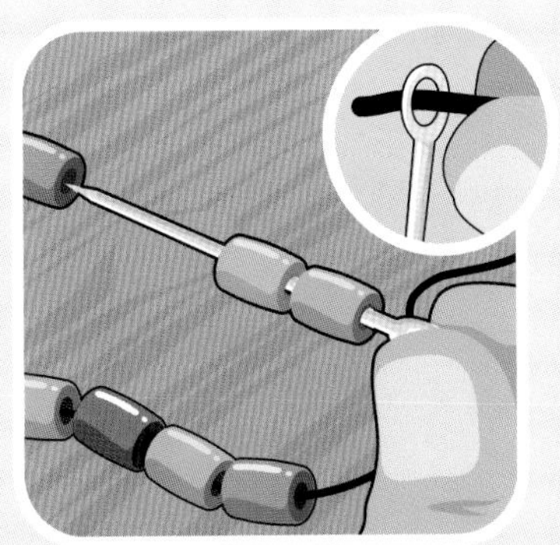
Thread needle; string a row of beads.

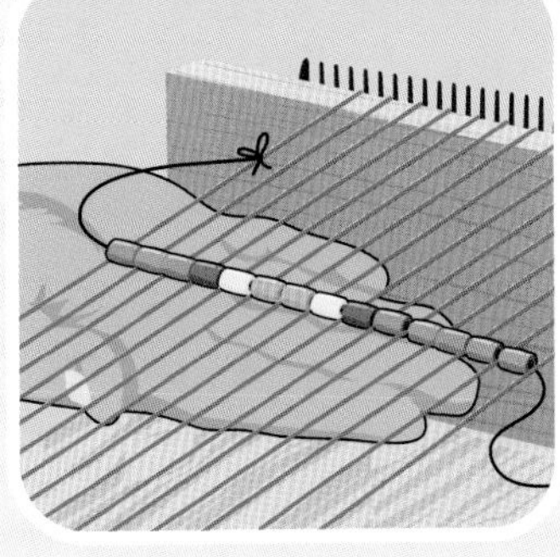
Line up beads under warp.

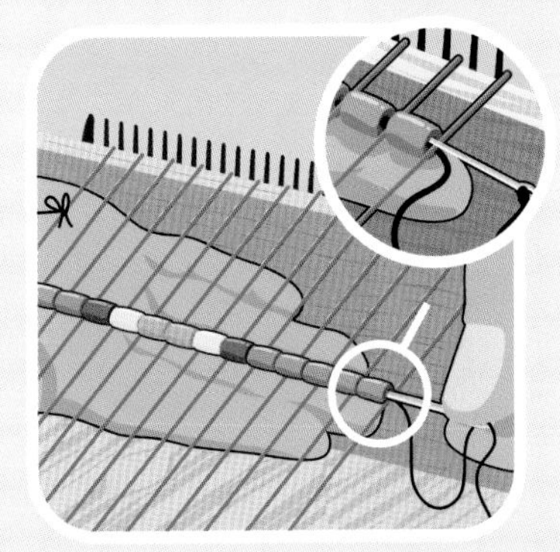
Thread back through bead row.

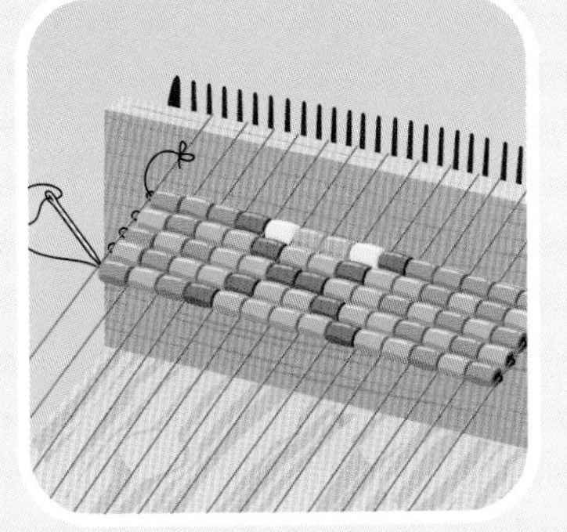
Continue beading.

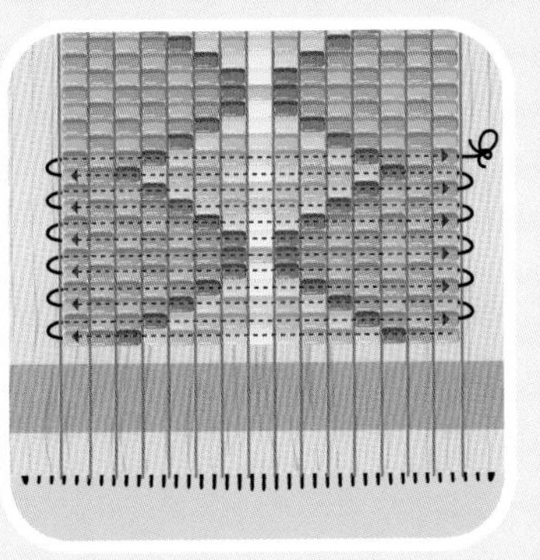
Weave through last few rows; tie.

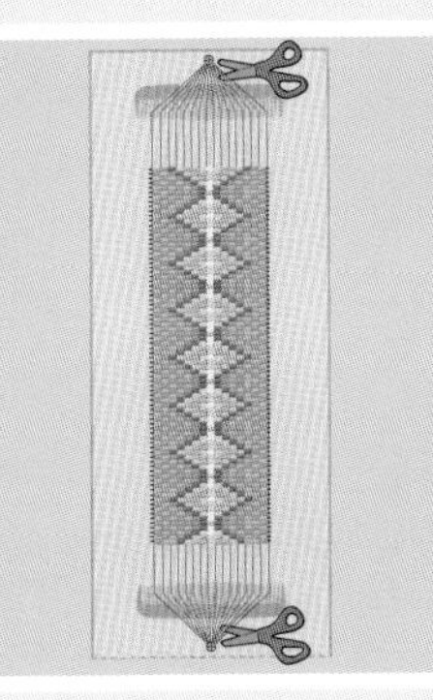
Snip warp ends off loom.

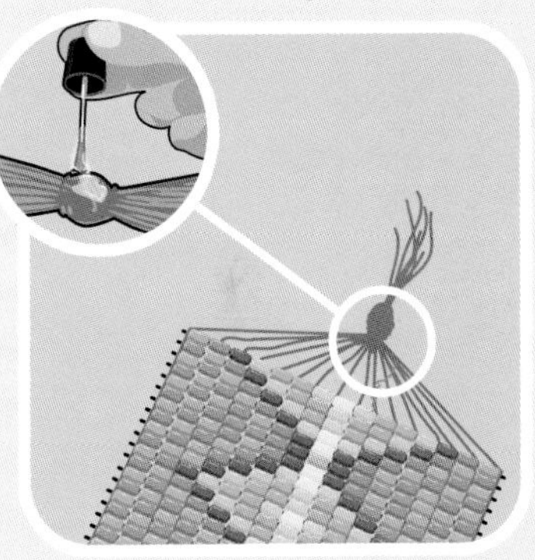
Tie; secure with nail polish.

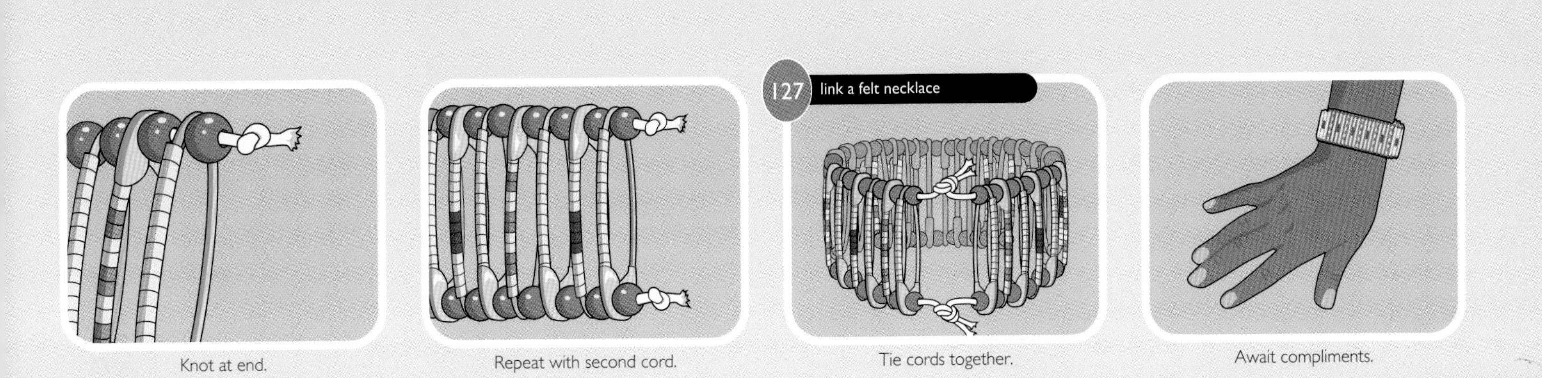

Knot at end.

Repeat with second cord.

Tie cords together.

Await compliments.

81 shrink photos for jewelry

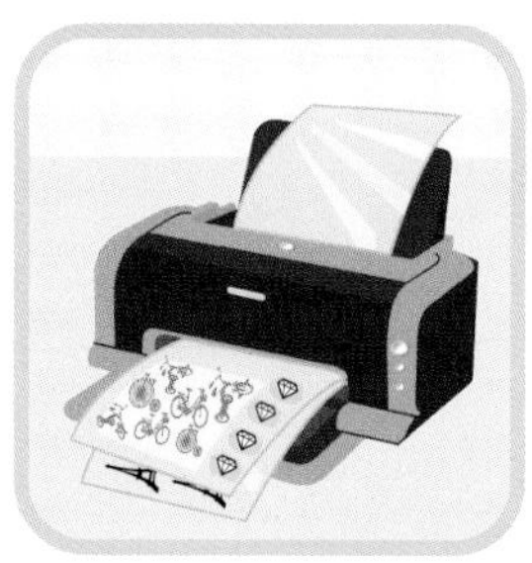

Print onto shrink plastic.

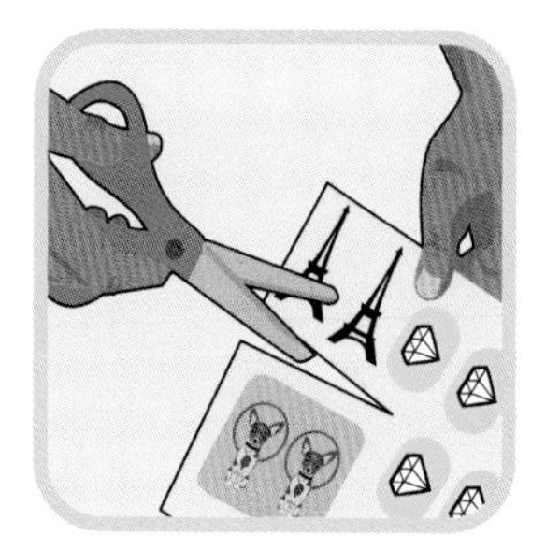

Cut out images.

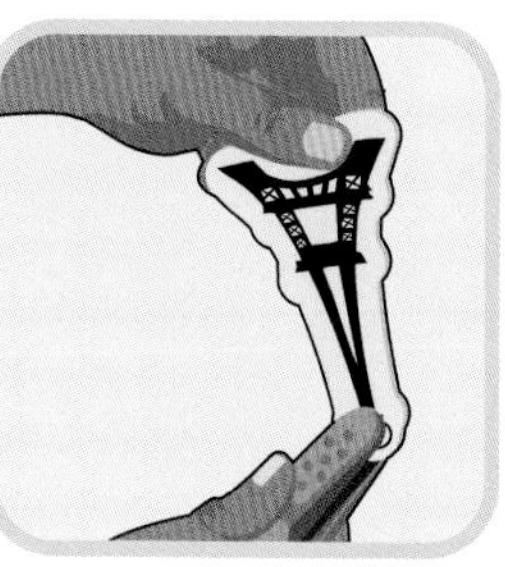

Punch holes as needed.

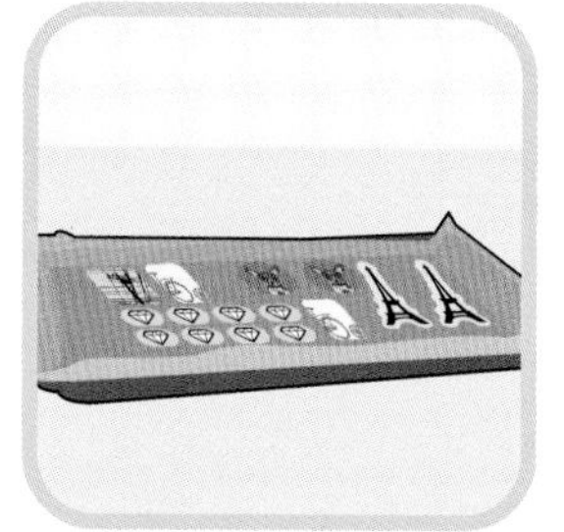

Set on parchment in pan.

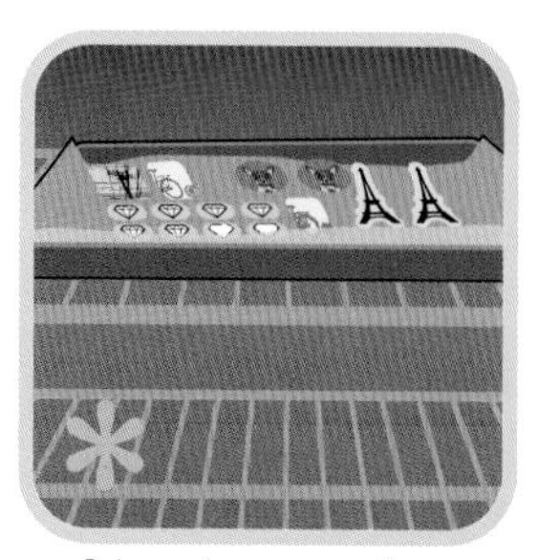

Bake as shown on package.

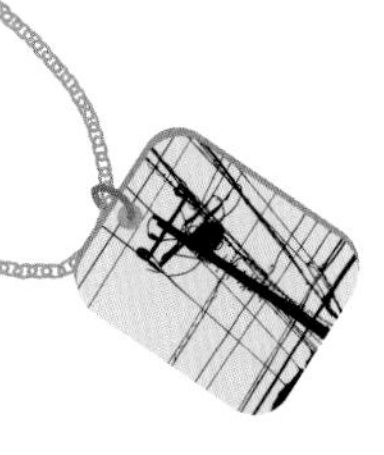

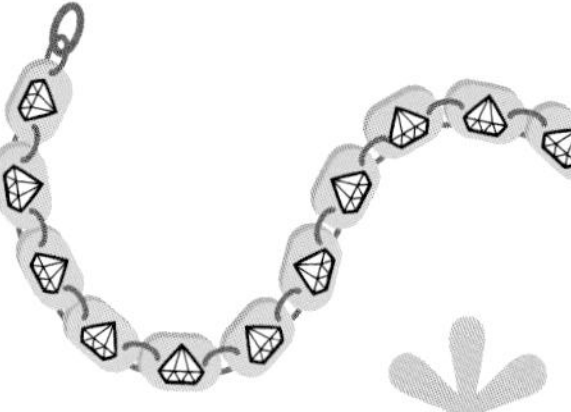

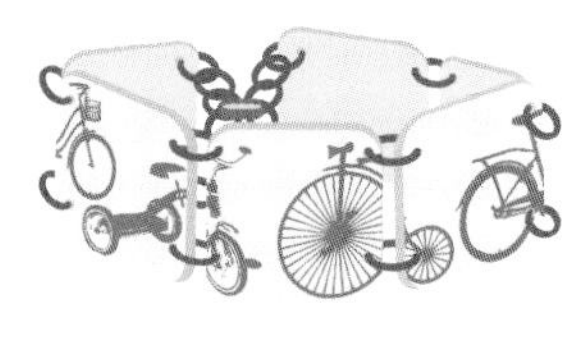

* The shrink plastic will come with instructions for baking. After the plastic cools, put jewelry findings in the holes you punched and rock your shrinky-bling!

82 decorate with a photo transfer

Enlarge an image on copier.

Tape the sheets together.

Apply clear acrylic gloss.

Press photocopy facedown.

Smooth with cardboard.

Let dry.

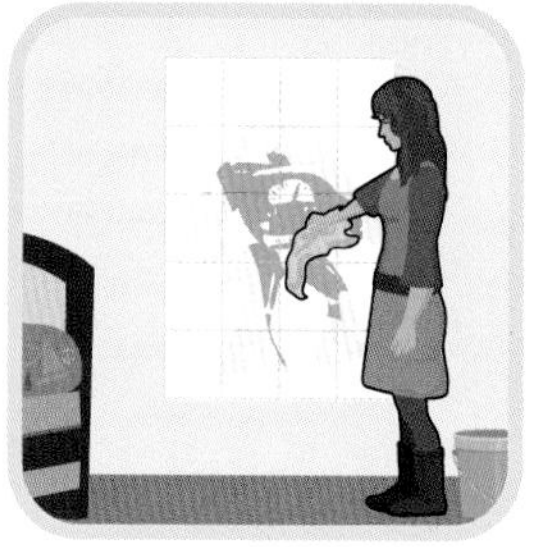

Dampen with wet towel.

Gently rub off wet paper.

Rub until image is clear.

Cover with acrylic to seal.

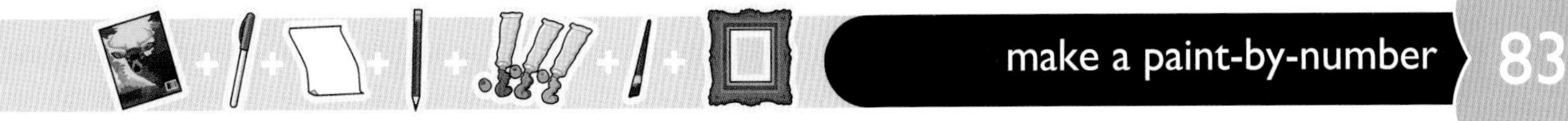

make a paint-by-number

Trace image in marker.

Trace onto blank paper.

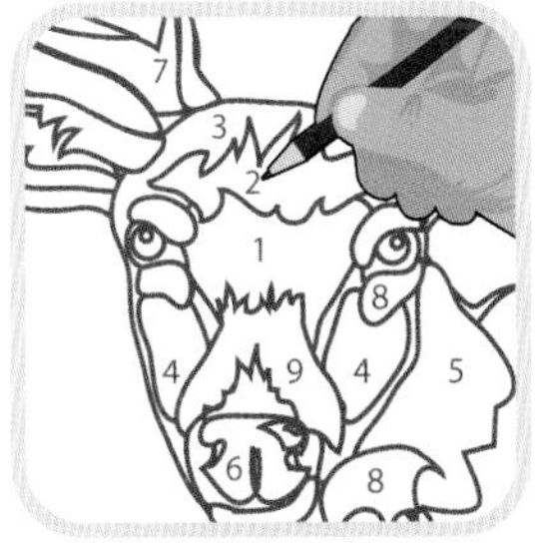

Number by shade.

Paint by numbers.

Display.

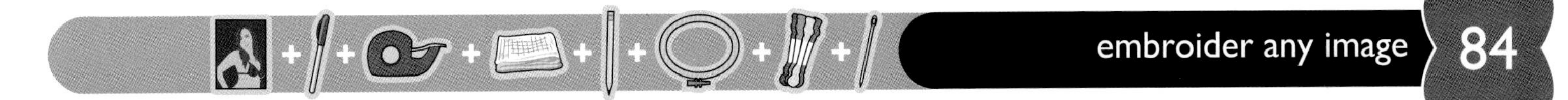

embroider any image

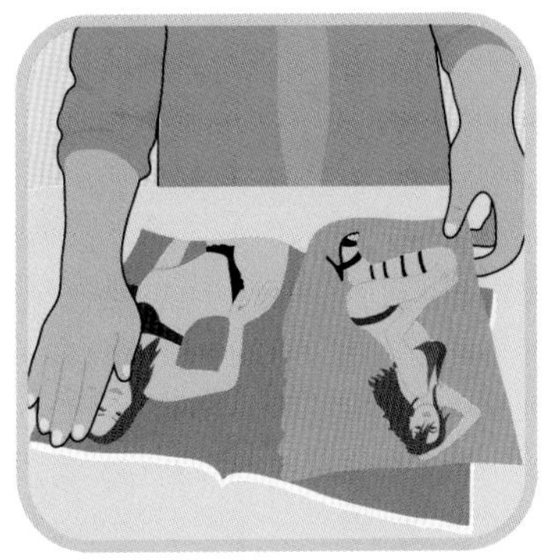

Tear out a picture.

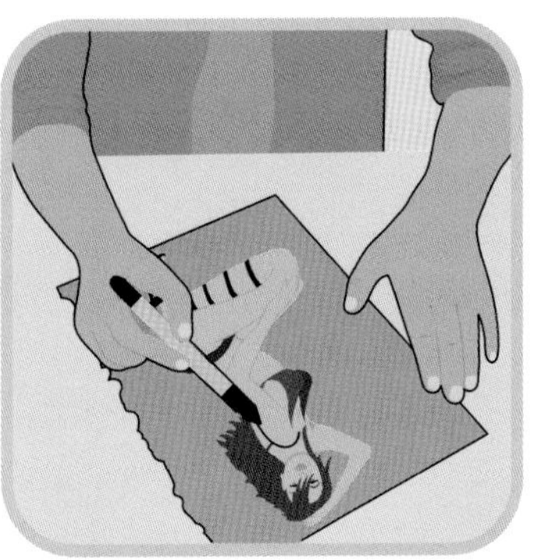

Outline in black marker.

Tape to a bright window.

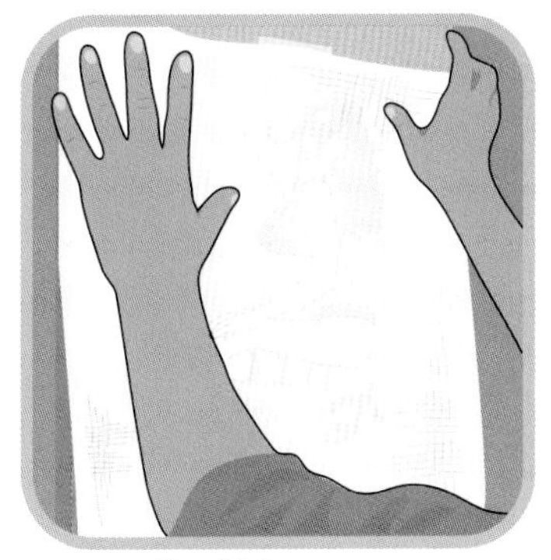

Tape fabric over page.

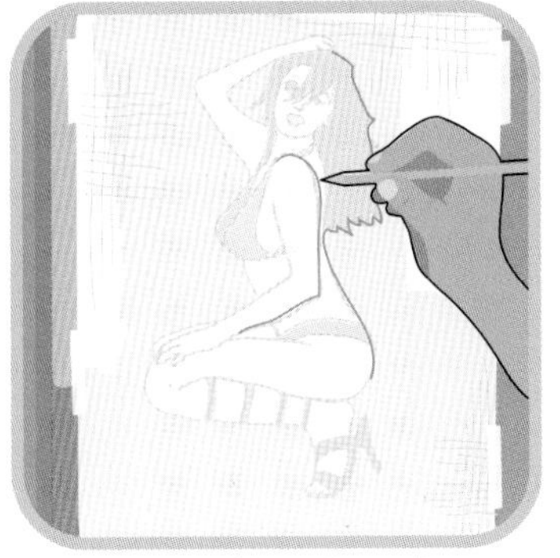

Trace on fabric with pencil.

Remove fabric and image.

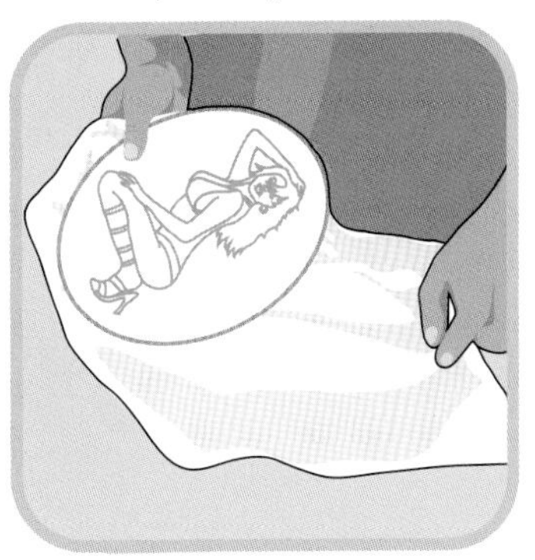

Place in embroidery hoop.

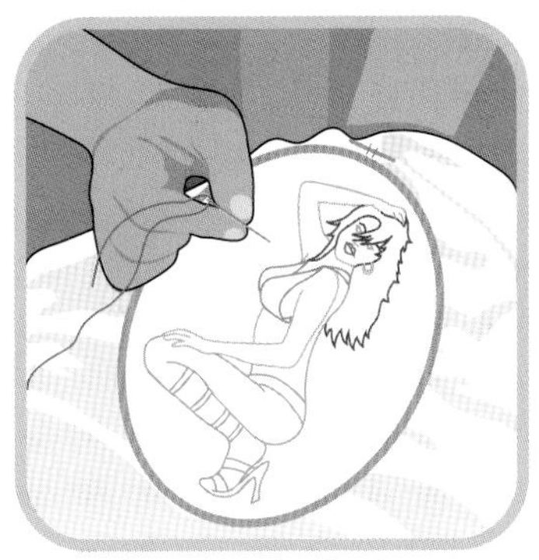

Stitch with embroidery floss.

85 create a coil pot

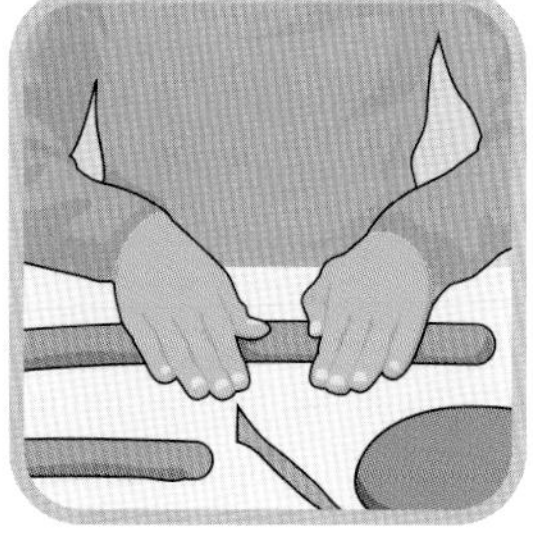
Roll clay coils; cut a base.

Fix coil to base; wrap coils.

Pinch coils together.

354 bake bread in a clay-pot oven

Keep coiling and pinching.

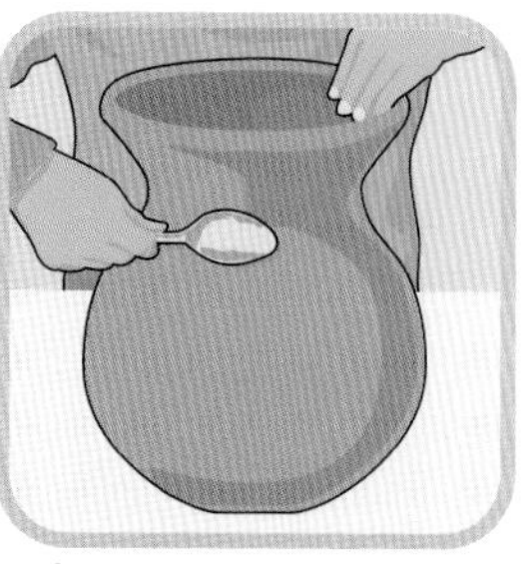
Smooth exterior with spoon.

86 mosaic a stepping stone

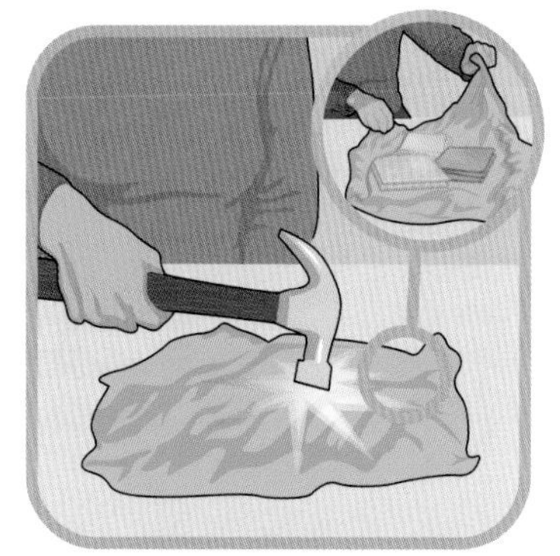
Wrap tiles in a towel; smash.

Arrange pieces into an image.

Prepare grout; lay on stone.

Press tile pieces into grout.

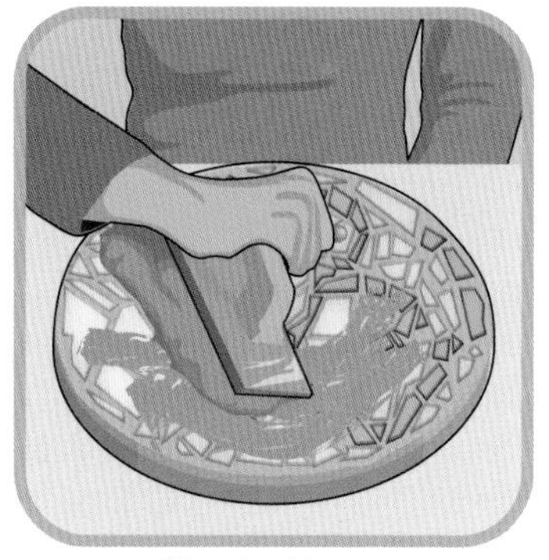
Fill cracks with grout.

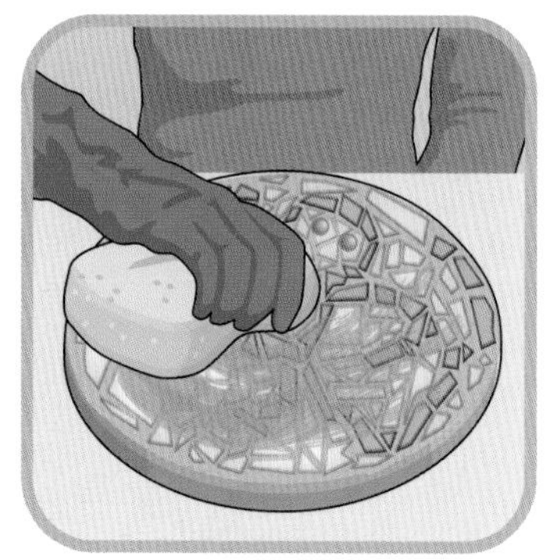
Wipe off surface.

Finish with sealant; let dry.

Install outdoors.

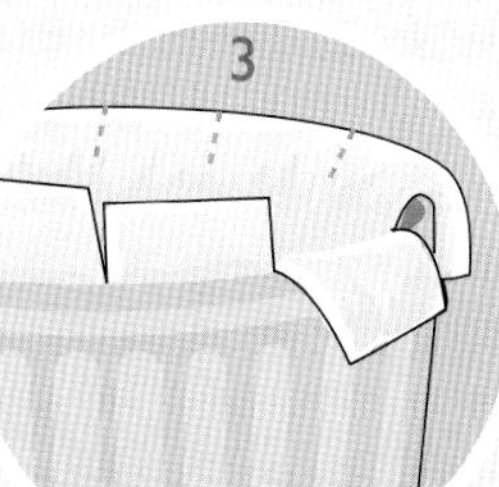

Slit insulation overhang; fold over sides.

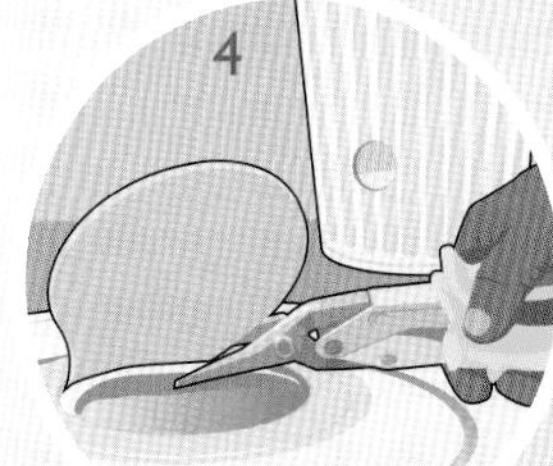

Cut vents in the lid and the can's side.

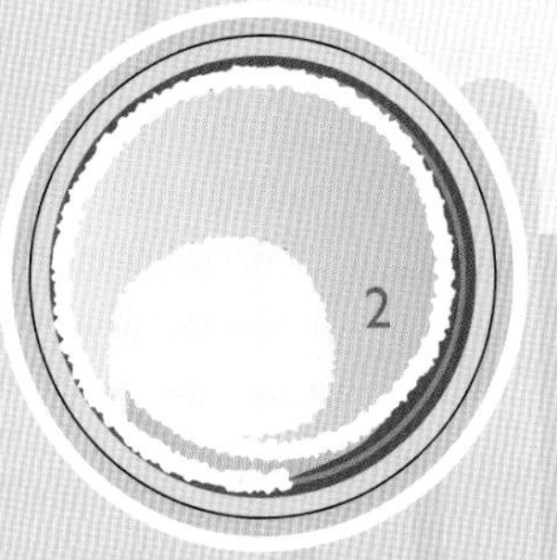

Line the can's interior with insulation, leaving an overhang at top.

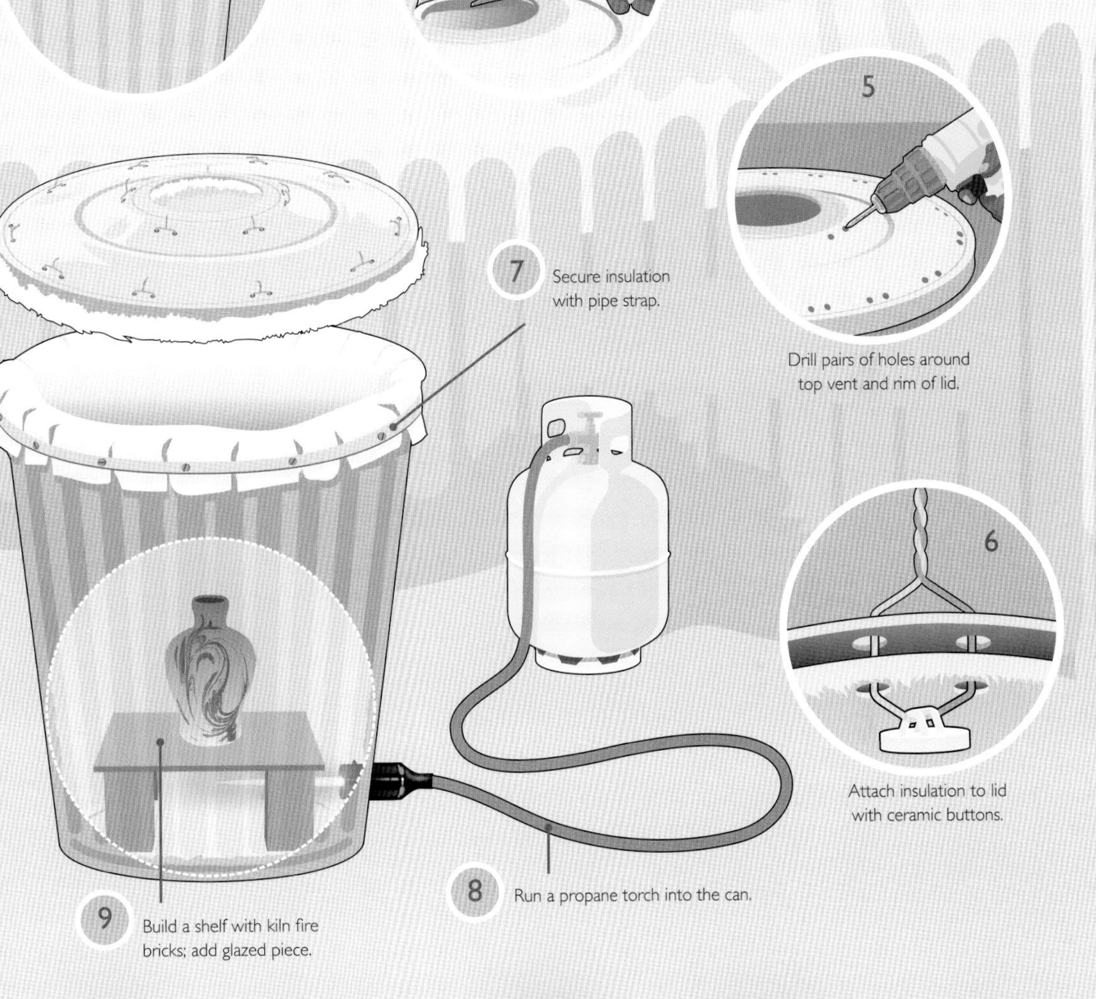

Trace and cut insulation to line the bottom of a garbage can.

This traditional Japanese firing technique requires smoke, flame, and a sense of adventure—special glazes and intense heat create effects that are beautiful, earthy, and unpredictable.

88
construct a thumb piano
×3
1
Snip large bobby pins in half.
2
Place four halves on block.
3
Lay one half across top.
4
Tape it in place.
5
Staple down crosswise pin.
6
Slide screw beneath keys.
7
Staple screw in place.
8
Play with thumbs.

play the glass harp
89
Tune your glasses to a piano for exact pitches.
⅞ full
¾ full
⅝ full
½ full
⅜ full
¼ full
⅛ full
1/16 full
carve a bone flute
90
1
Cut end off large bone.
2
Glue sandpaper to metal rod.
3
Grind to clean bone's interior.
4
File off a mouthpiece at end.
5
Gently drill finger holes.

style

91 face the world as a flapper

92 be a swell '40s bombshell

Lengthen and darken brows with a pencil.

Sweep on subtle eyeshadow.

Start with a dark foundation; apply light powder.

Exaggerate top lip; use bold lipstick.

93 get a mind-blowing '60s look

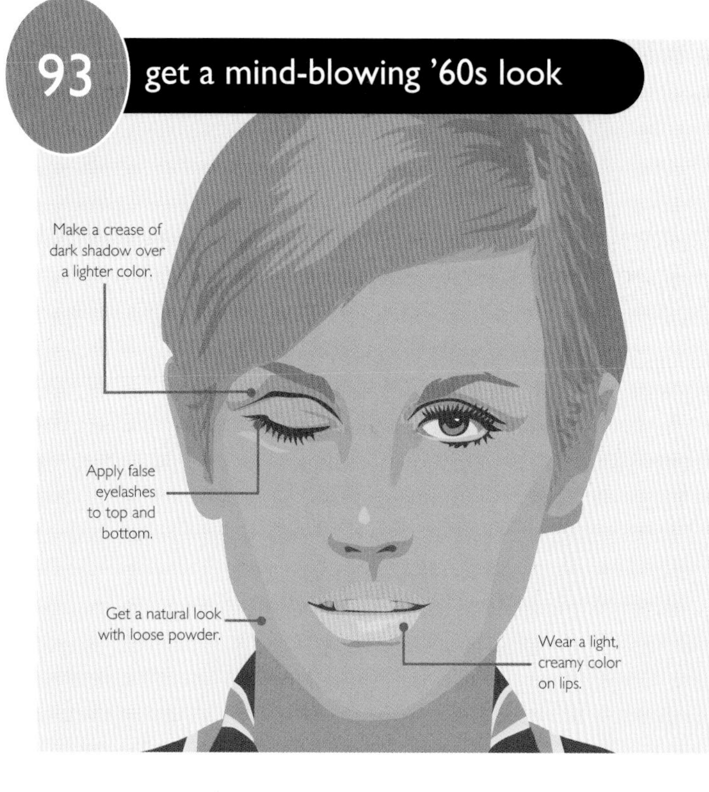

94 sport groovy '70s style

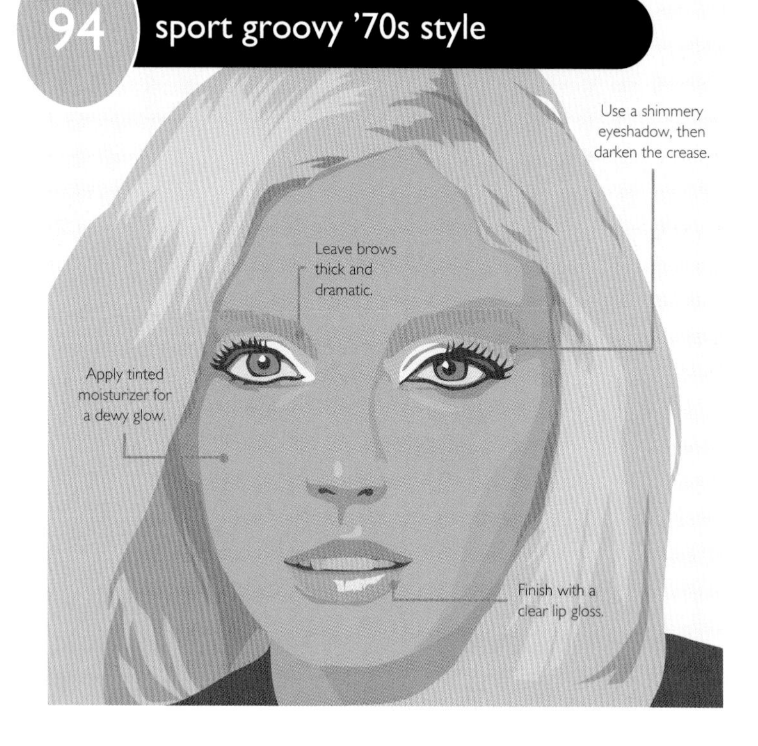

apply mineral makeup

95

Tap powder into lid.

Dip large brush.

Tap to remove excess.

Apply in widening circles.

Spritz with water to set.

contour cheekbones

96

Apply cream foundation.

Suck in cheeks.

Brush along sunken line.

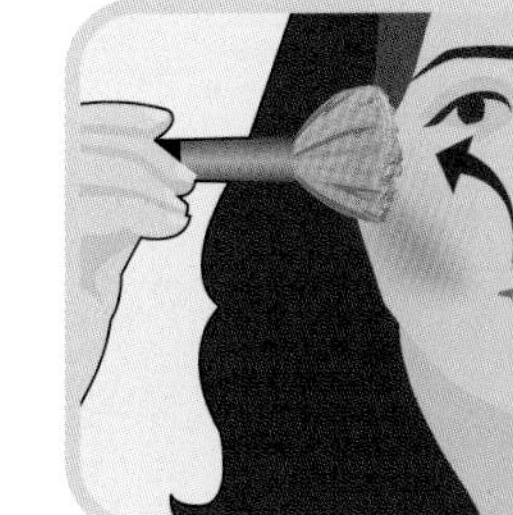
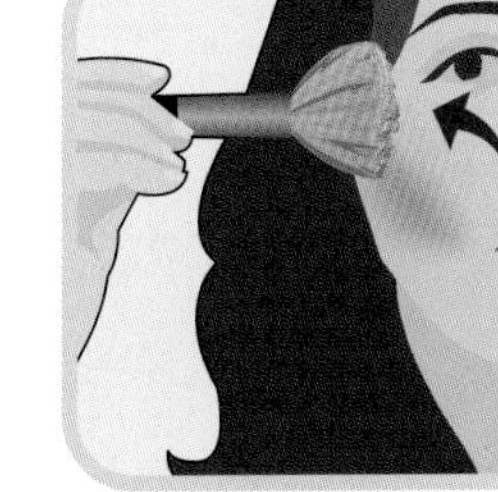
Sweep highlighter above.

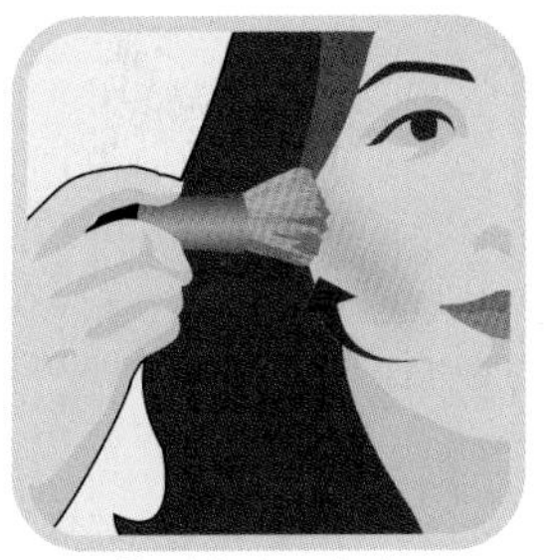
Blend the two shades.

draw on dramatic eyes

97

Warm eye pencil briefly.

Blow on pencil to set.

Apply lavishly.

Rock like an Egyptian.

98 customize lip gloss

Cut up old lipsticks.

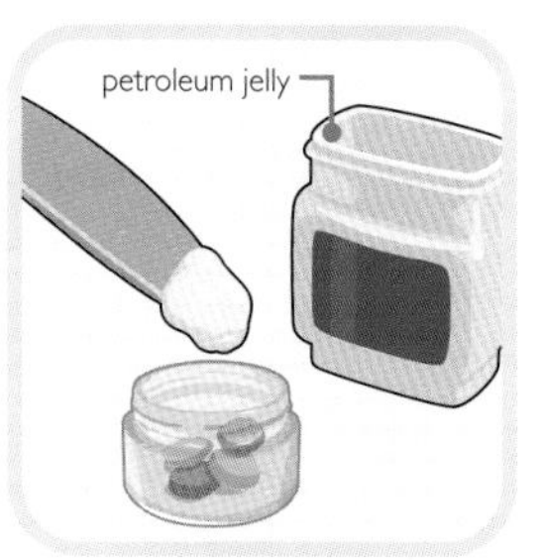

Combine in jar; close lid.

Submerge in hot water.

Stir to blend.

Cool before applying.

99 blend my own toner

Cut into thin slices.

Combine fruit and alcohol.

Blend until pureed.

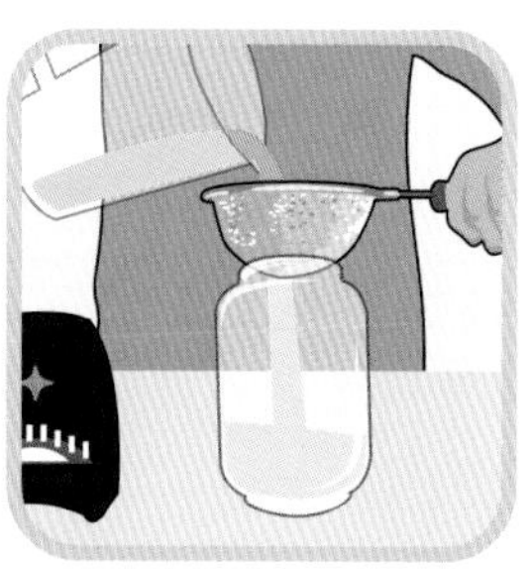
Strain; reserve liquid.

Store in refrigerator.

100 whip up facial moisturizer

* To personalize, add:
2 tsp lemon juice (acne-prone skin
2 tsp evening primrose oil (dry skin
1 tsp witch hazel (oily skin)

210 lift wax from a carpet

Melt in a double boiler.

Stir in distilled water.

Stir until completely blended.

*
Pour into container; cool before using.

make custom beauty products 101

know my cosmetics' shelf lives 102

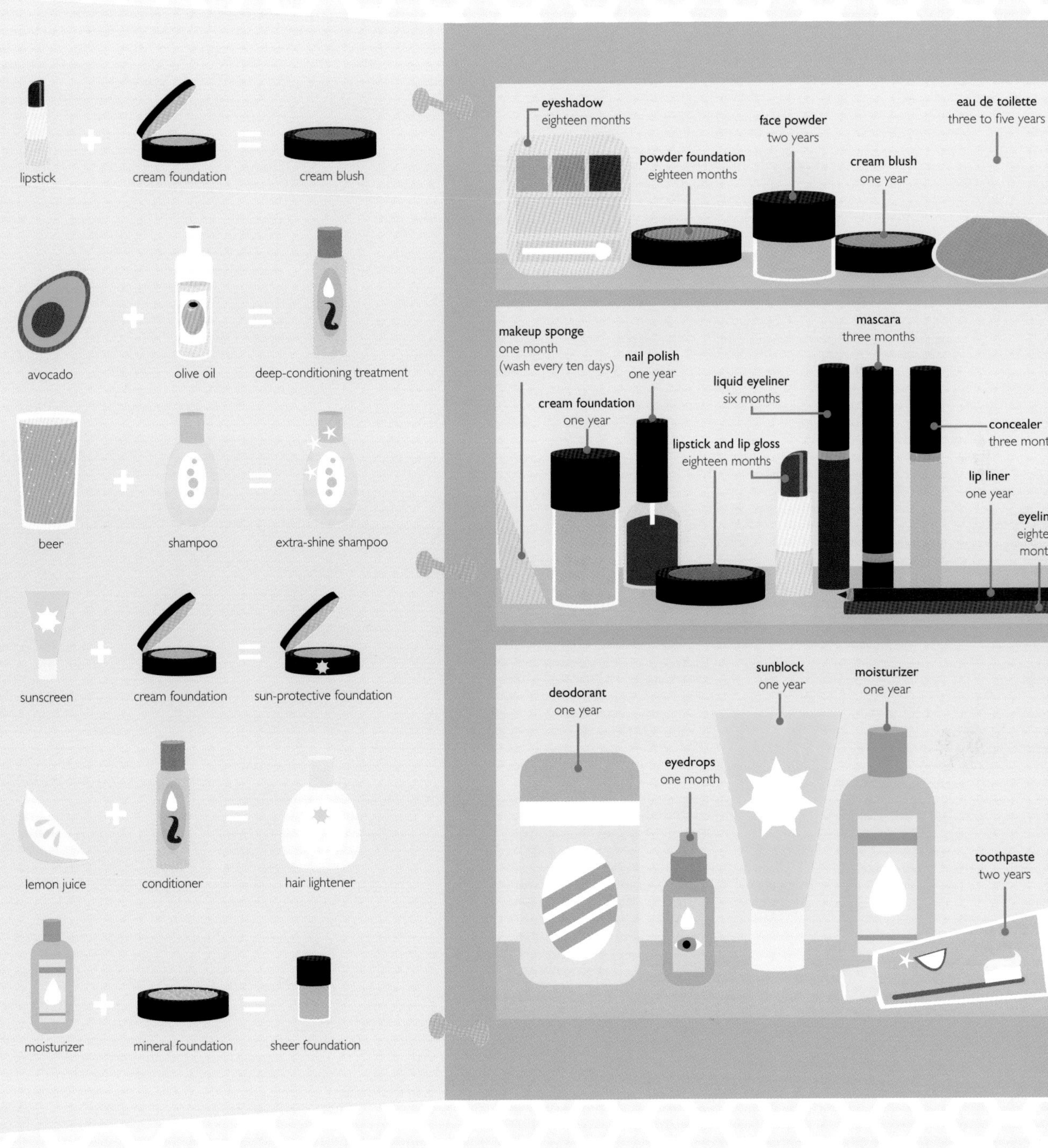

103 pick sunglasses to suit my face

If you want to be made in your shades, identify your face shape below, then pick one of the corresponding frame styles.

104 fix bent glasses

Note bent glasses.

Tighten arm screws.

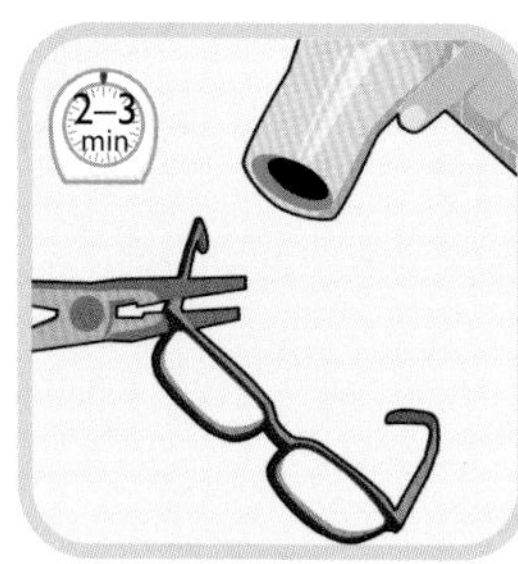

Heat bent arm.

Grip with pliers; straighten.

Test in front of mirror.

105 divide a unibrow

Find the best brow shape.

Pluck slowly and carefully.

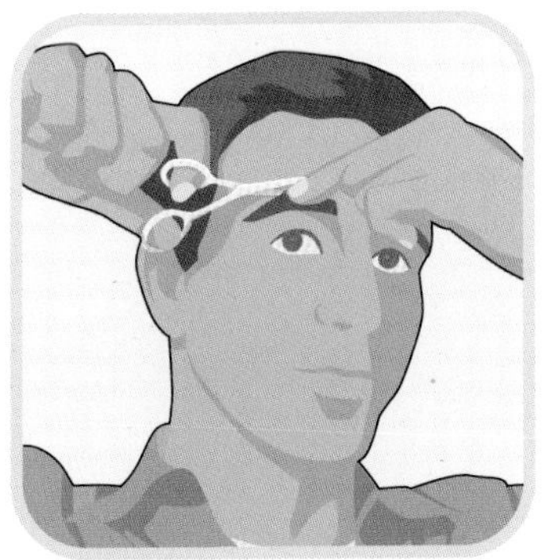

Trim long hairs with scissors.

Soothe skin with aloe.

Find the end points for your new brows by holding a pencil alongside your nose. Pluck from the edge of the pencil toward the bridge of your nose.

106 wax body hair at home

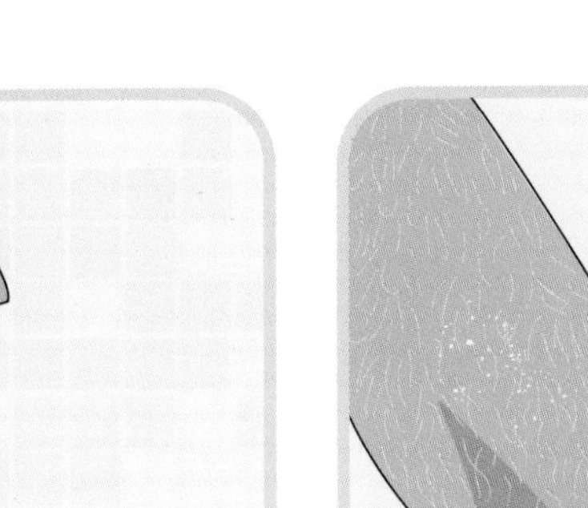

Begin after a hot shower.

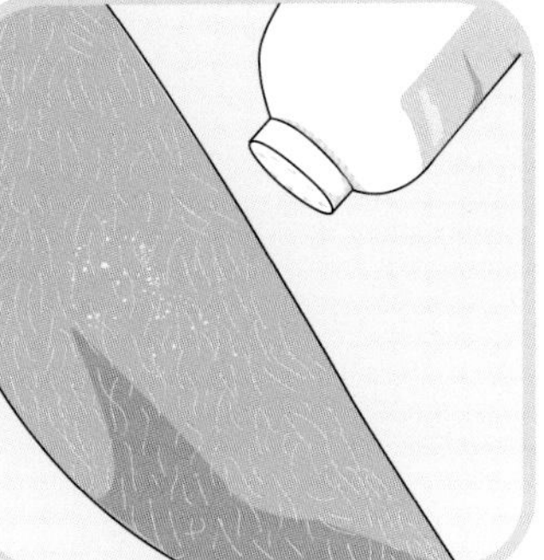

Rub talcum powder into skin.

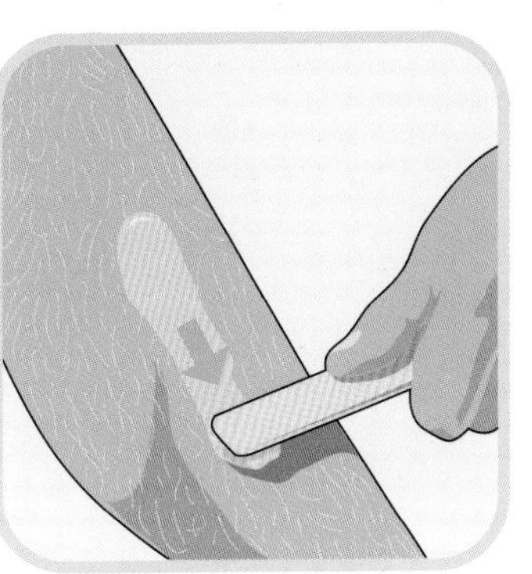

Add wax along hair growth.

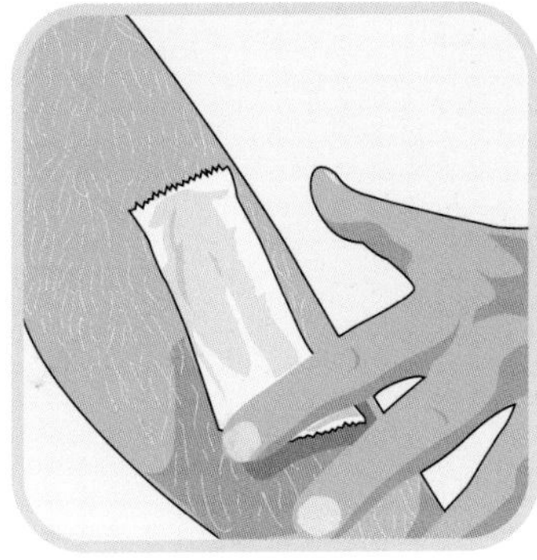

Smooth on cloth strip.

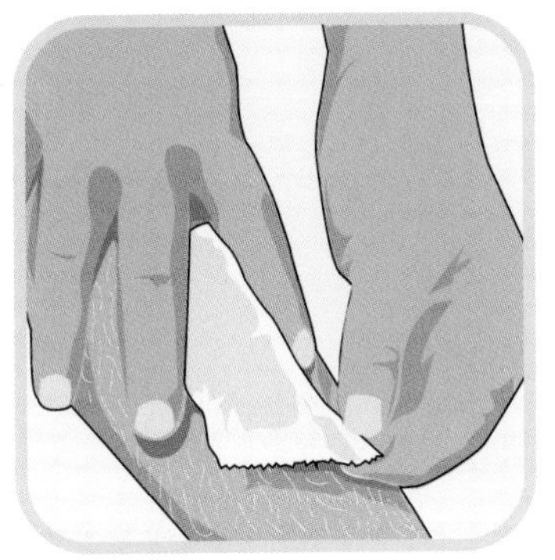

Grasp end of strip.

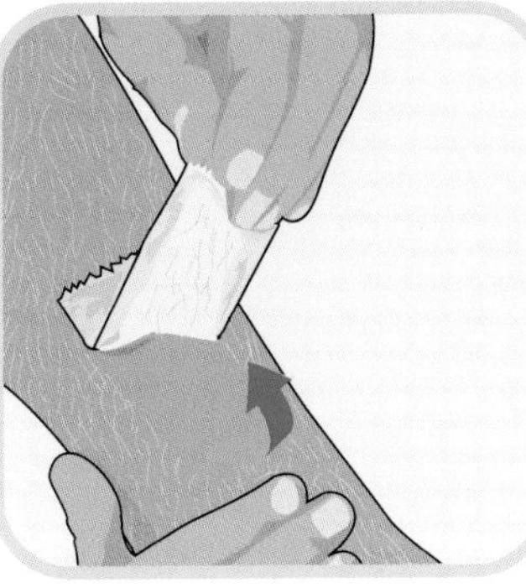

Pull in direction opposite of growth.

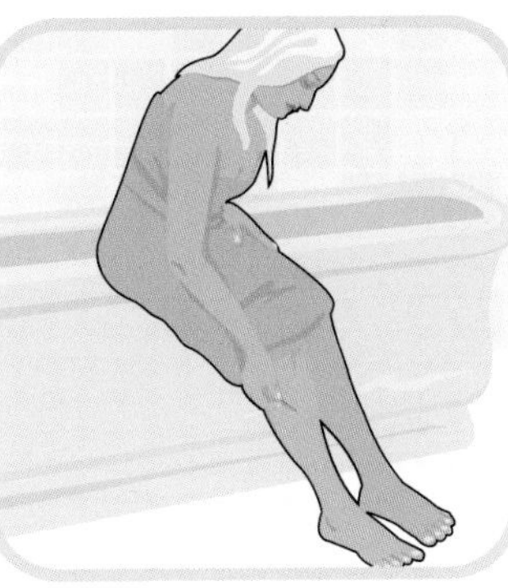

Wax rest of leg; tweeze stray hairs.

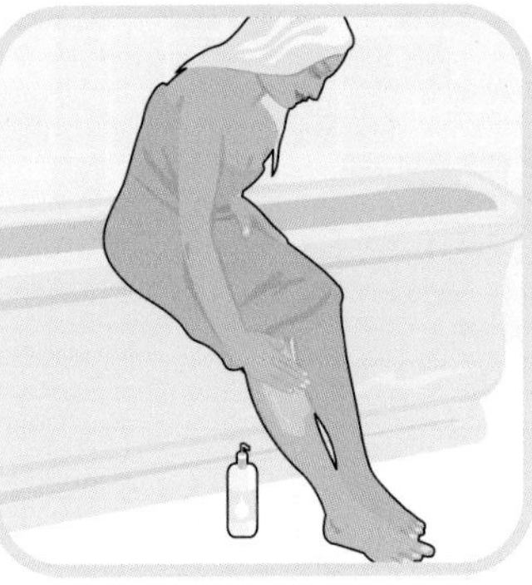

Apply soothing balm.

107 give a simple haircut

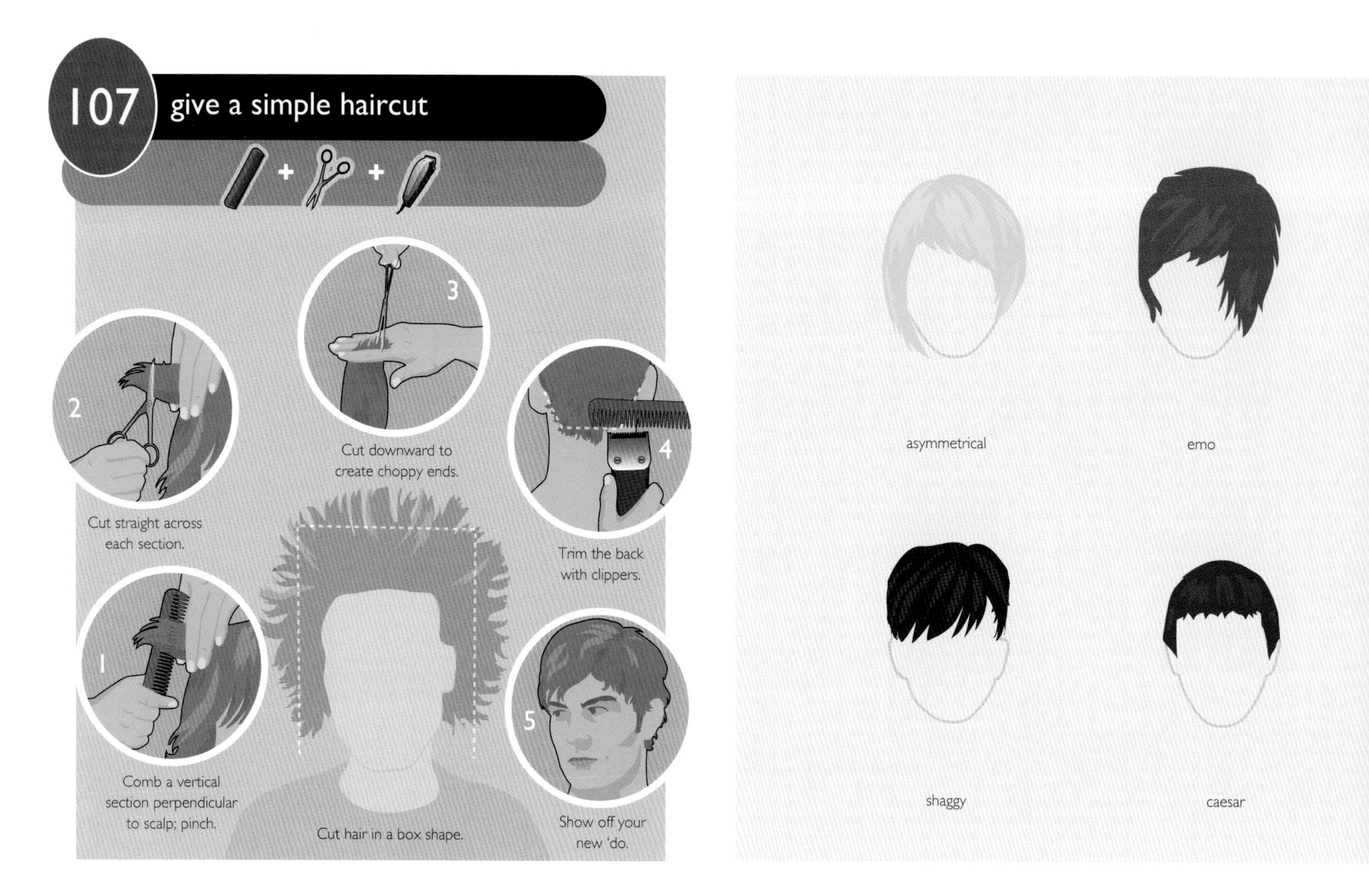

Comb a vertical section perpendicular to scalp; pinch.

Cut straight across each section.

Cut downward to create choppy ends.

Trim the back with clippers.

Cut hair in a box shape.

Show off your new 'do.

asymmetrical

emo

shaggy

caesar

108 get home highlights

Cut strips of tinfoil.

Clip hair; pull out one section.

Weave comb handle through section.

351 roast veggies in foil

Put foil beneath top fringe.

choppy

blunt

pixie

'80s

bettie page

side-swept

wispy

parted

feathered

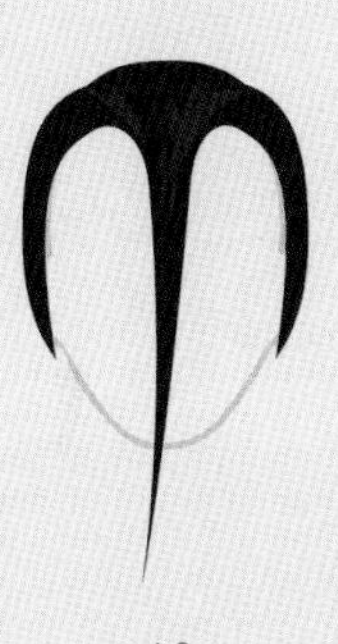

misfits

Brush bleaching dye onto hair.

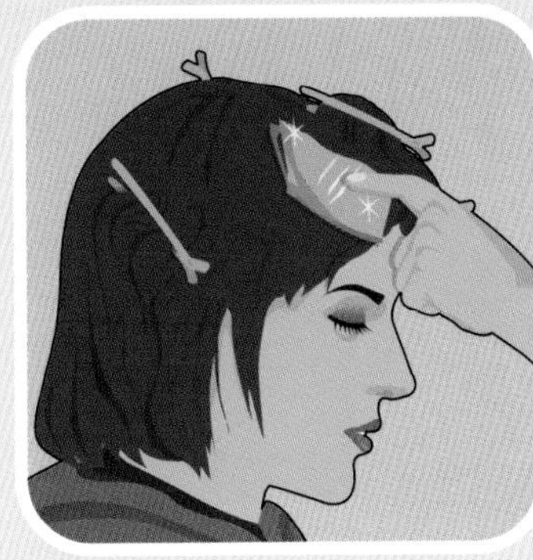

Fold up the foil.

Repeat on more sections of hair.

Let set. Remove foil and wash hair.

110 braid my own headband

Secure top section of hair.

Separate a side section; braid.

Braid a section on the other side.

183 braid a denim rug

Wrap one across crown; pin.

Wrap and pin second braid.

Charming!

111 sport a fauxhawk

Wash, then dry hair.

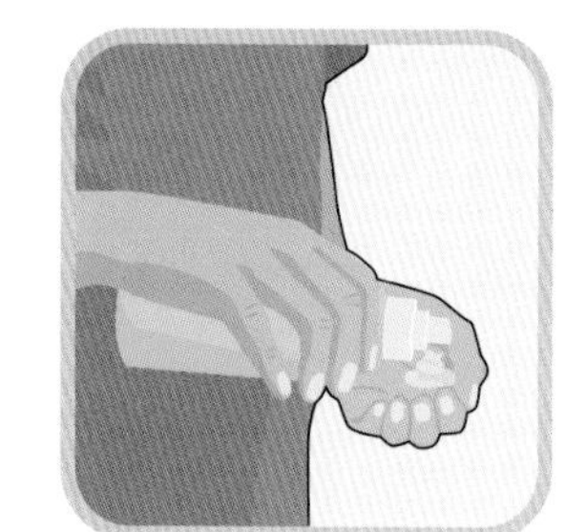
Coat palms with gel.

Pull hair up and forward.

Hold with hair spray.

Enjoy your awesomeness.

112 style a ducktail

Apply pomade liberally.

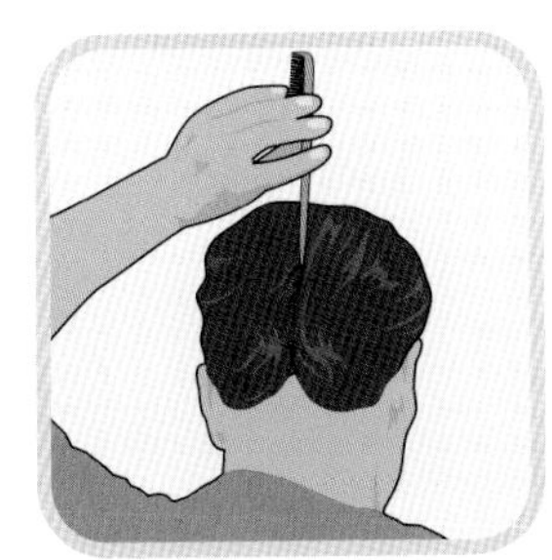
Part hair in back.

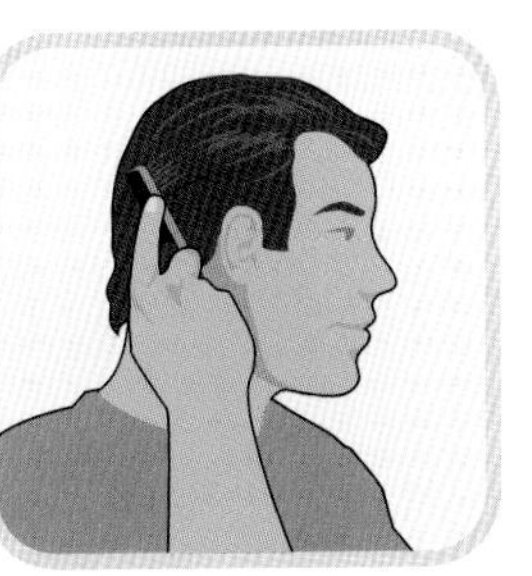
Comb sides straight back.

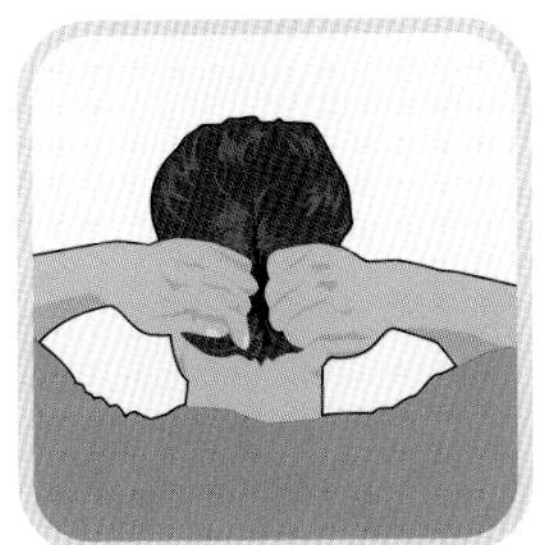
Fold ends into back part.

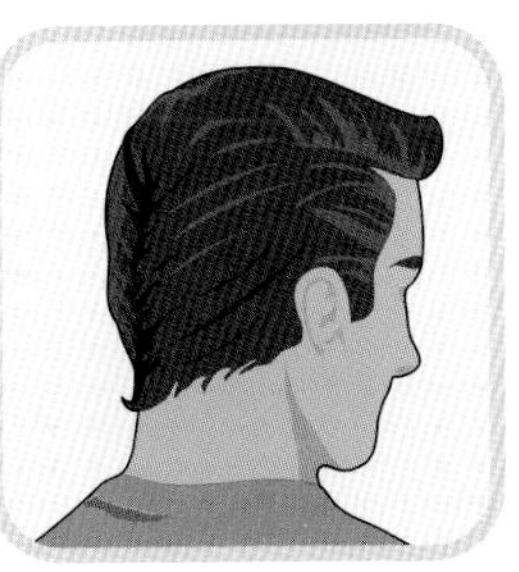
Lookin' sharp!

113 roll up luscious waves

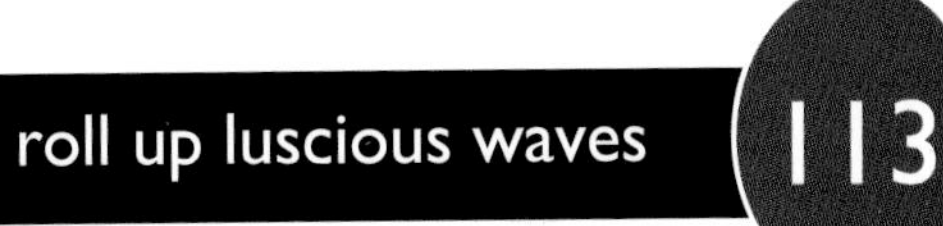

Separate hair into sections.

Clip sections as needed.

Mist with hair spray; comb.

Wrap hair around curlers.

Mist with hair spray.

Set with a blow-dryer.

Let hair cool; remove rollers.

Comb wtih fingers to style.

114 bundle up in winter fabrics

bouclé

dobby

chenille

jacquard

herringbone

houndstooth

115 step out in spring textiles

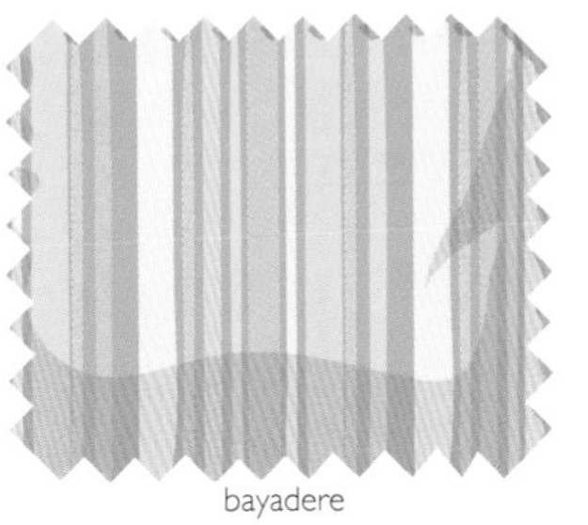
bayadere

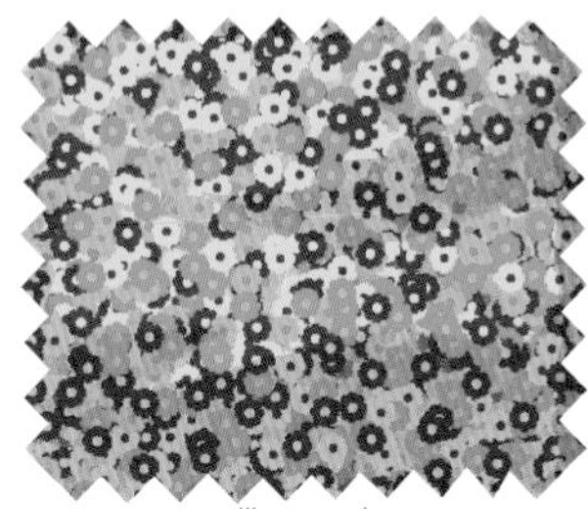
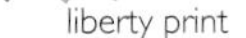
liberty print

eyelet

leno weave

jamdani

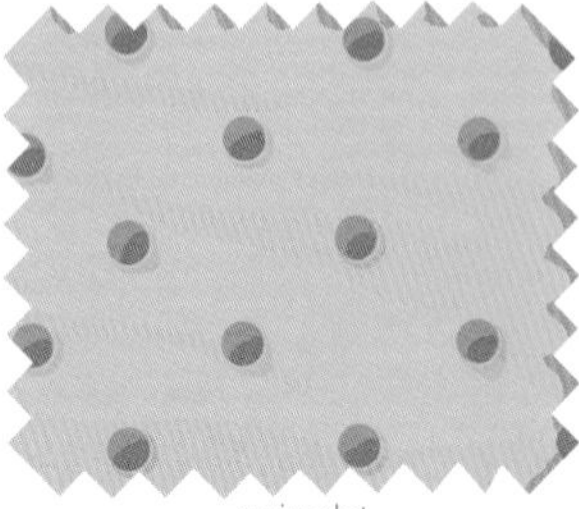
swiss dot

116 stay cool in summer material

117 cozy up in autumn fabrics

withstand a heat wave 256

batik

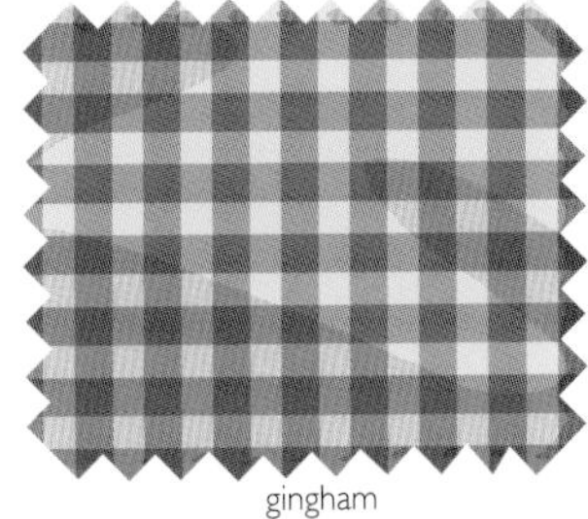
gingham

ikat

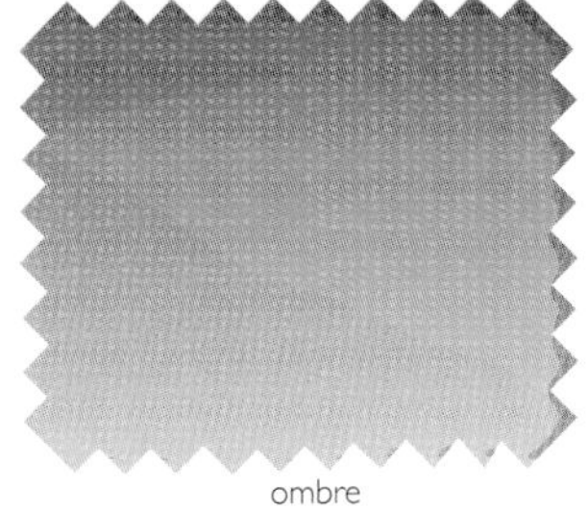
ombre

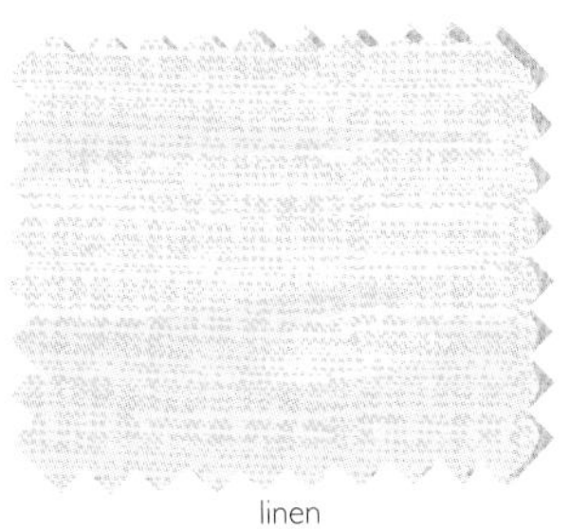
linen

seersucker

damask

corduroy

pointelle

paisley

tartan

toile

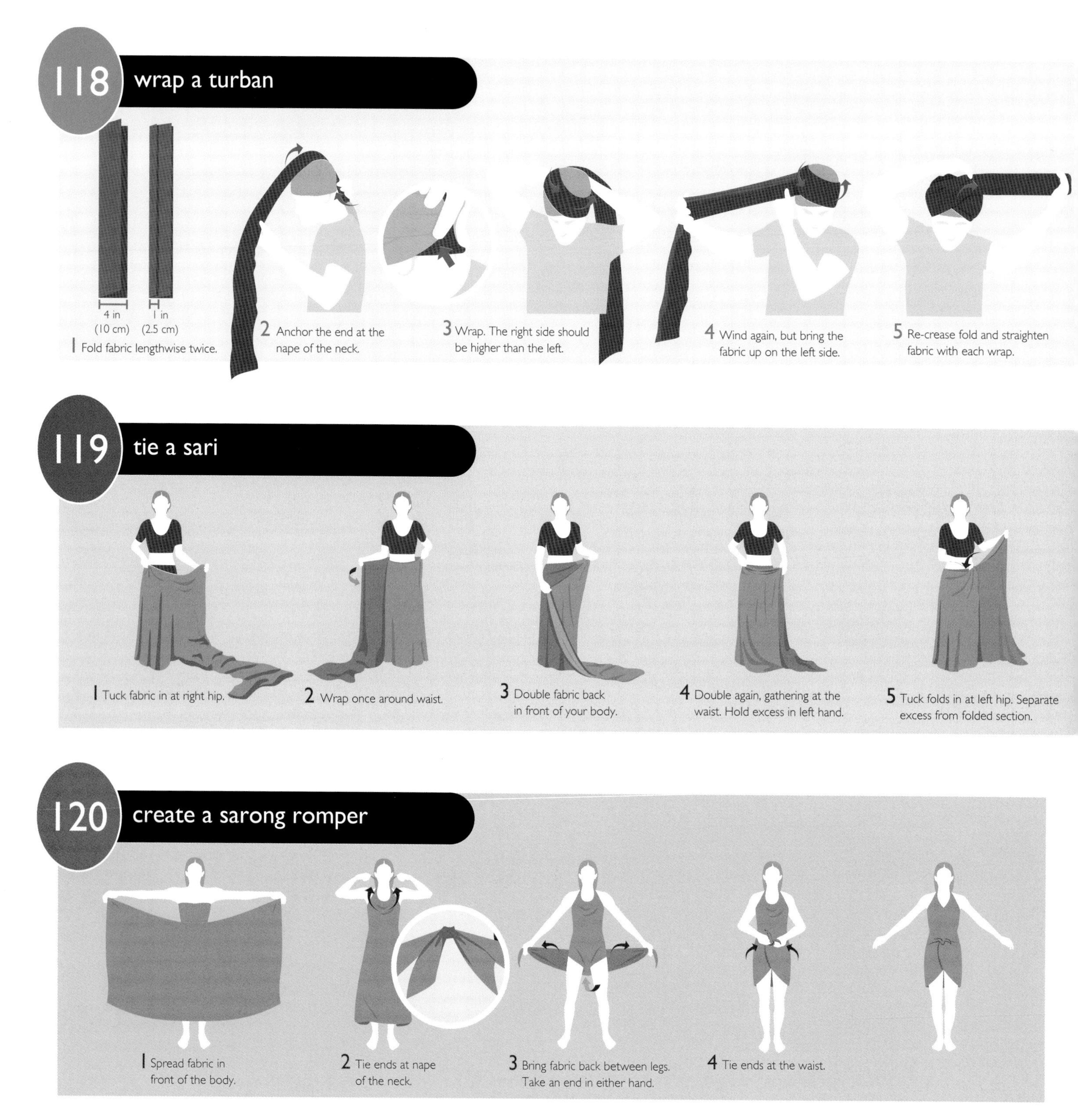
118
wrap a turban
4 in (10 cm)
1 in (2.5 cm)
1 Fold fabric lengthwise twice.
2 Anchor the end at the nape of the neck.
3 Wrap. The right side should be higher than the left.
4 Wind again, but bring the fabric up on the left side.
5 Re-crease fold and straighten fabric with each wrap.
119
tie a sari
1 Tuck fabric in at right hip.
2 Wrap once around waist.
3 Double fabric back in front of your body.
4 Double again, gathering at the waist. Hold excess in left hand.
5 Tuck folds in at left hip. Separate excess from folded section.
120
create a sarong romper
1 Spread fabric in front of the body.
2 Tie ends at nape of the neck.
3 Bring fabric back between legs. Take an end in either hand.
4 Tie ends at the waist.

6 When 1 ft (30 cm) of fabric remains, fold it back on itself.

7 Tuck in the folded fabric.

8 Pull out the fabric from the nape of neck. Spread over front.

9 Fold back over crown; tuck in at the back.

6 With right hand, grab folded fabric. Drape over left shoulder.

7 Knot the middle of the draped fabric.

8 Tuck the knotted fabric into the waist.

9 Bring the excess fabric around the back.

10 Drape the excess over the left shoulder.

121 rock a sarong halter dress

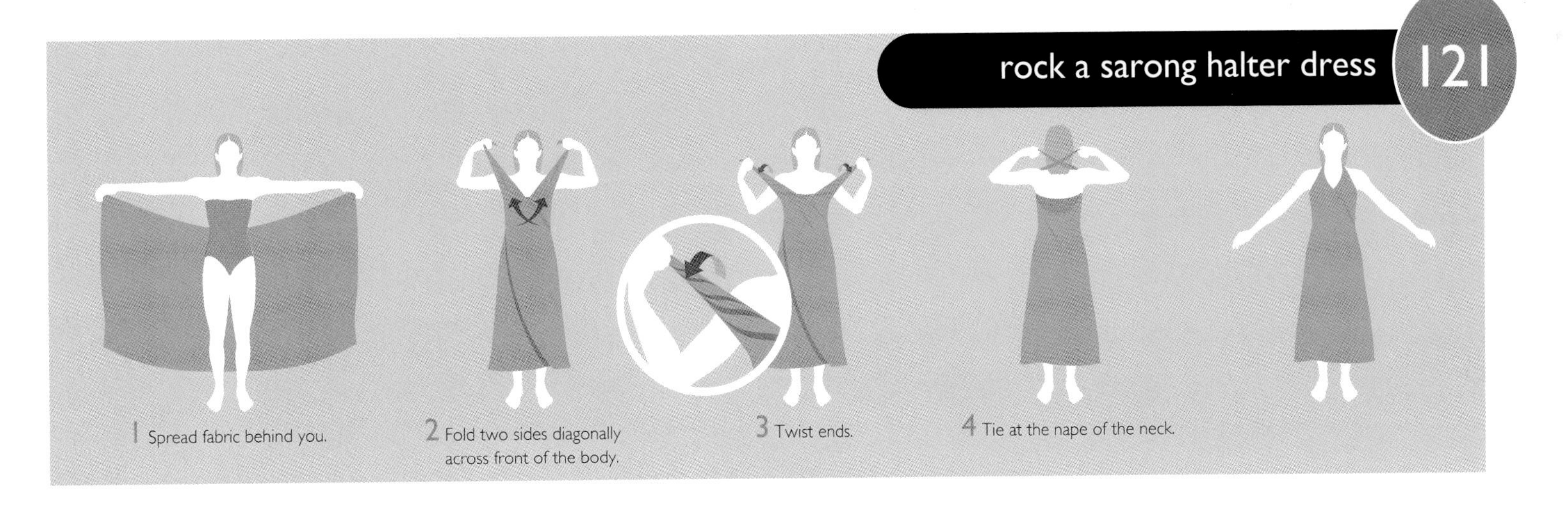

1 Spread fabric behind you.

2 Fold two sides diagonally across front of the body.

3 Twist ends.

4 Tie at the nape of the neck.

122 hem a pair of pants

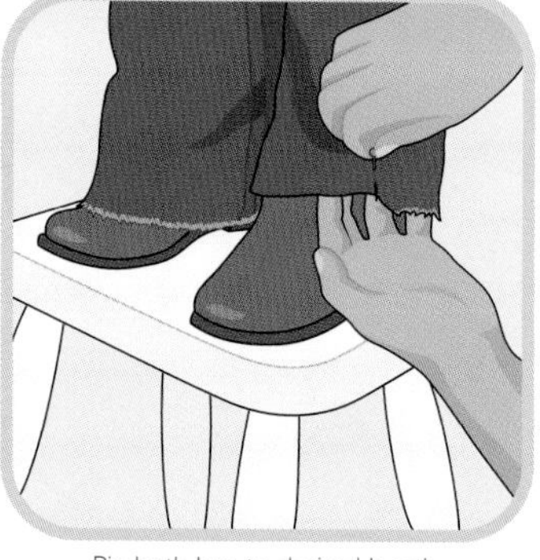
Pin both legs to desired length.

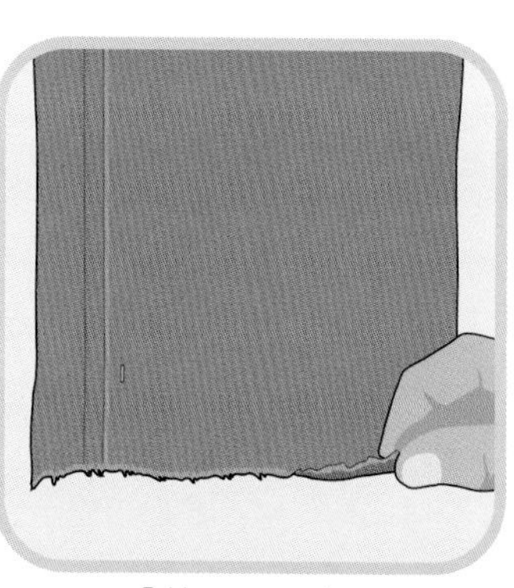
Fold over raw edge.

Iron fold in place.

Fold up to point marked with pin.

Iron fold in place.

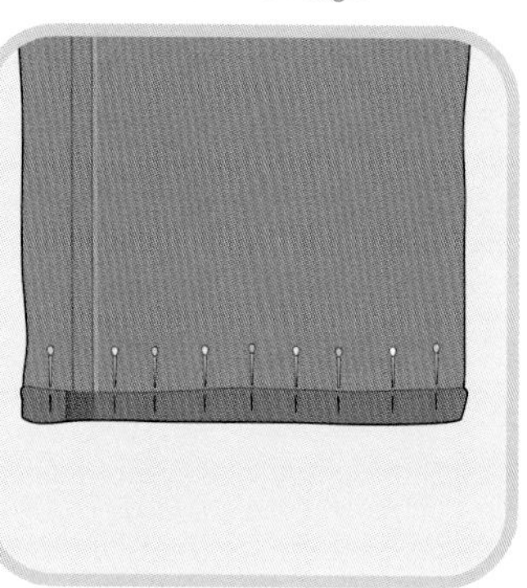
Pin.

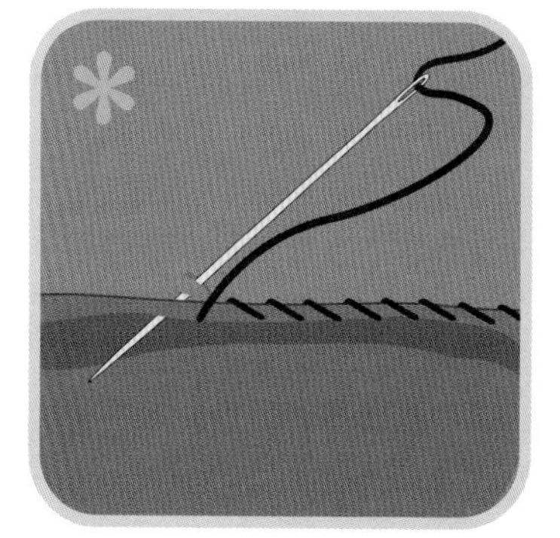
Sew hem using blind stitch.

Repeat process on other pant leg.

A blind stitch is only visible from the "wrong" side of the seam. Only a tiny bit of thread is apparent on the "right" side.

123 shorten a zipper

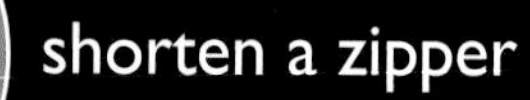

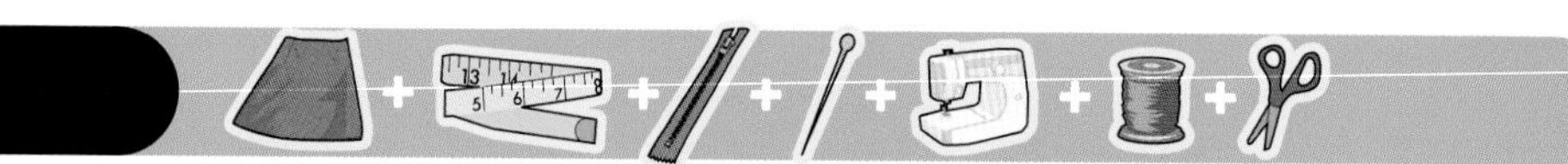

Determine ideal length.

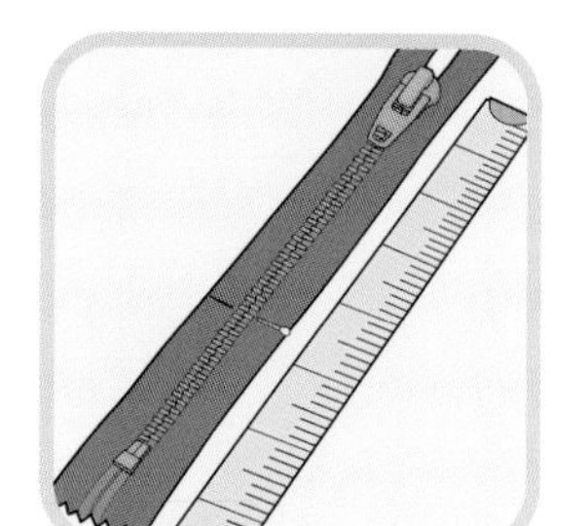
Mark desired length.

Set machine to zigzag.

Overstitch to make stopper.

Trim beneath stitching.

sew in a zipper 124

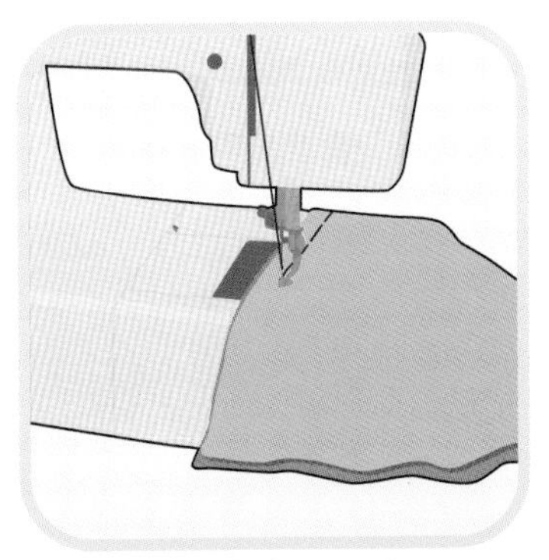
Baste fabric pieces together.

Press seam open.

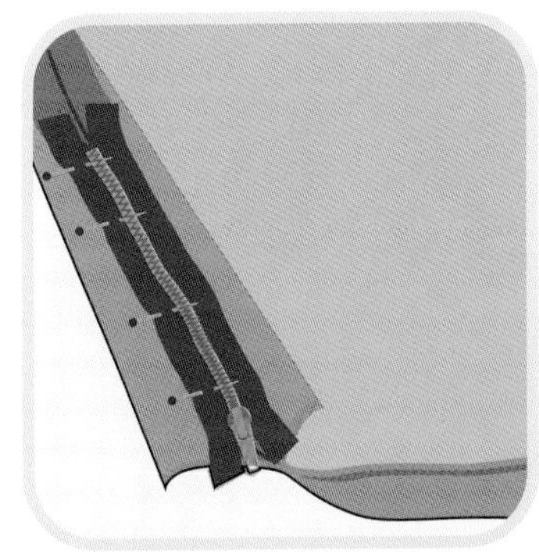
Pin zipper to one side of seam.

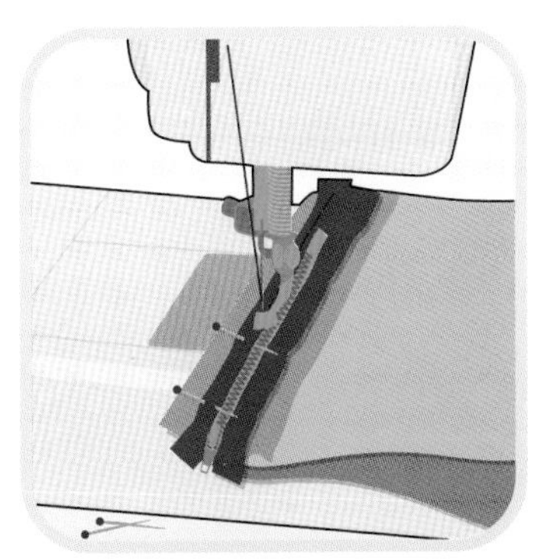
Baste pinned side onto fabric.

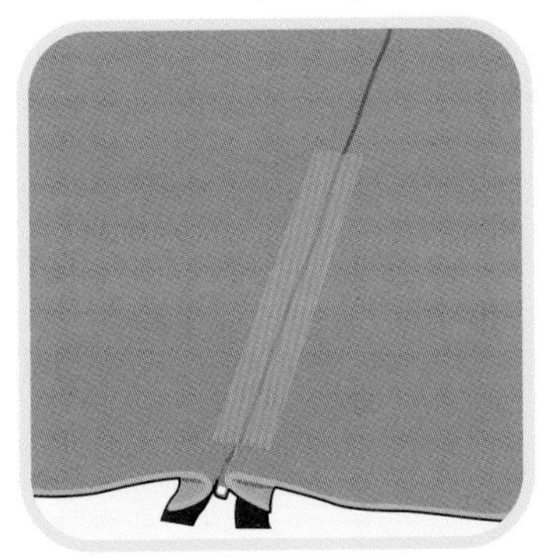
Affix clear tape as a guide.

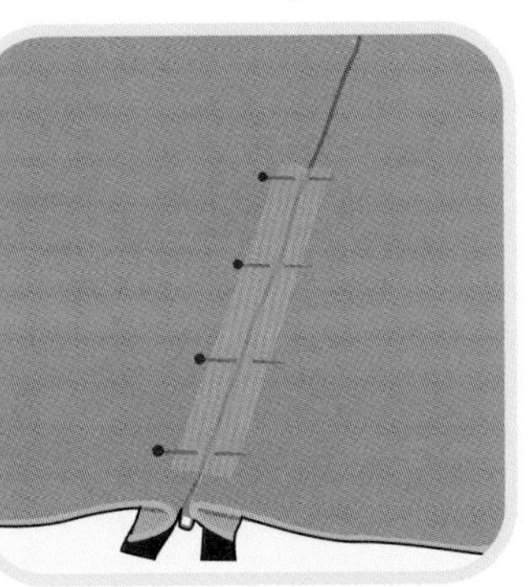
Secure fabric with pins.

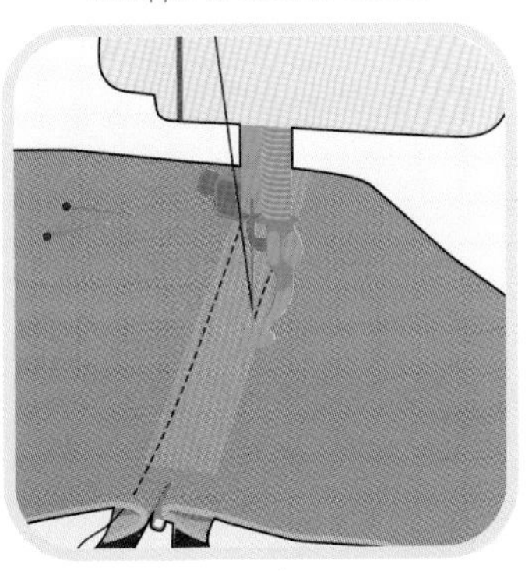
Sew zipper along tape.

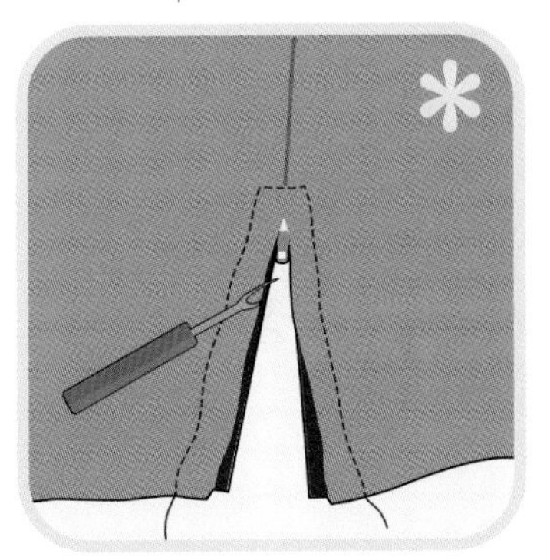
Remove tape; cut fabric seam open.

* The basting you're doing here has nothing to do with turkeys. It means putting a quickie stitch in place to hold fabric together. Remove the basting stitch when you're done with your project.

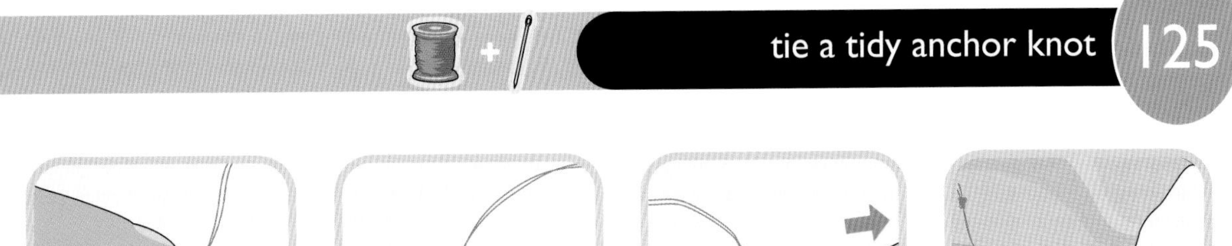

tie a tidy anchor knot 125

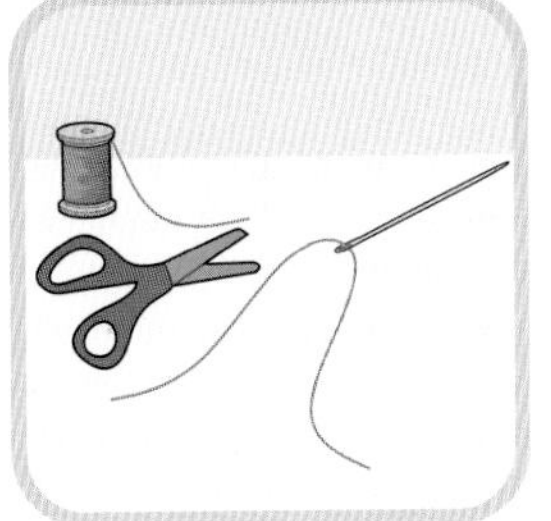
Thread and center needle.

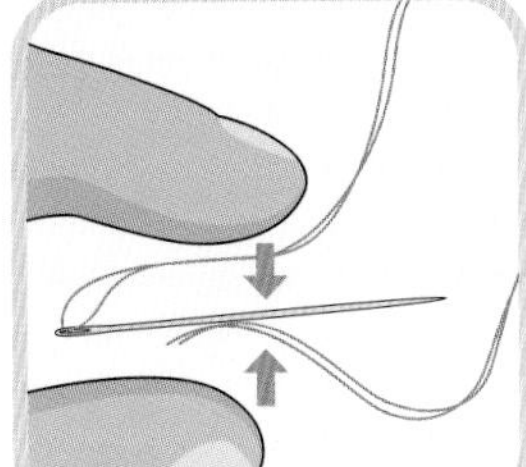
Pinch thread to needle.

Loop three times.

Pinch loops; pull needle out.

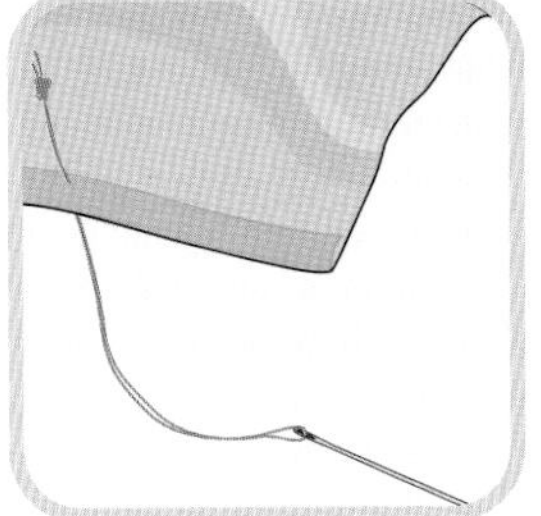
Use knot as anchor.

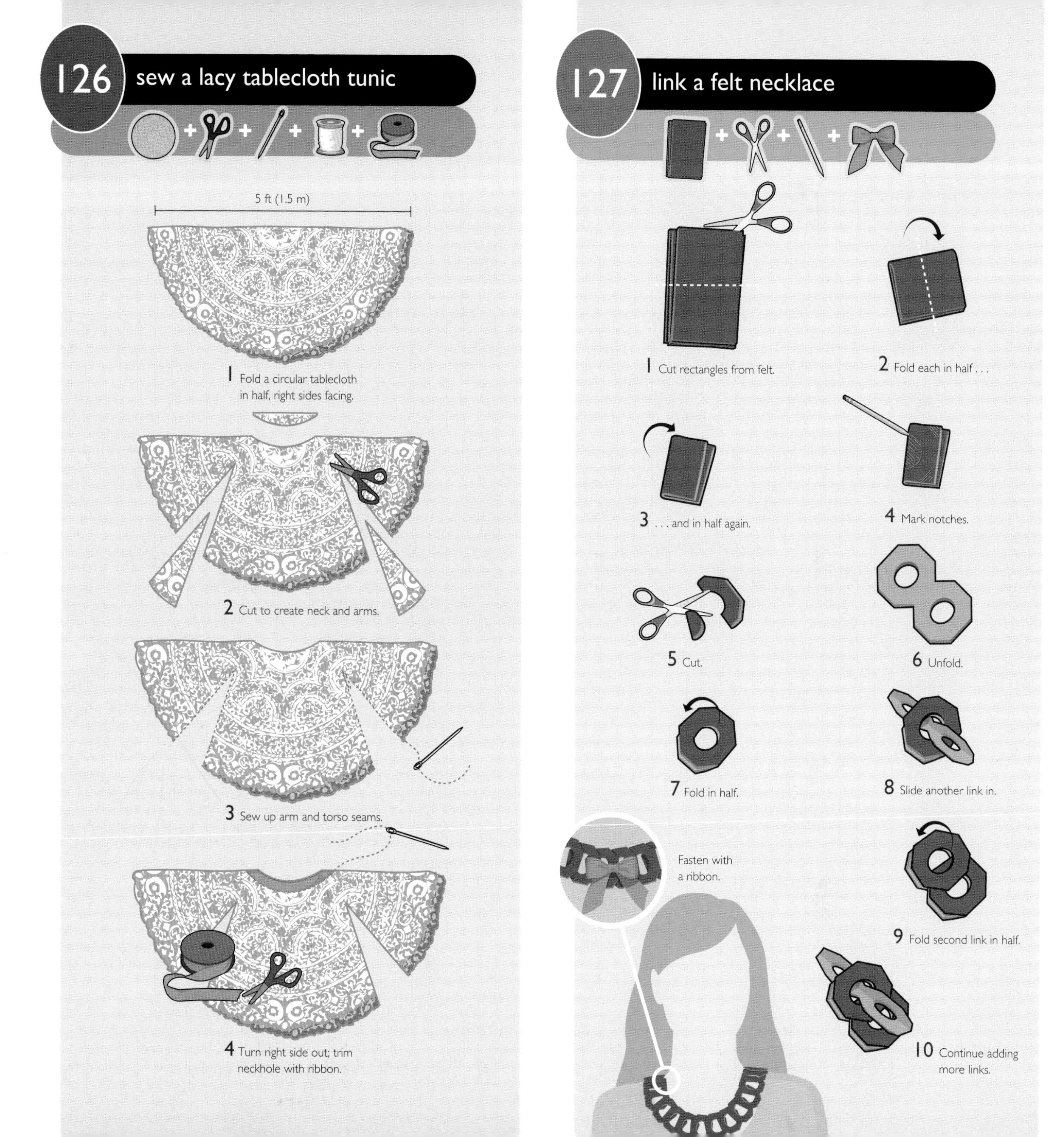
126
sew a lacy tablecloth tunic
5 ft (1.5 m)
1 Fold a circular tablecloth in half, right sides facing.
2 Cut to create neck and arms.
3 Sew up arm and torso seams.
4 Turn right side out; trim neckhole with ribbon.
127
link a felt necklace
1 Cut rectangles from felt.
2 Fold each in half . . .
3 . . . and in half again.
4 Mark notches.
5 Cut.
6 Unfold.
7 Fold in half.
8 Slide another link in.
Fasten with a ribbon.
9 Fold second link in half.
10 Continue adding more links.

romper

teddy

babydoll

bodysuit

merry widow

waist cincher

corset

bustier

garter belt

camisole

whip up edible undies 162

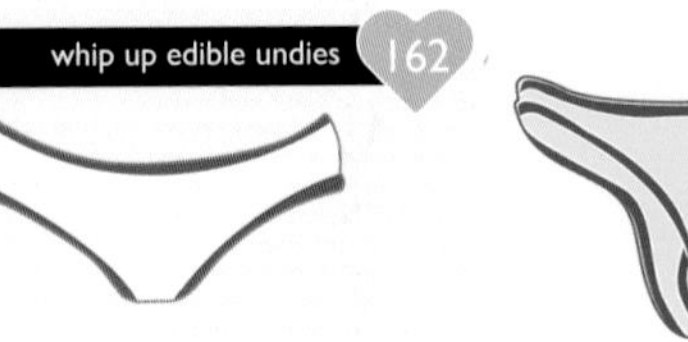
bikini

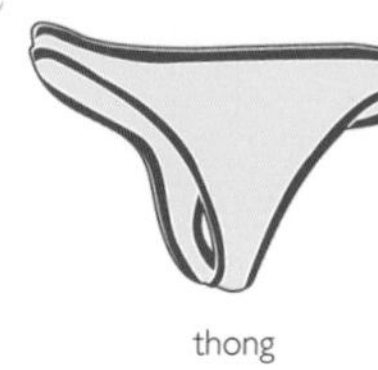
thong

g-string

boy shorts

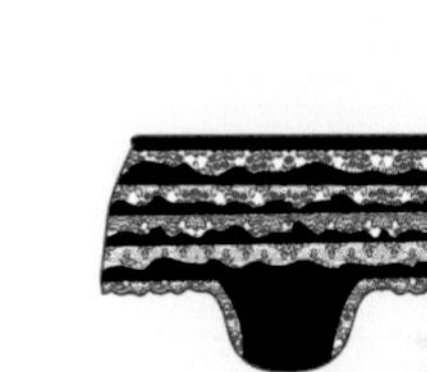
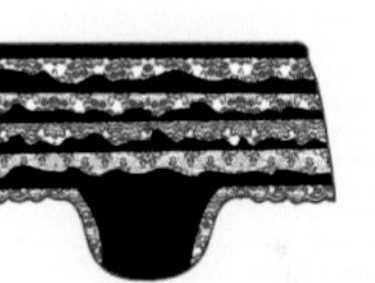
rhumba panties

peignoir

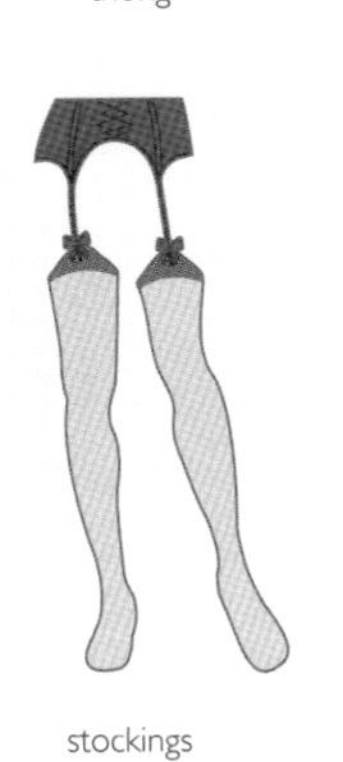
stockings

thigh-highs

catsuit

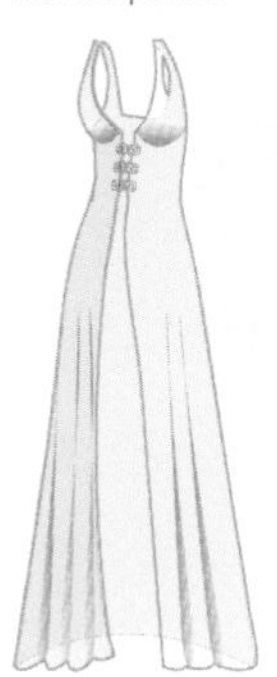
negligee

129 tie a dapper ascot

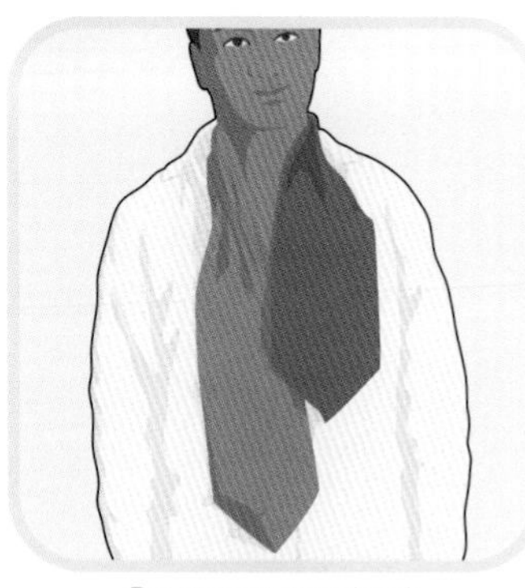
Drape ascot around neck.

Loop right end under left.

Cross right end back over.

Slide up inside, against neck.

Bring right end down.

Adjust to cover knot.

Tuck into shirt.

Fluff it out. Dandy!

130 identify men's furnishings

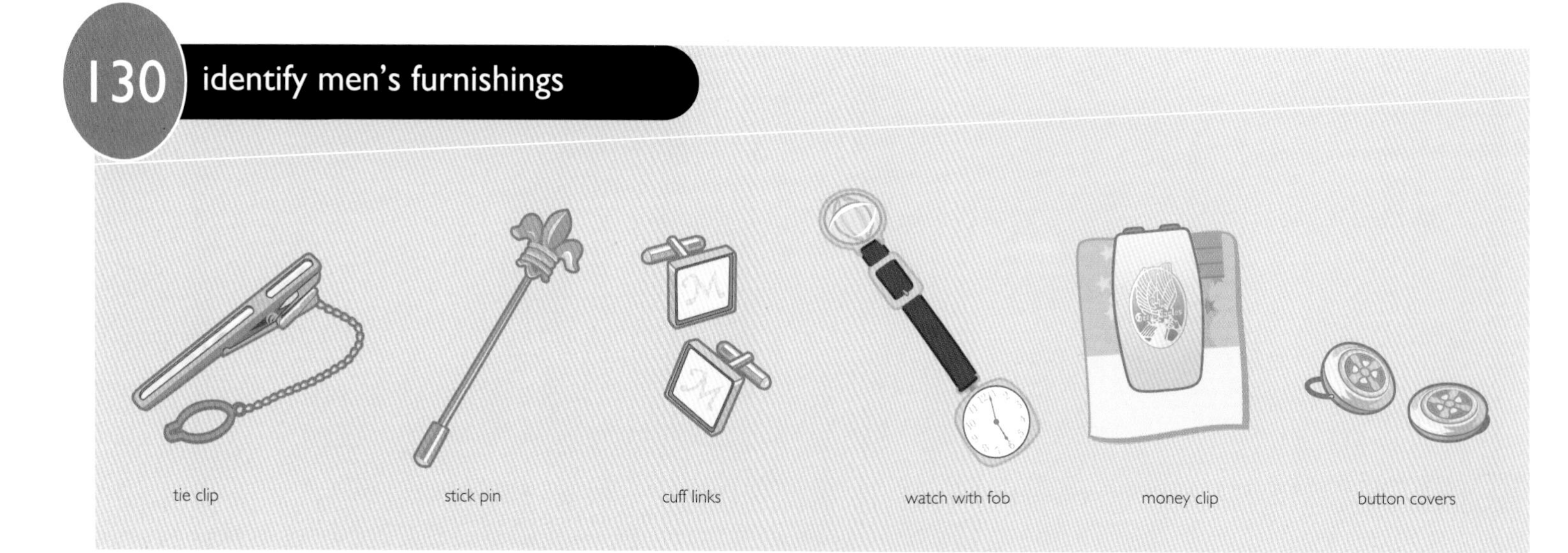

accessorize with a seat belt

131

Remove belt and buckle.

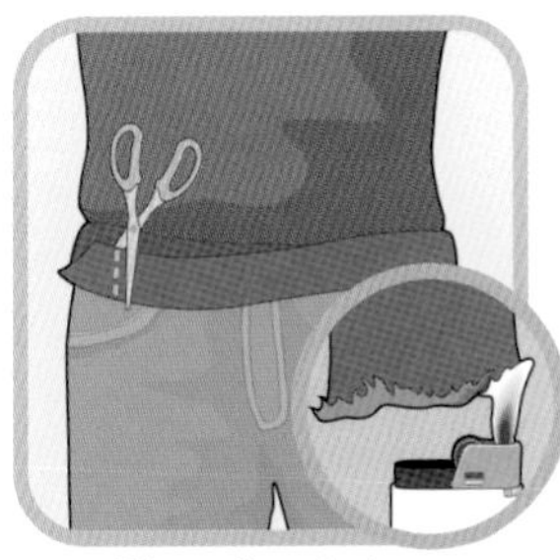
Trim to fit; seal edges.

Loop belt into loose buckle.

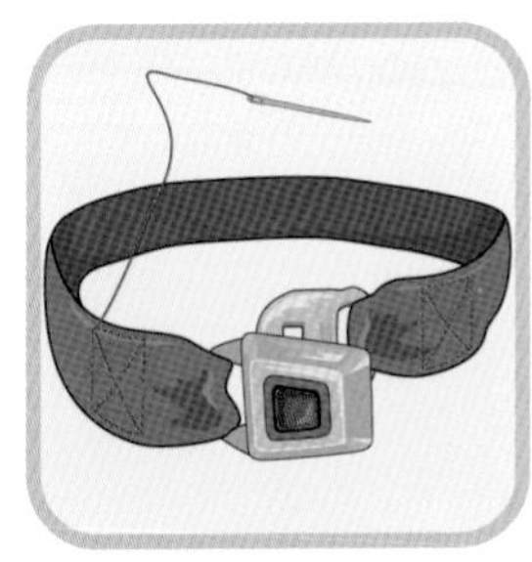
Secure with thread.

Buckle up!

carry a cassette-tape money clip

132

Unscrew; detach tape halves.

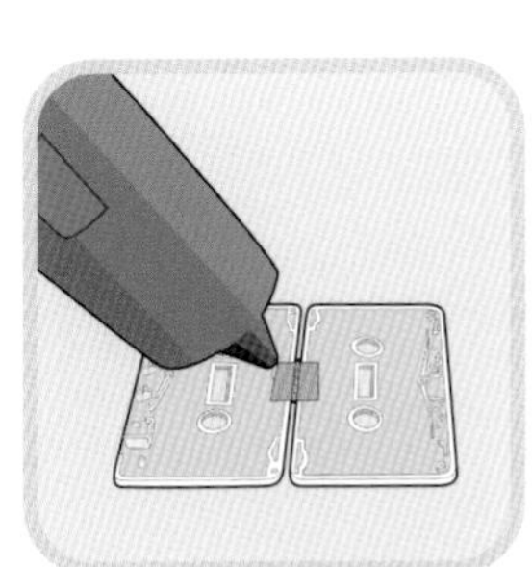
Glue in a hinge.

Attach paper clips.

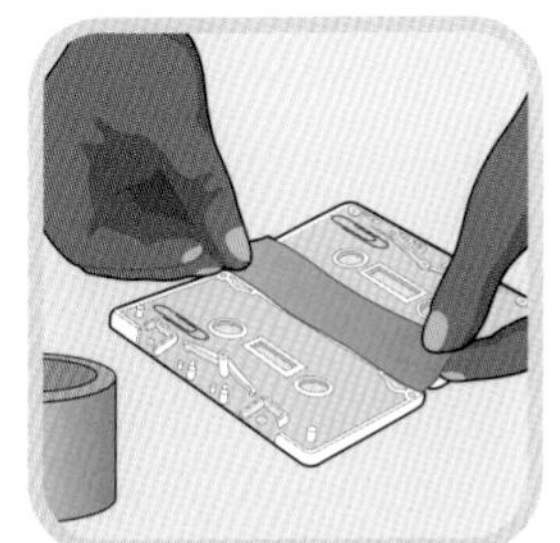
Add strip of duct tape.

Clip in cash and cards.

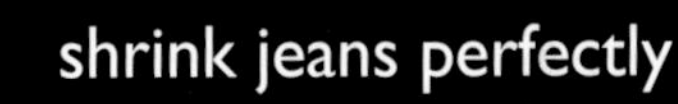

shrink jeans perfectly

133

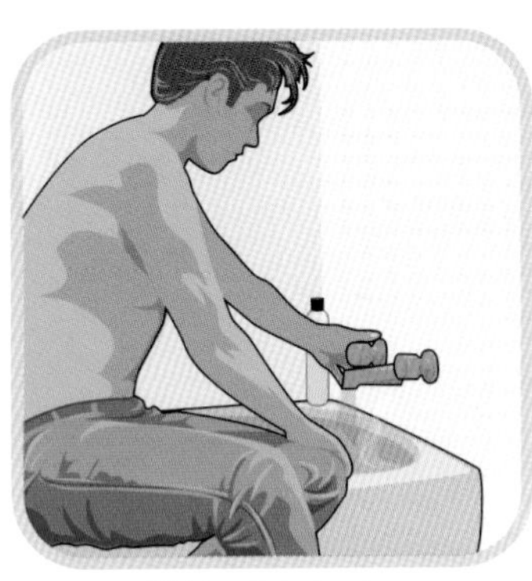
Fill tub with hot water.

Get in, wearing your jeans.

Stay in tub until water cools.

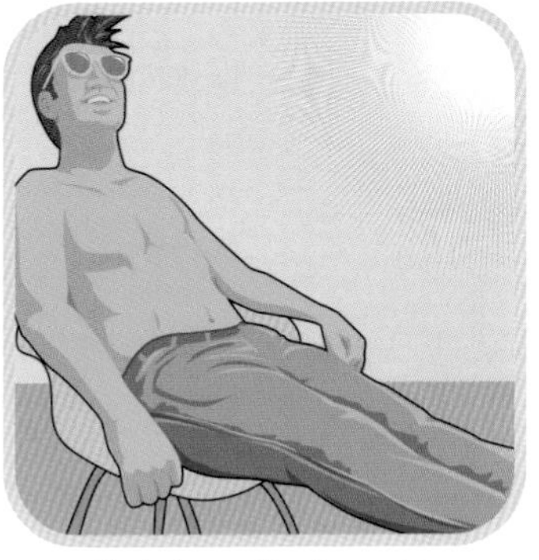
Air-dry jeans without removing.

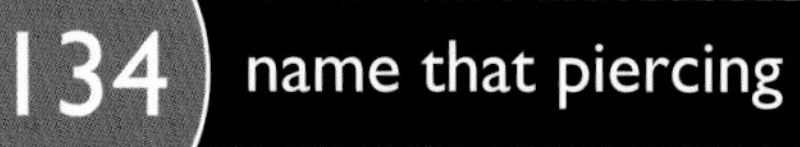

134 name that piercing

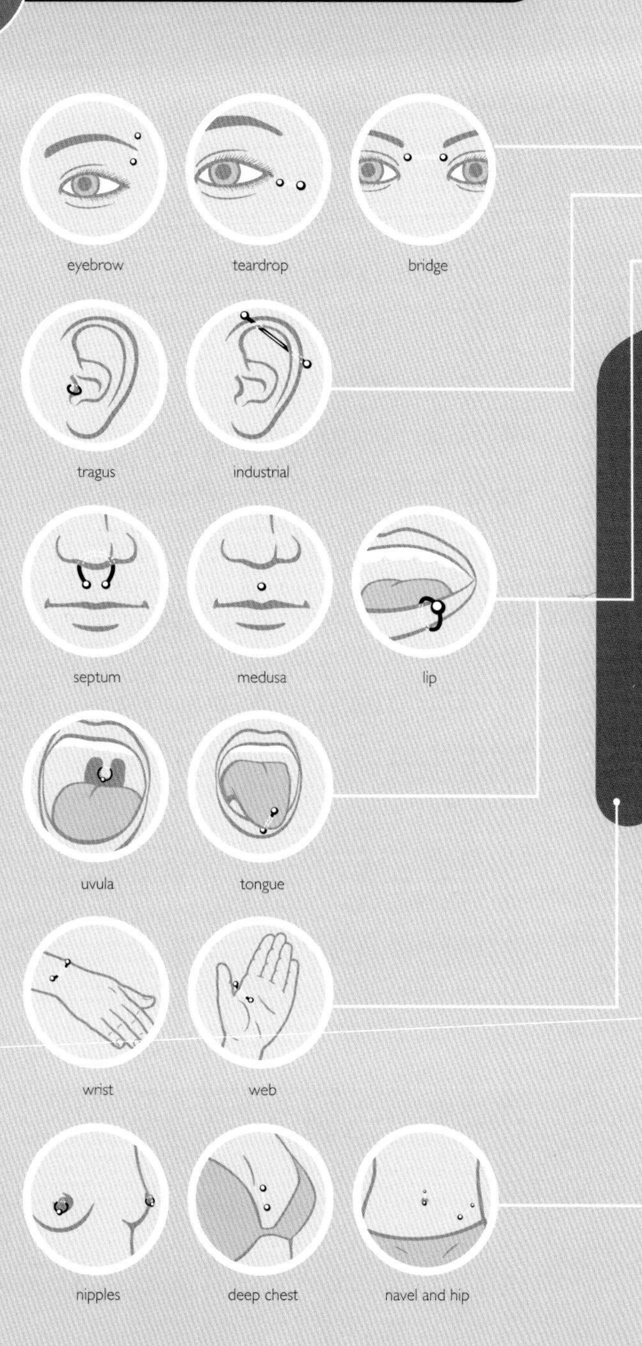

eyelet

plug

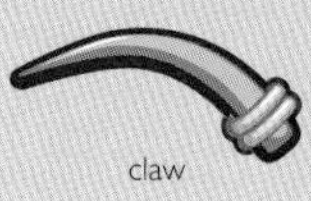

claw

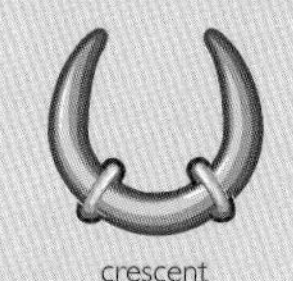

crescent

open spiral

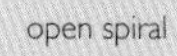

open ring

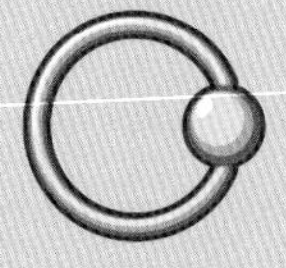

captive bead ring

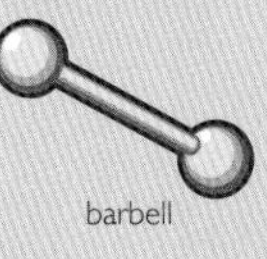

barbell

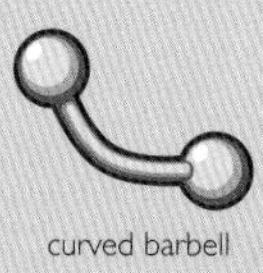

curved barbell

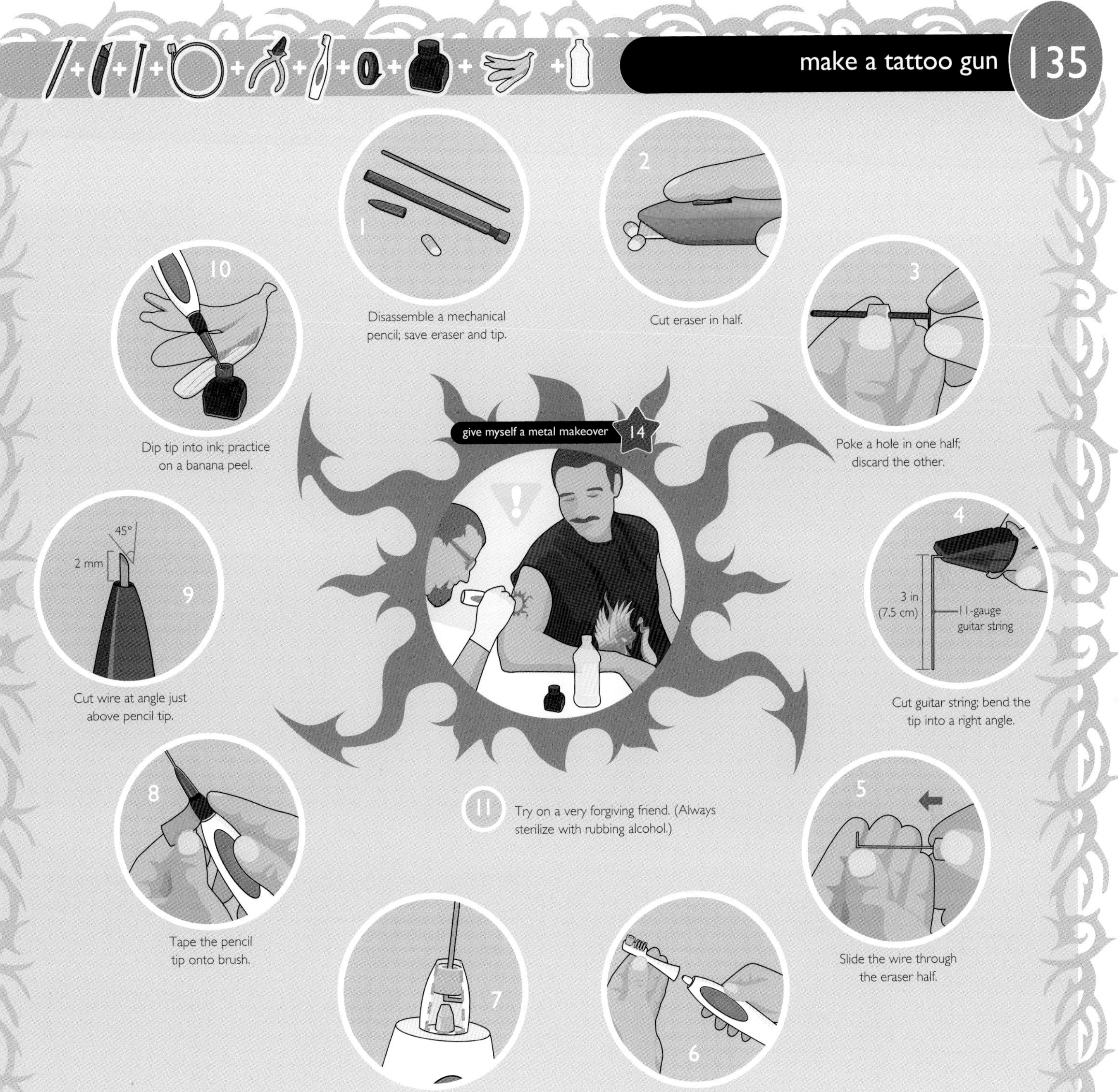
1
Disassemble a mechanical pencil; save eraser and tip.
2
Cut eraser in half.
3
Poke a hole in one half; discard the other.
4
3 in (7.5 cm)
11-gauge guitar string
Cut guitar string; bend the tip into a right angle.
5
Slide the wire through the eraser half.
6
Remove tip of toothbrush.
7
Push firmly onto metal shaft.
8
Tape the pencil tip onto brush.
9
45°
2 mm
Cut wire at angle just above pencil tip.
10
Dip tip into ink; practice on a banana peel.
give myself a metal makeover 14
11
Try on a very forgiving friend. (Always sterilize with rubbing alcohol.)

136 spot harajuku styles

Tokyo's Harajuku neighborhood is a mecca for fashionable youngsters, who flock there on Sundays to hang out, show off, and get inspired. Check out a few of the many styles on display.

Draw shape on cardboard; cut out.
Fan hair over template.
Saturate with gel.
Blow-dry.
Secure with hair spray.
bosozoku
long coat with kanji script
pompadour
headband
mechanic's jumpsuit
visual kei
brightly colored spiked hair
lipstick and heavy eyeliner for both genders
fur or feathers
futuristic clothing
decora
hair ornaments
pink lipstick
brightly colored, layered clothing
stuffed animals as accessories

love

138
charm love into my life
Entice new love into your life with these romantic charms.
Light a pair of candles as you visualize your dream lover.
Attract love with a vase of vanilla blossom, daffodil, and bachelor's button.
Spill your worries to worry dolls, then tuck the dolls under a pillow.
Place a love amulet under the bed.
Use a love potion as perfume or bath oil.
Carry moonstone, rose quartz, and jade with you.

mix up a love potion 139

one part lavender oil

one part sandalwood

one part ylang-ylang

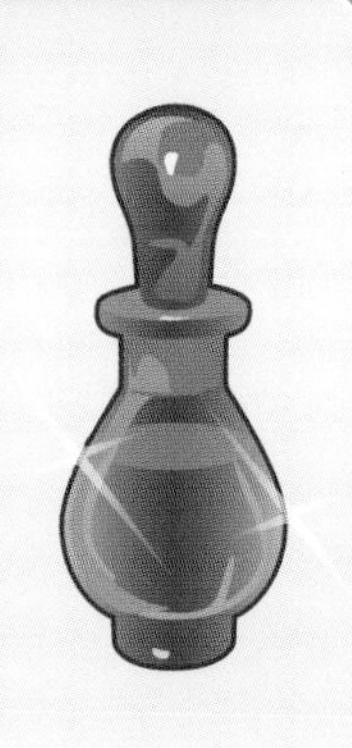

bathe my way to love 140

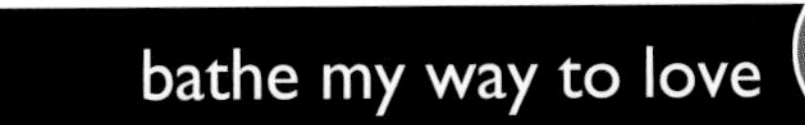

Describe your perfect lover.

Light a pink candle.

Place the list by the candle.

Draw a warm bath.

Add several drops of love potion.

Soak, visualizing your lover.

Blow out the candle.

Stash list; await your lover.

141 shop for love at the market

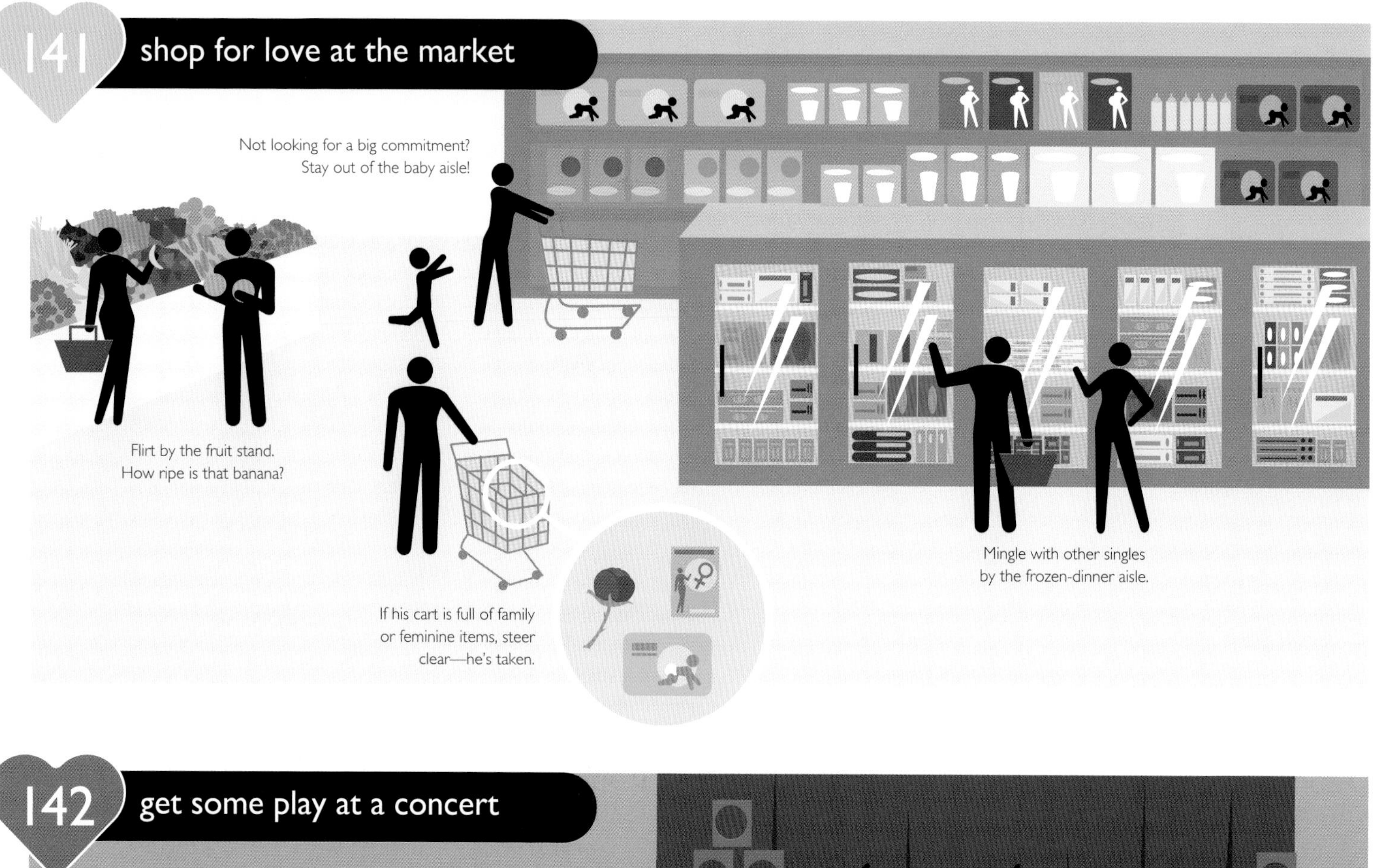

142 get some play at a concert

spot someone at the gym
143
Chat it up at the water fountain. Bonus points: let a cutie cut in line.
Strike a pose or two for a fetching stretching partner.
Work out next to someone who really gets your heart pumping.
Ask a hottie to spot—and then wow with your skills.

get carried away on the bus
144
Bring a prop that could spark a conversation.
Are you lost, or just hoping to be found? Ask for help from someone who's both good-looking and transit-savvy.
Oops! Good thing that handsome stranger was there to break your fall.

145 ace a school crush

Sit close to your crush.

Discuss interesting ideas in class.

Pass notes during lecture.

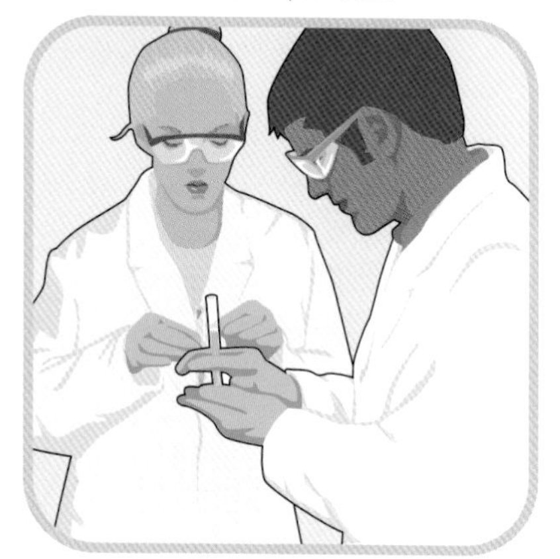
Team up on projects.

Study together outside school.

Celebrate after an exam.

146 win over a work crush

Make yourself noticeable.

Play footsie during meetings.

Have fun "working late."

Tryst in closets with locks.

Know the policy on dating.

find puppy love at a dog park 147

craft a dog bed 237

Go at busy times.

Spot the right dog (owner).

Share your toys.

Chat up the hot owner.

Respect dog-park etiquette.

Follow a routine.

score at a kids' soccer game 148

Dress to impress.

Share snacks and drinks.

Seek out single parents.

Get excited, within reason.

Plan to hang out post-game.

149 have fun with my partner

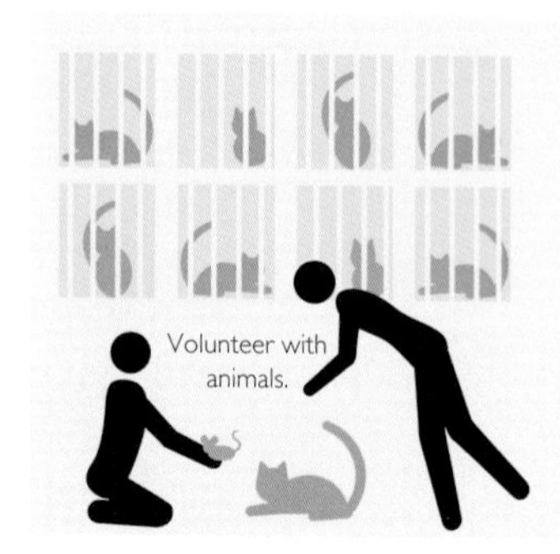

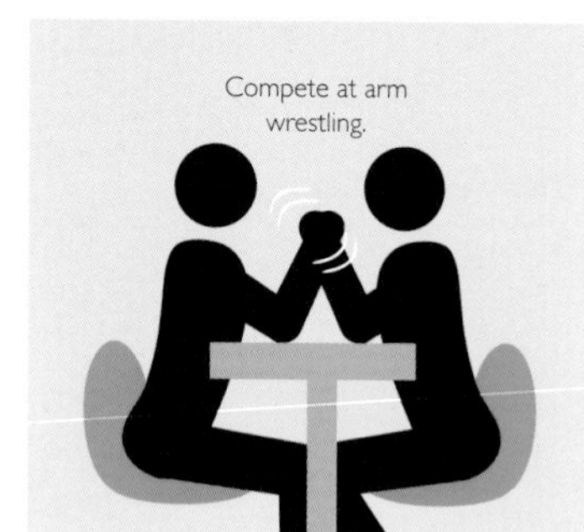

4 ace the balloon darts

Visit an aquarium.

jump mount a pony 386

Ride horseback.

Compete at a poetry slam.

Play cards.

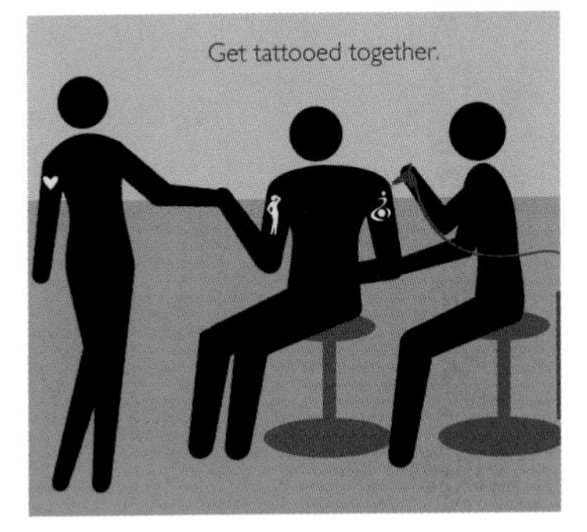
Get tattooed together.

Row a boat.

Cook a meal.

Construct a giant robot.

Hit the open road.

Relax at a spa.

Ride a tandem bike.

Take a wild ride.

Learn to fence.

Play a duet.

Build a snowman.

Go rock climbing.

150 shoot a hot home movie

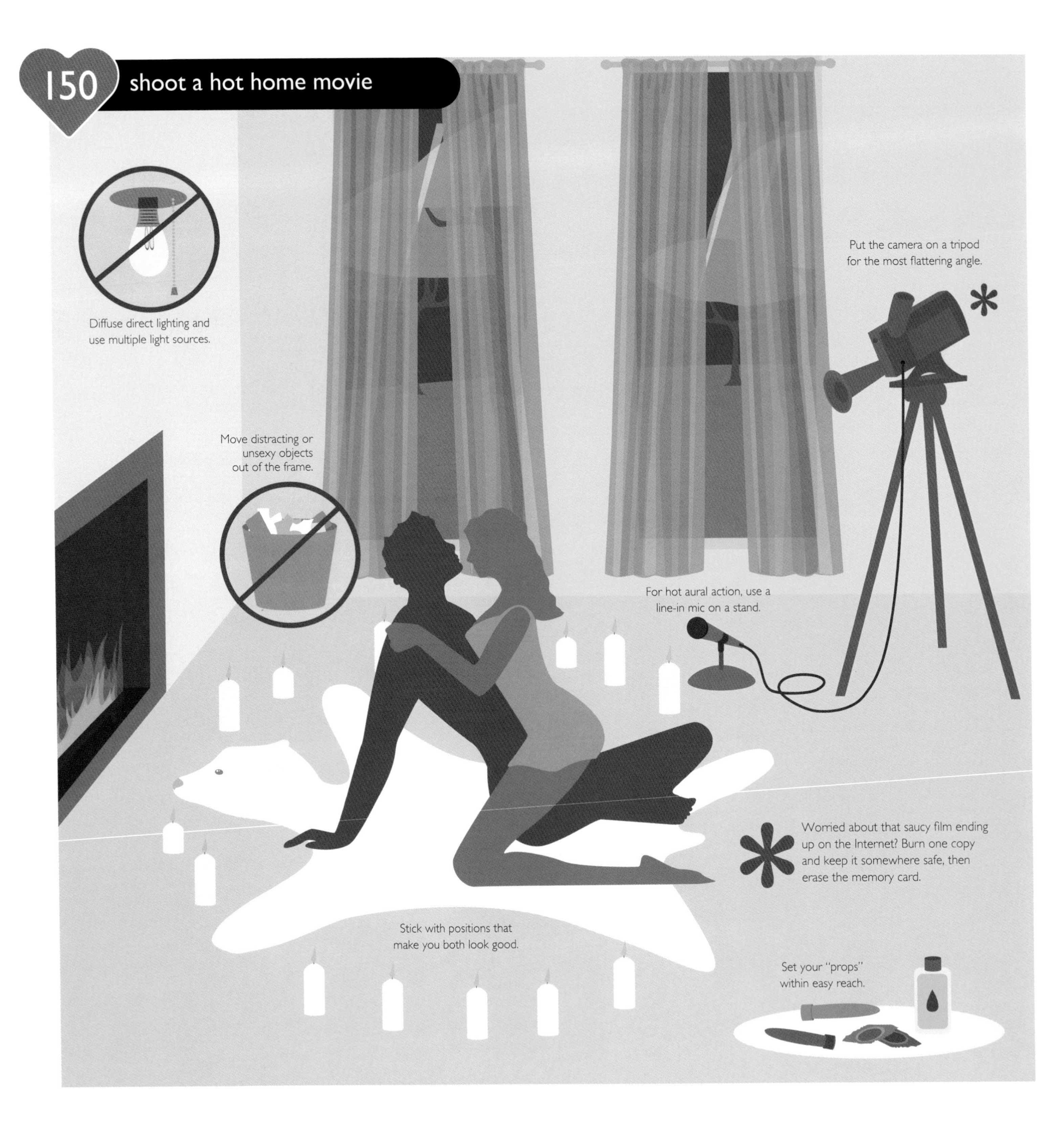

151 snap a racy cell-phone shot

Remove distracting objects.

Light yourself from the front.

Shoot from shoulder level.

wax body hair at home 106

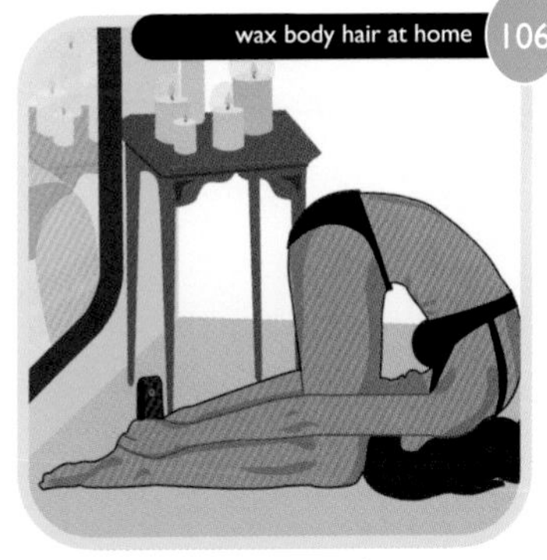
Try interesting angles.

152 play the sexy stranger game

Agree on fantasy details.

Dress the part.

Arrive separately.

Flirt with your eyes.

Send your lover a drink.

Introduce yourself.

Leave together.

Complete the seduction at a hotel.

153 rock the boat

Remove potential hazards.

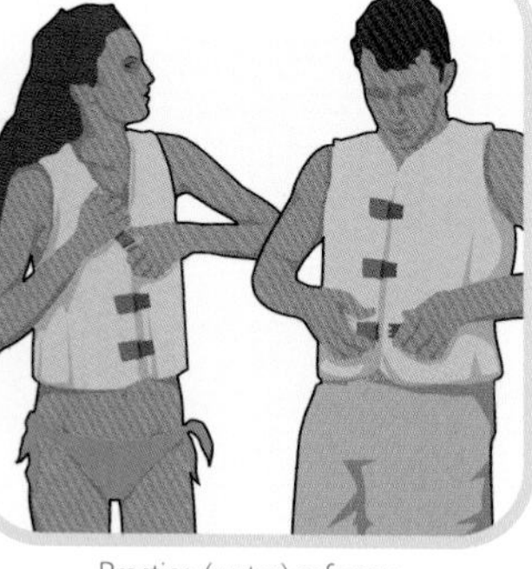

Practice (water) safe sex.

Paddle far from shore; drop anchor.

Don't rock too hard!

154 master parallel parking

Be sure car is clean.

Park in a secluded spot.

Turn off all lights.

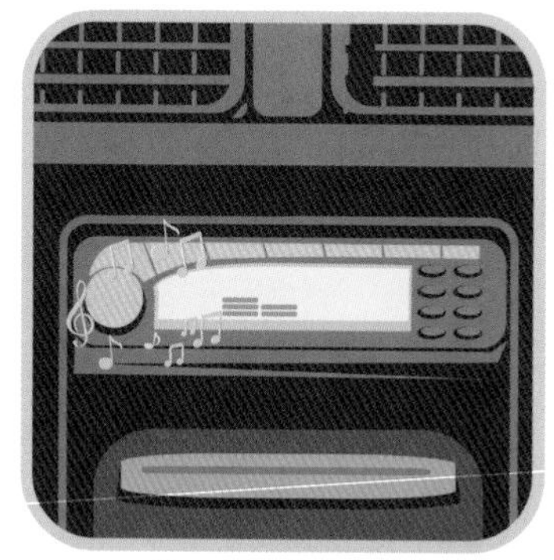

Turn radio down low.

Move to the passenger seat.

Recline the seat slightly.

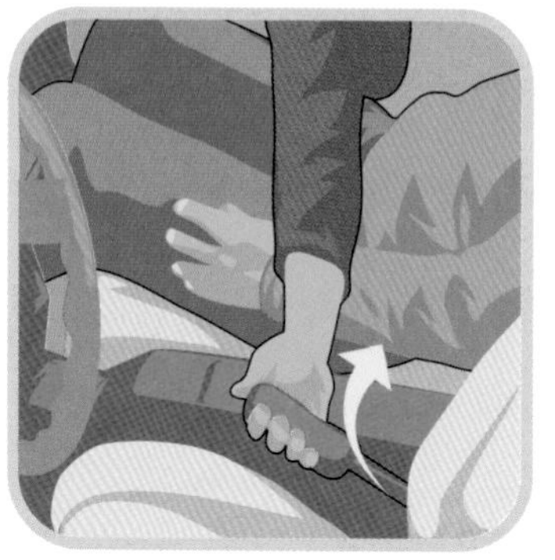

Apply the parking brake.

Time to park 'n' ride.

make love in an elevator 155

Dress for ease of access.

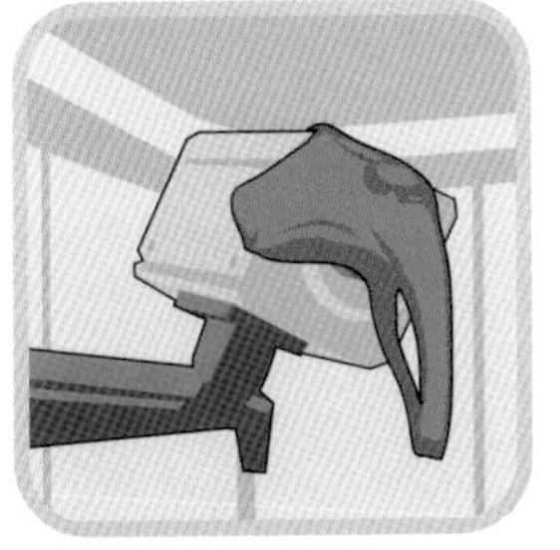

Cover the security camera.

Hit "stop" between floors.

Be efficient.

Exit nonchalantly.

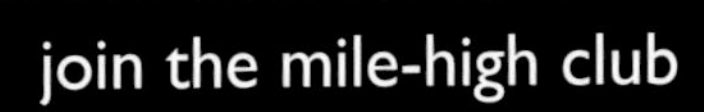

join the mile-high club 156

Nighttime flights are best.

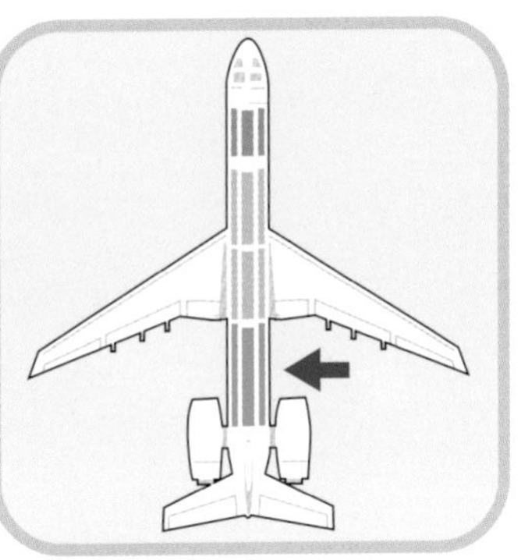

Get a seat at the back.

Get started under a blanket.

Get up while everyone's distracted.

Lock the door behind you.

Lay down paper towels.

Be quick.

Welcome to the club!

157 select aphrodisiacs

Due to their stimulating flavors (or their suggestive shapes), these foods have been decreed aphrodisiacs by various cultures around the world—while others have been deemed decidedly unsexy. Bon appétit!

banana

avocado

ginger

pine nuts

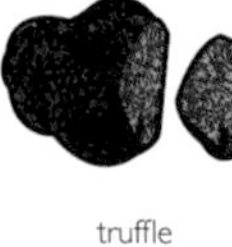
truffle

carrots

honey

almonds

garlic

sweet basil

sparrow

pineapple

chocolates

figs

skink

watercress

lentils

tofu

158 serve sushi extra-raw

159 eat sweets off my sweetie

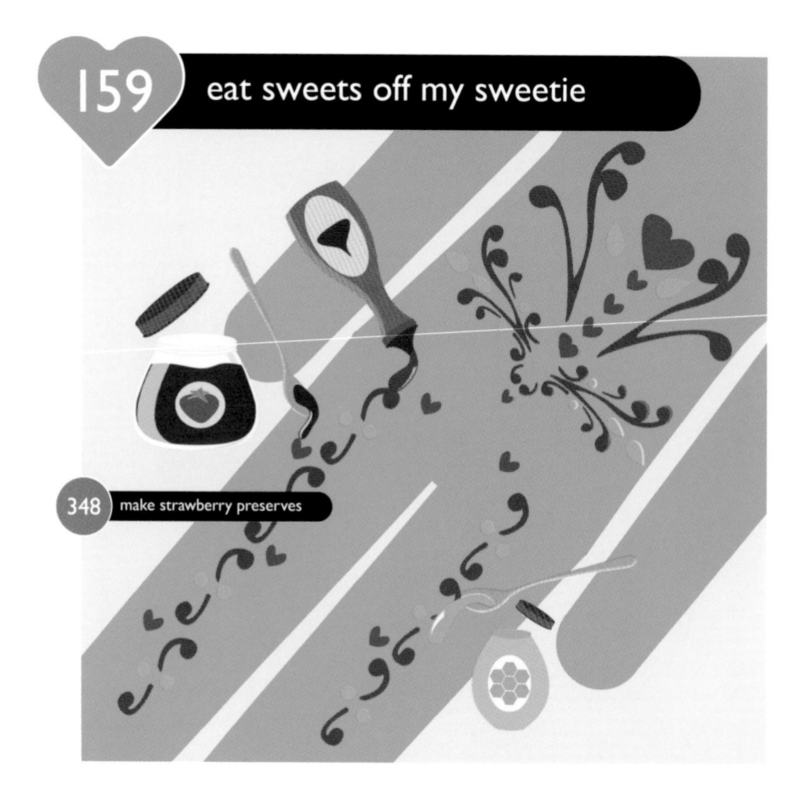

348 make strawberry preserves

get my just desserts 160

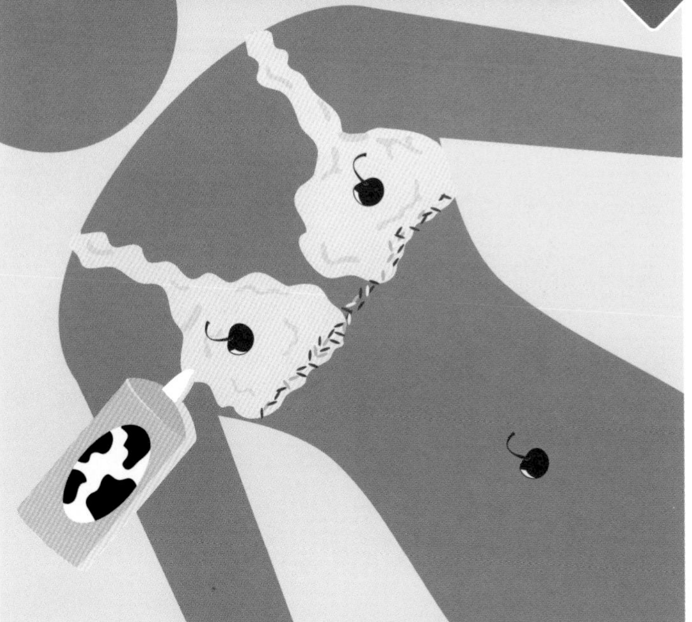

have a delicious morning after 161

whip up edible undies 162

Cover pot and simmer.

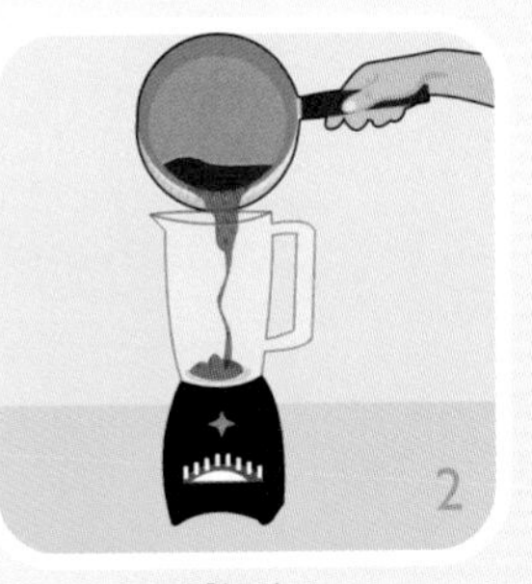

Blend.

Pour onto lined baking pan.

Bake.

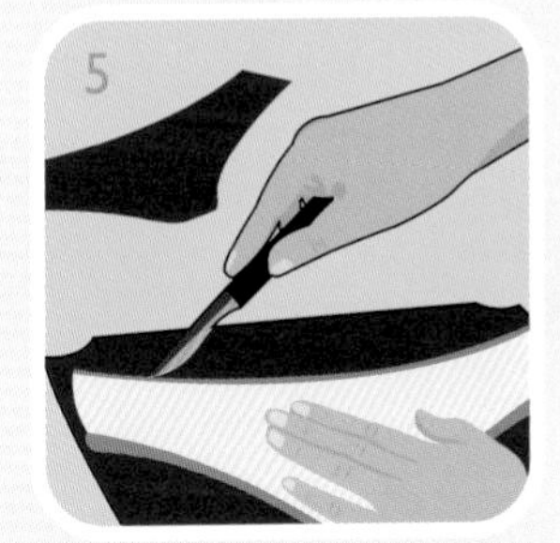

Use panties as a template.

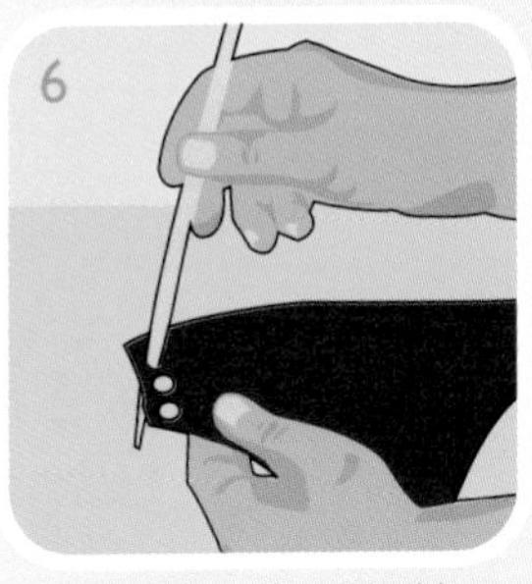

Poke holes with a chopstick.

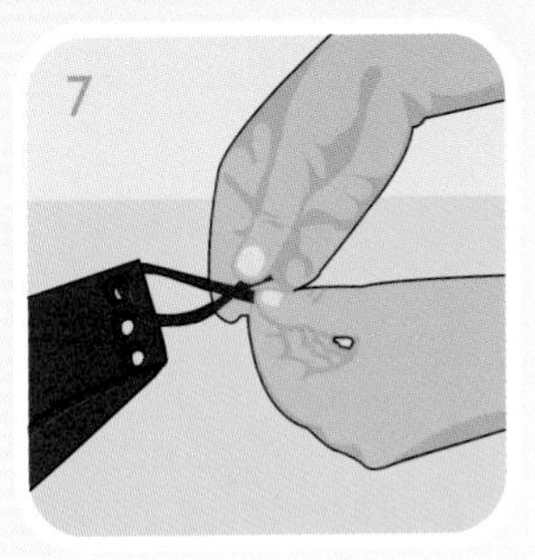

Tie together with licorice.

Serve at once!

163 pick perfect bridal attire

No matter what your shape or style, there is an ensemble out there to make you look fantastic on your special day.

sheath
tall; slim-hipped

ball gown
hourglass figure

mermaid
slender figure

empire waist
small-breasted; petite

sweetheart neckline
large bustline

bateau neck
ample bosom

high collar
all body types

low v-neck
small bustline

chapel train
elegant event

watteau train
informal; outdoors

cathedral train
formal wedding

royal train
state event

fingertip veil
any dress

waltz veil
dress without train

mantilla veil
simple gown

cathedral veil
formal gown

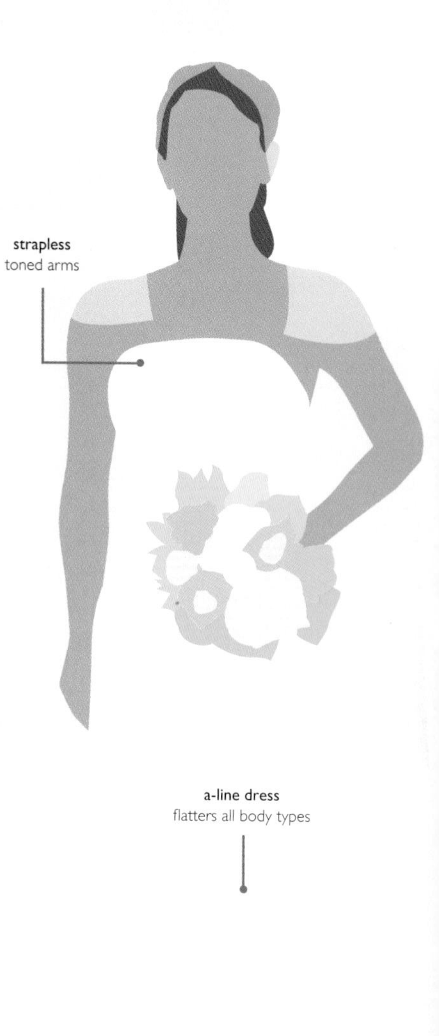

strapless
toned arms

a-line dress
flatters all body types

brush train
appropriate for any wedding

Having trouble figuring out what to wear? Your venue and style of ceremony will narrow those choices for you.

euro tie
wear with wing or spread collar

vest
add personality with colors or fabrics

tuxedo jacket
single- or double-breasted

dinner jacket
outdoors; evening

cutaway tux
formal daytime

mandarin jacket
informal; modern

tailcoat tuxedo
formal evening

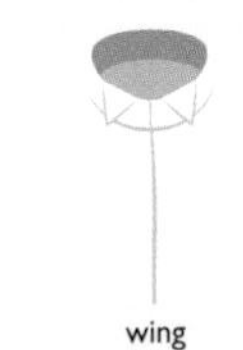

wing
tuxedo; bow tie

tunnel
tuxedo; no tie

banded
mandarin jacket; no tie

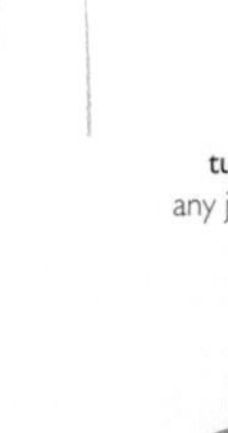

turndown
any jacket or tie

necktie
dinner jacket

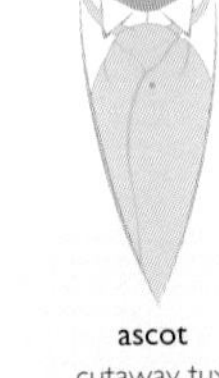

ascot
cutaway tux

bolo
turndown collar

bow tie
tuxedo; wing collar

cuff links
add personality and flair

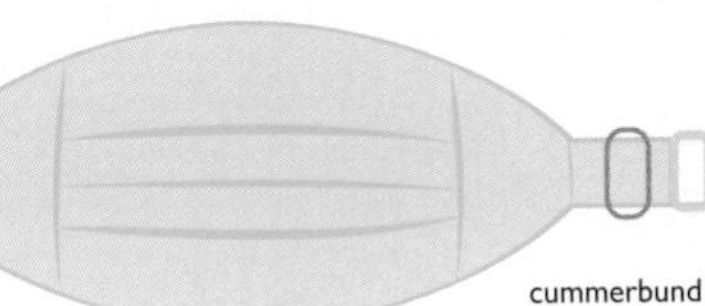
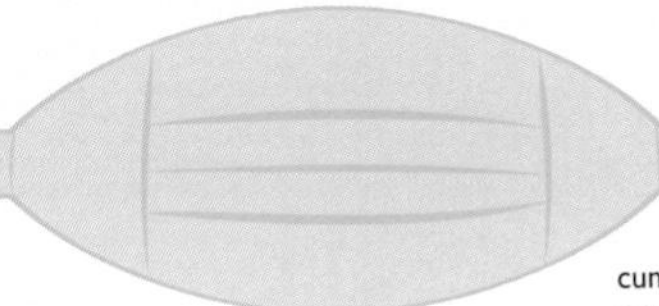

cummerbund
match color to bridesmaids' dresses

165
say "i do" italian-style
Guests shower the newlyweds with Jordan almonds.
A lump of iron in the groom's pocket wards off evil spirits.
A rip in the veil brings good luck.
The pieces of a broken glass symbolize the many happy years ahead.

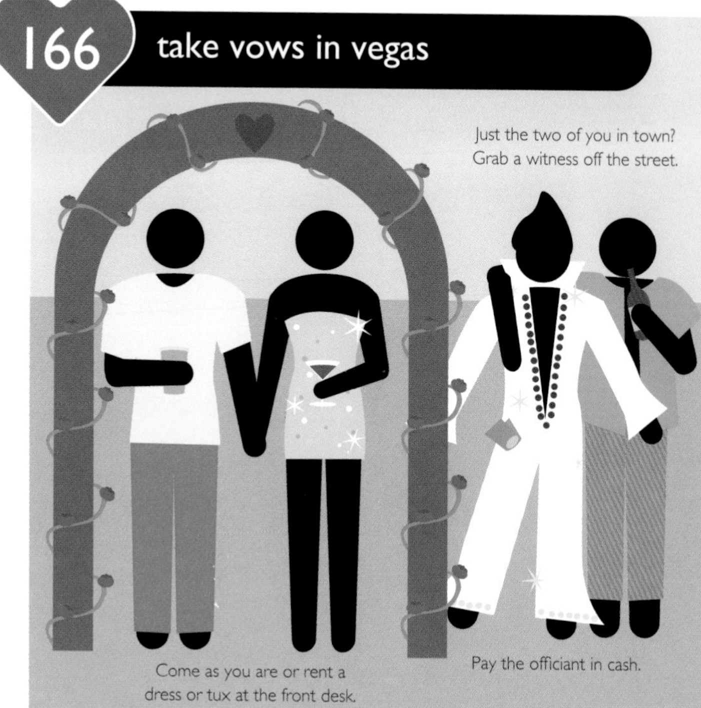
166
take vows in vegas
Just the two of you in town? Grab a witness off the street.
Come as you are or rent a dress or tux at the front desk.
Pay the officiant in cash.

167
get married in the philippines
Children bear wedding rings and coins.
An heirloom rosary is part of the bride's bouquet.
A veil and cord are placed on bride and groom to symbolize their unity.

168
forge a bond in china
Hire lucky lion dancers to entertain guests.
The couples' parents are honored with a tea ceremony.
The couple makes a vow by drinking from wine glasses tied together with string.
306 celebrate chinese new year's day

hear wedding bells in cuba
169
Family or friends hold lace above the couple during the ceremony.
A rowdy party procession follows the couple to and from the courthouse.
Serve a delicious roast pig.
get hitched in hungary
170
Arrive in a horse-drawn carriage.
Hire a Roma band to play a spirited serenade.
Bride and groom exchange coins and handkerchiefs.
The bride carries a bouquet for herself and one for Saint Stephen.
She breaks an egg to ensure that they have healthy kids.

tie the knot in nigeria
171
The father of the bride offers prayers and ceremonial kola nuts.
The groom hides among the guests, and the bride seeks him out.
She offers him palm wine; when he drinks, they are married.
plan a persian wedding
172
Female friends and relatives grind sugar onto fabric held above the couple.
The wedding spread includes candelabras, special foods, and a mirror that reflects the couple.

nest

173 fix a sagging door

Prop door with magazines.

Unscrew bracket from wall.

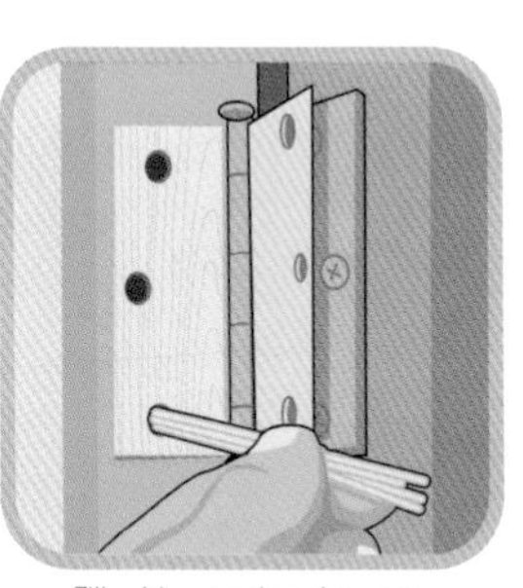
Fill with wooden skewers.

Trim skewers flush to wall.

Screw bracket into wall.

A door sags when the wood around a hinge screw gets worn down or stripped. Fill the hole with bamboo skewers to give the screw something solid to thread into.

174 replace a broken tile

Cut along grout with grinder.

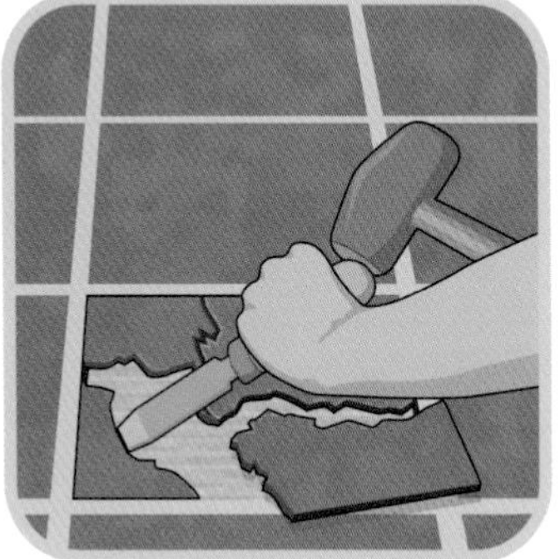
Chip away the broken tile.

Scrape the mortar bed clean.

Apply mortar to new tile.

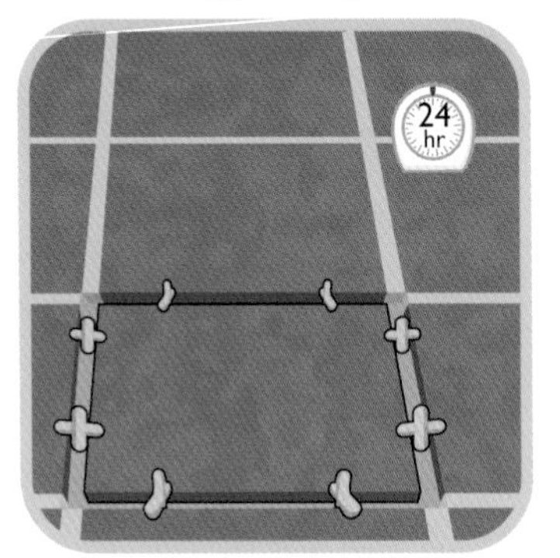

Add spacers; let dry.

Secure tile with grout.

Wipe off excess; let dry overnight.

Apply sealant.

mend a cracked shingle 175

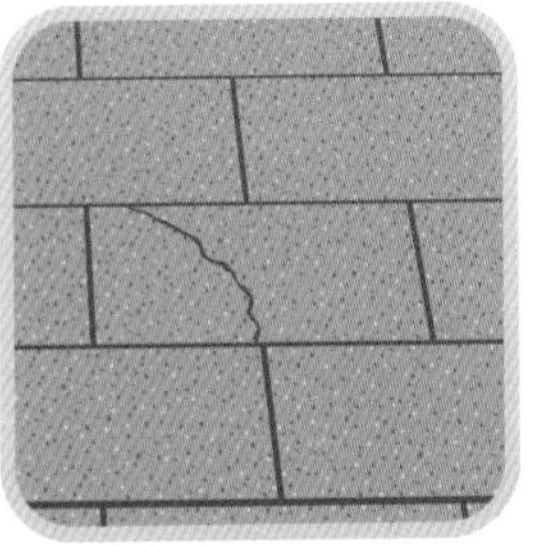

Locate cracked tile.

Apply roofing cement.

Put cement on crack.

Sprinkle gravel from gutter.

Nail down edges.

patch a rain gutter 176

Clean crack in gutter.

Spread roofing cement.

Add fiberglass mesh.

Layer a piece of plastic.

Cover with roofing cement.

seal with caulk 177

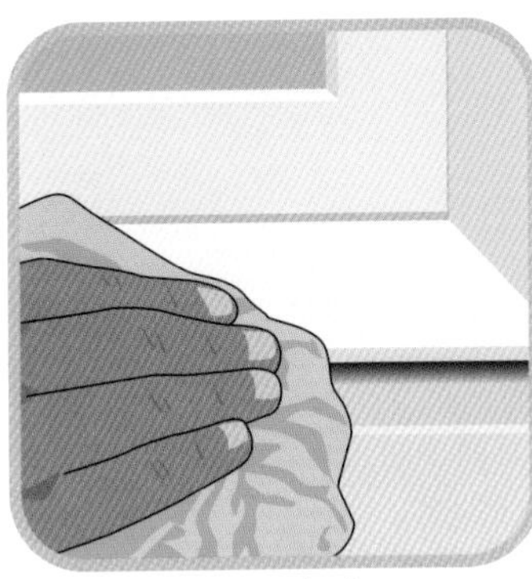

Clean area; let dry.

Apply more caulk than needed.

frost a perfectly smooth cake 358

Smooth with wet finger.

Smooth with damp sponge.

178 strip and splice wires

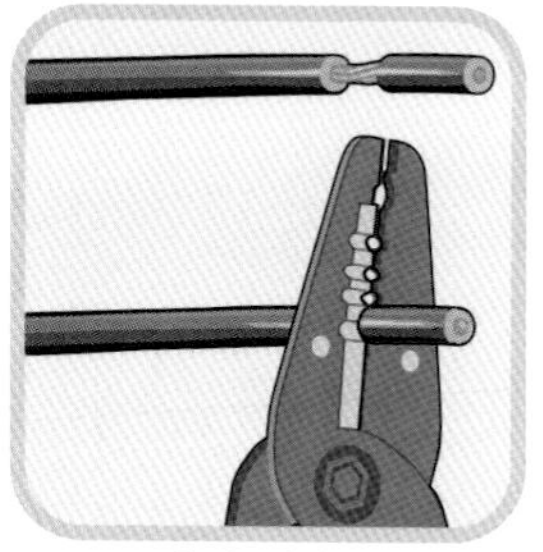
Cut insulation.

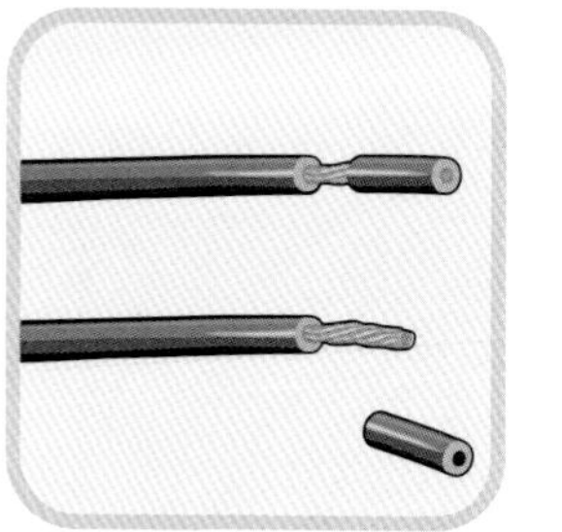
Remove, exposing wires.

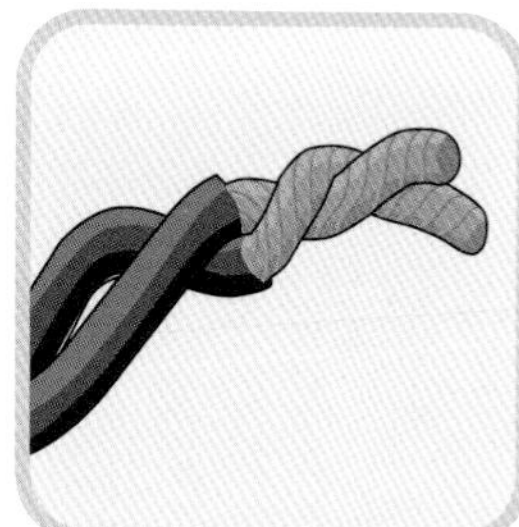
Twist wires together.

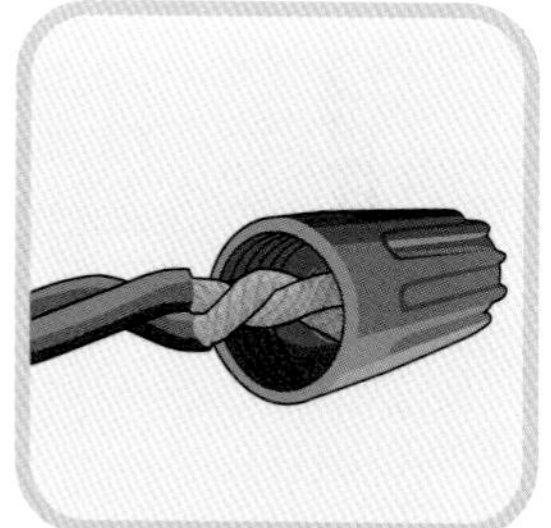
Push into wire nut.

Seal with electrical tape.

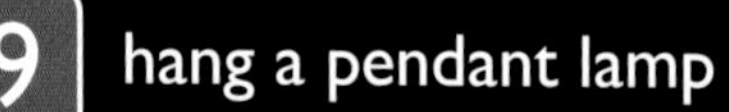

179 hang a pendant lamp

Shut off appropriate fuse.

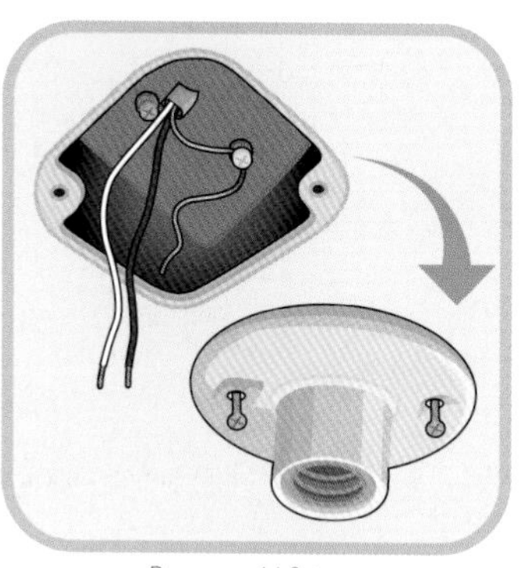
Remove old fixture.

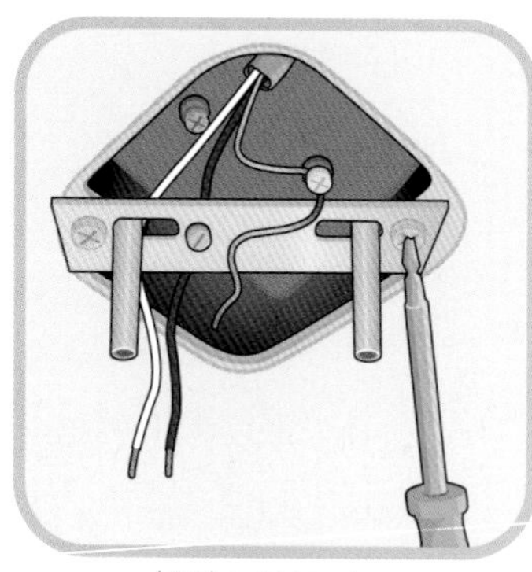
Attach a canopy strap.

Measure desired length; add extra; cut.

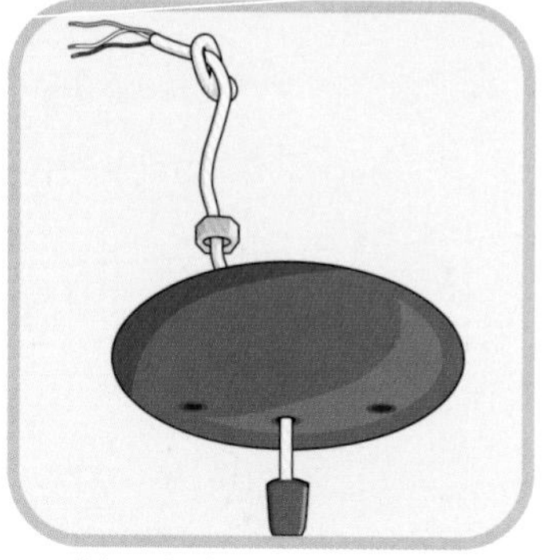
Bring wire through base; knot end.

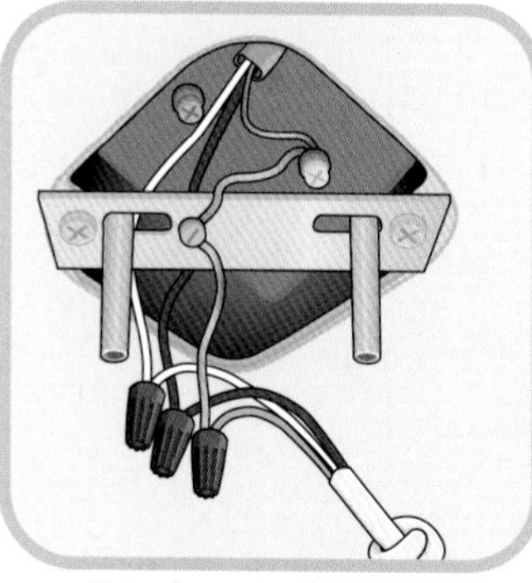
Strip wires and splice together.

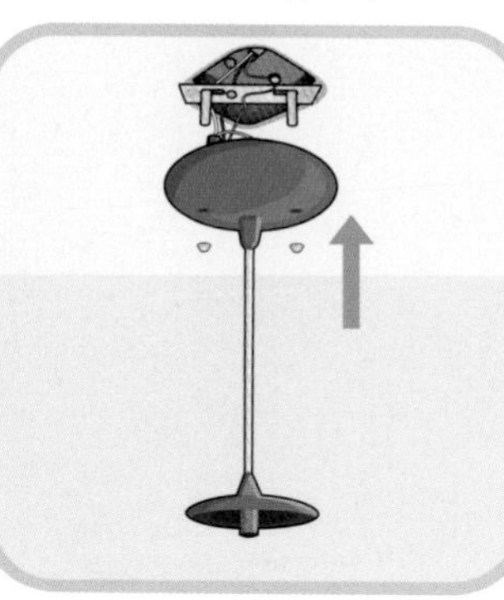
Attach lamp base to canopy strap.

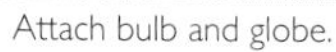
Attach bulb and globe.

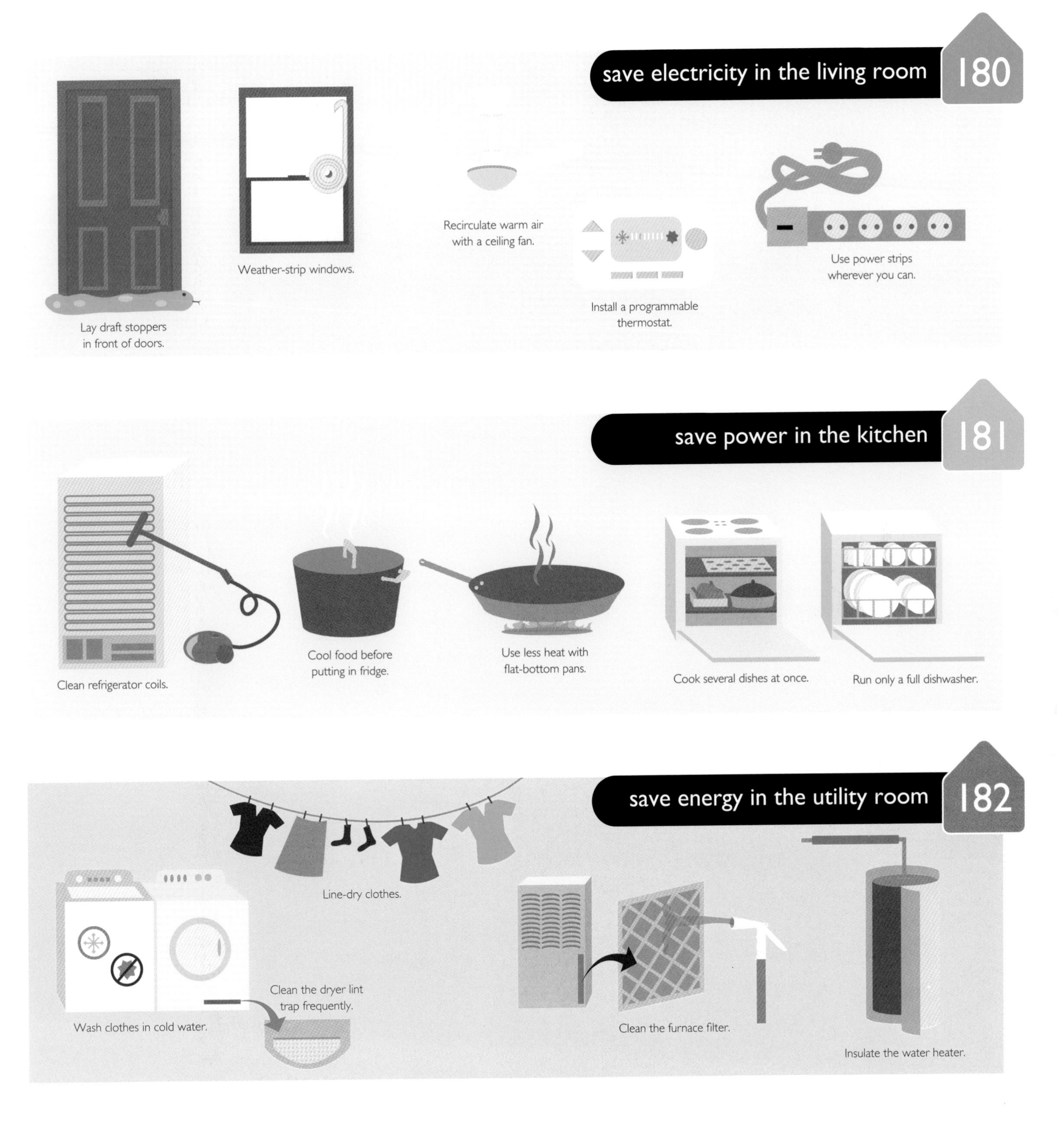
save electricity in the living room
180
Lay draft stoppers in front of doors.
Weather-strip windows.
Recirculate warm air with a ceiling fan.
Install a programmable thermostat.
Use power strips wherever you can.
save power in the kitchen
181
Clean refrigerator coils.
Cool food before putting in fridge.
Use less heat with flat-bottom pans.
Cook several dishes at once.
Run only a full dishwasher.
save energy in the utility room
182
Line-dry clothes.
Wash clothes in cold water.
Clean the dryer lint trap frequently.
Clean the furnace filter.
Insulate the water heater.

183 braid a denim rug

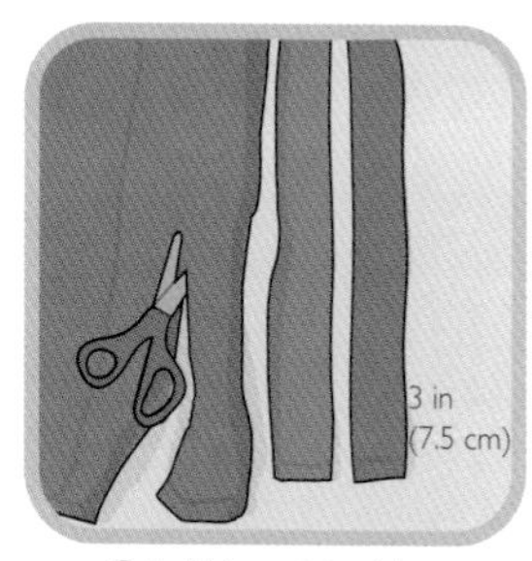

Cut old jeans into strips.

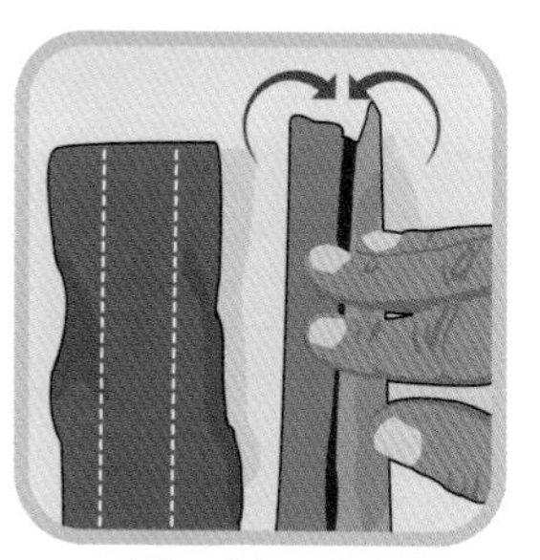
Fold each into thirds.

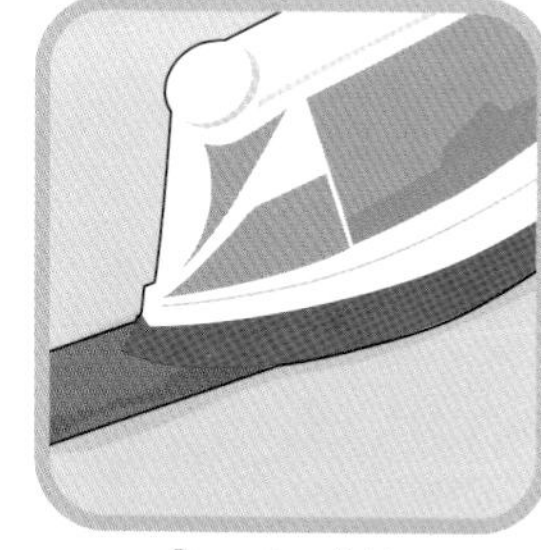
Press along folds.

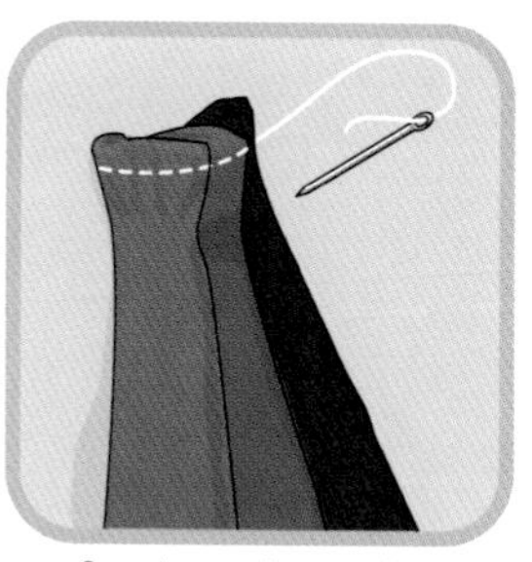
Sew three strips together.

Braid.

Add new strips.

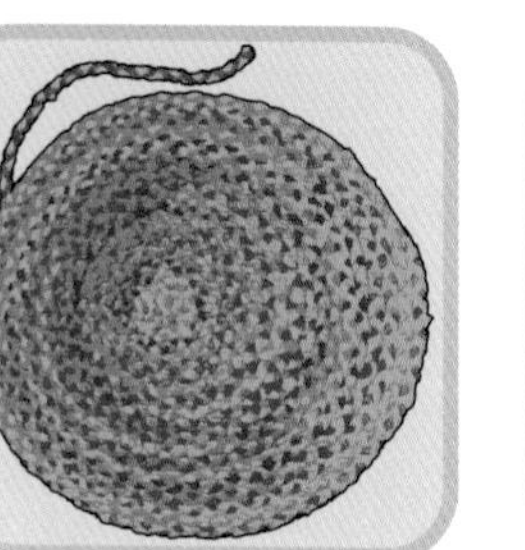
Coil braid to desired size.

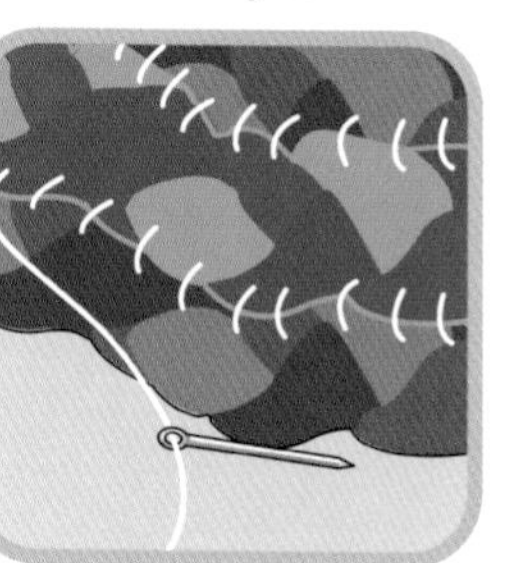
Stitch coils together.

Sew a small patch.

Cover end with patch.

When you come to the end of a strip, sew in a new one with a diagonal seam. This way, the seam won't look lumpy when folded in thirds. Don't have a pile of old jeans to cut up? Try using old wool blankets—it makes for a supersoft alternative.

184 hang a shimmering paillette wall

Paint corkboard wall color.

Let dry; snap chalk-line grid.

Attach corkboard to wall.

Nail paillettes onto grid.

Bask in the shimmer.

185 construct a cardboard chaise

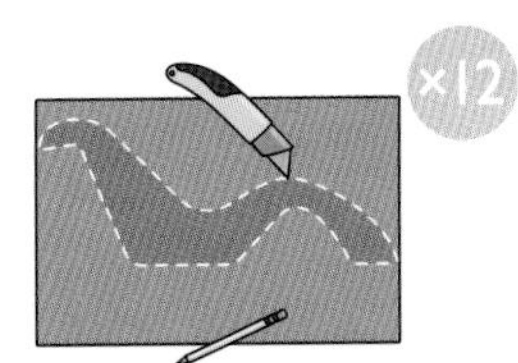

1 Cut chair shape from cardboard.

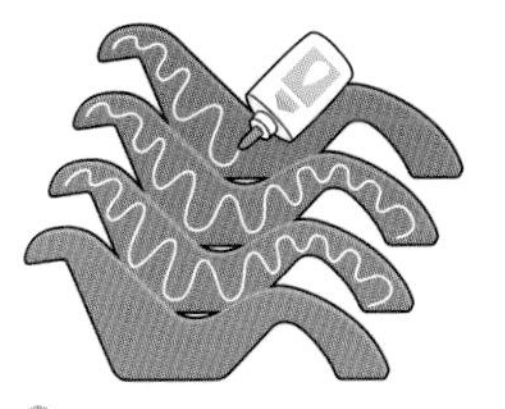

2 Glue into three sets of four.

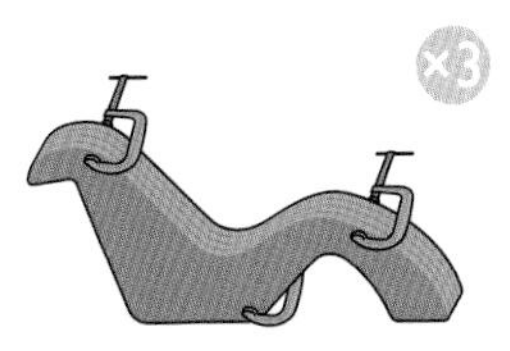

3 Clamp each set individually; let dry.

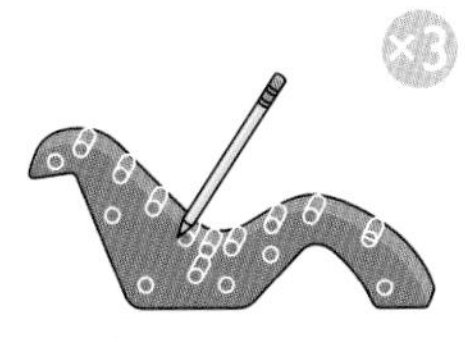

4 Mark notches and holes in each set.

5 Cut each set.

6 Line up sets; glue cardboard tubes.

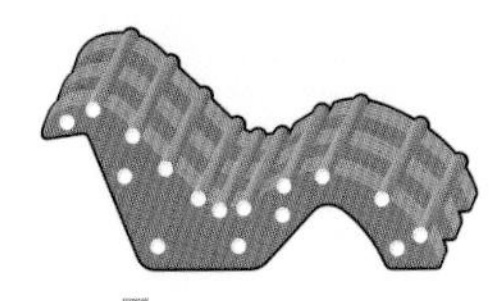

7 Let dry.

186 re-cover a kitchen chair

Remove chair's seat.

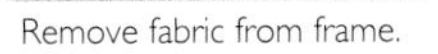

Remove fabric from frame.

Trace old fabric; cut.

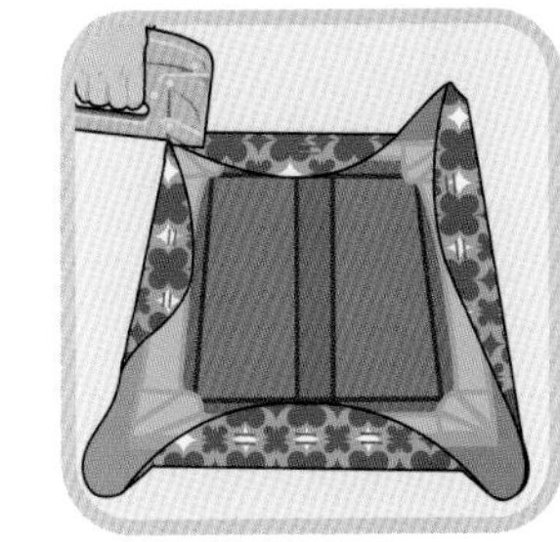

Staple fabric tautly to frame.

Reattach seat.

231 freeze a snack for my octopus
187 paint a mural using a grid
Draw image and trace a grid over it.
Draw a grid on the wall.
Copy each box onto the wall.
Erase the grid.

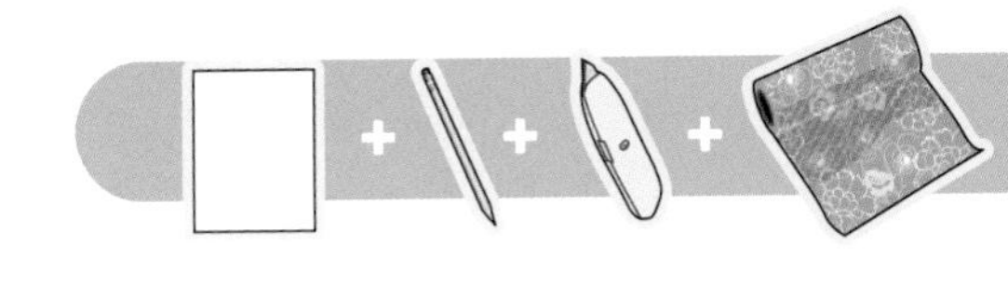

design a wallpaper headboard 188

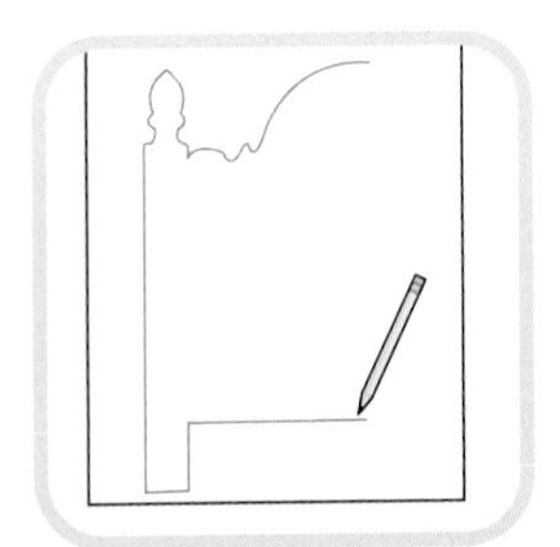
Draw half the design.

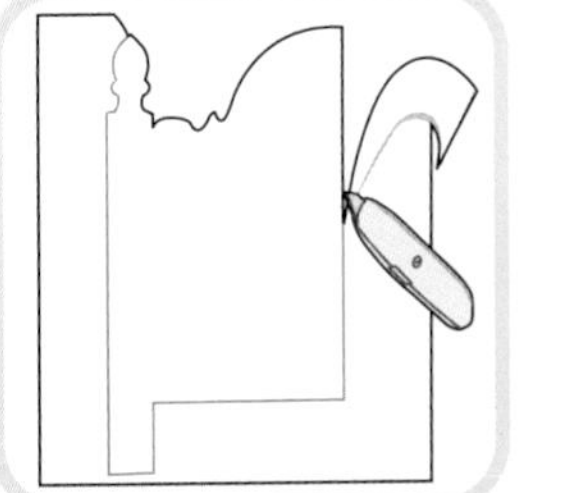
Cut to make template.

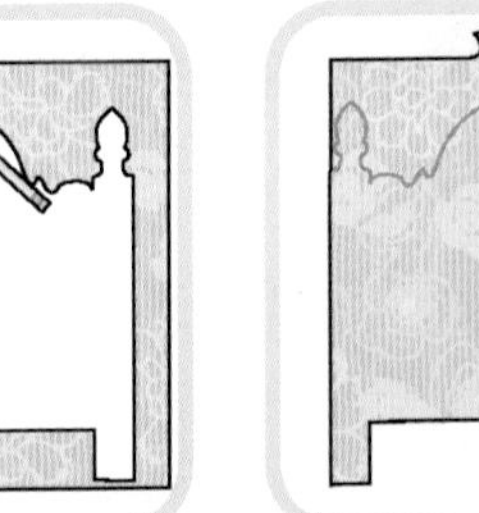
Trace twice onto wallpaper.

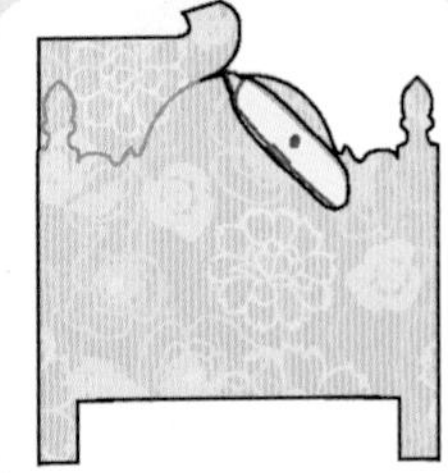
Cut out.

Hang wallpaper as directed.

Want to go a little crazy in bed design?
Use these silhouettes for inspiration.

cut out wallpaper silhouettes 189

To make wallpaper art, draw a grid over a source image, or use one of these examples. Mark a larger grid on the back of a sheet of wallpaper and copy the original frame by frame, just as you would do with the wall mural. Then cut out and hang as directed.

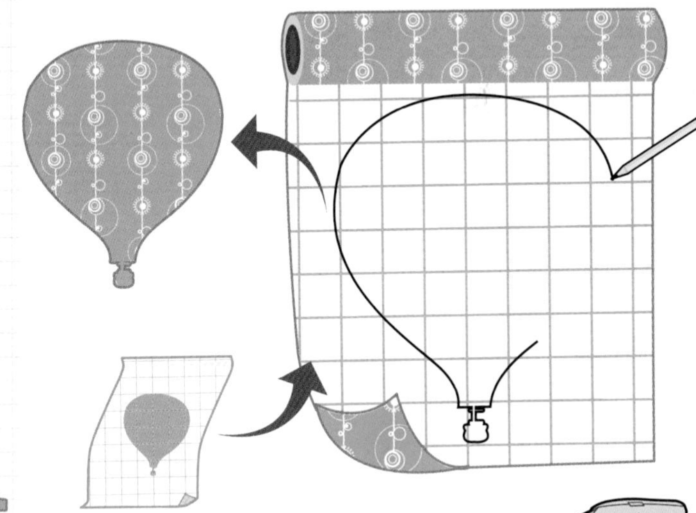

190 decoupage a tabletop

Clean intended surface.

Cut out images.

Arrange cut-outs.

Glue cut-outs.

Smooth as you go; let dry.

Layer varnish; let dry.

Sand varnish.

Varnish. Once dry, sand.

* Continue varnishing and sanding until the surface is completely smooth.

191 create a faux-bois effect

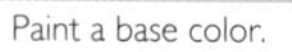
Paint a base color.

Add water to detail color.

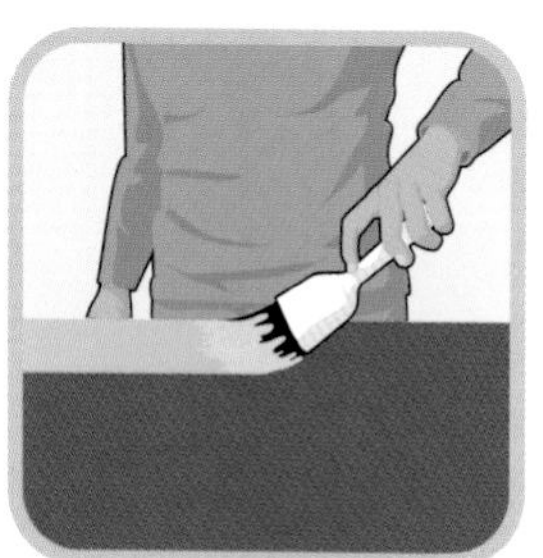
Paint a strip of detail color.

Drag, rock faux-bois tool.

Repeat to vary pattern.

192 etch a glass platter

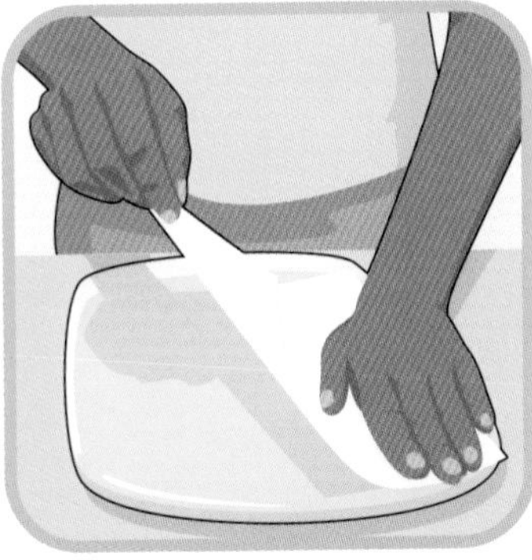

Lay contact paper on glass.

Set carbon paper on top.

Tape and trace an image.

Remove carbon paper.

Cut image from contact paper.

Spread etching cream on design; wait.

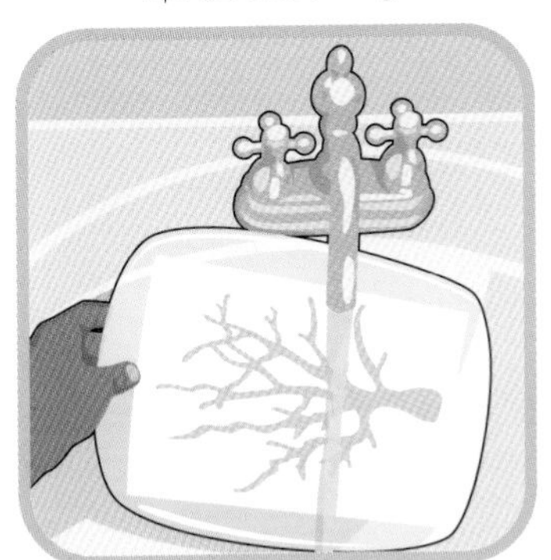

Rinse.

Remove contact paper; rinse again.

193 gild an old picture frame

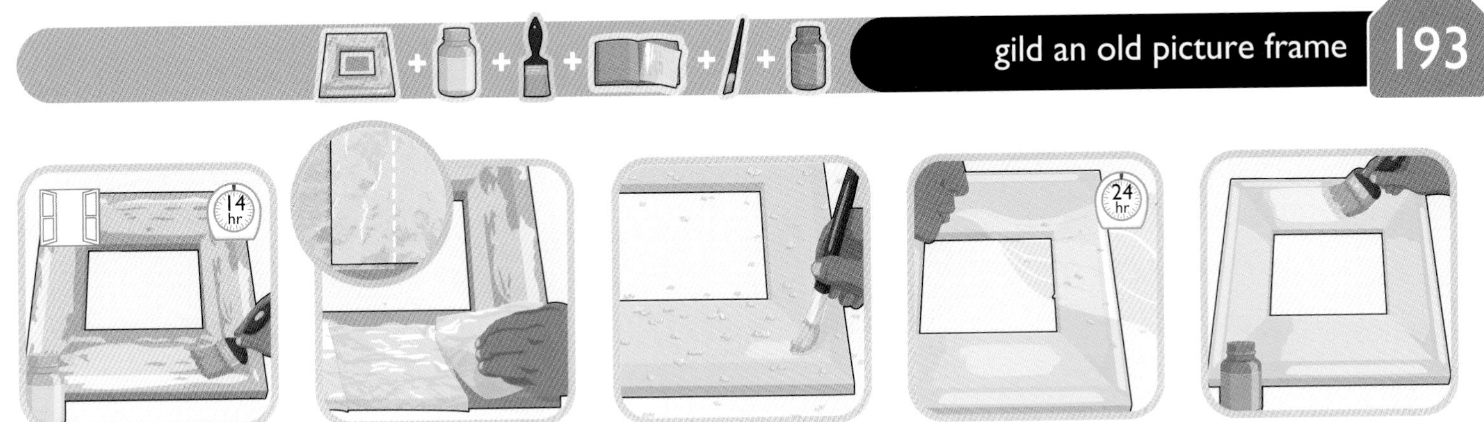

Paint gold-leaf sizing; let dry.

Overlap gold sheets.

Burnish with a soft brush.

Blow away excess; wait.

Seal with polyurethane.

194 hang fabric wallpaper

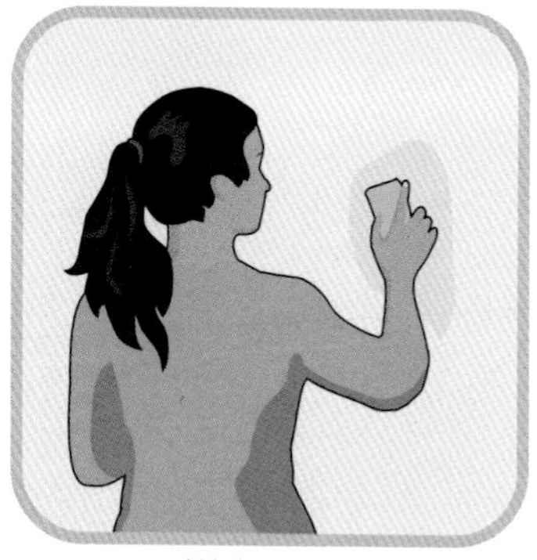
Wash the wall.

Measure.

Measure fabric; leave overhang.

Pour liquid starch in a pan.

Brush starch up wall.

Smooth fabric, leaving excess at edges.

Cover wall; let dry.

Trim off excess fabric.

When you're ready for a decor makeover, just peel the fabric off the wall—it won't leave residue or harm the paint.

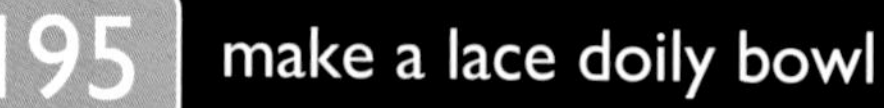

195 make a lace doily bowl

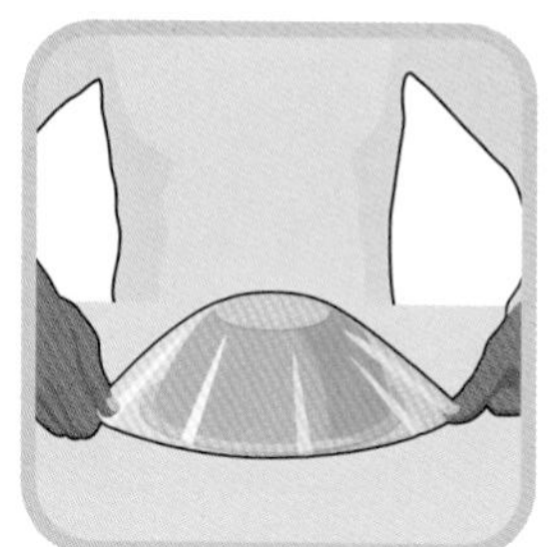
Cover bowl with cling wrap.

Center doily on bowl.

Brush on liquid starch; dry.

Remove bowl; starch inside.

Let dry; display.

pick a pillow 196

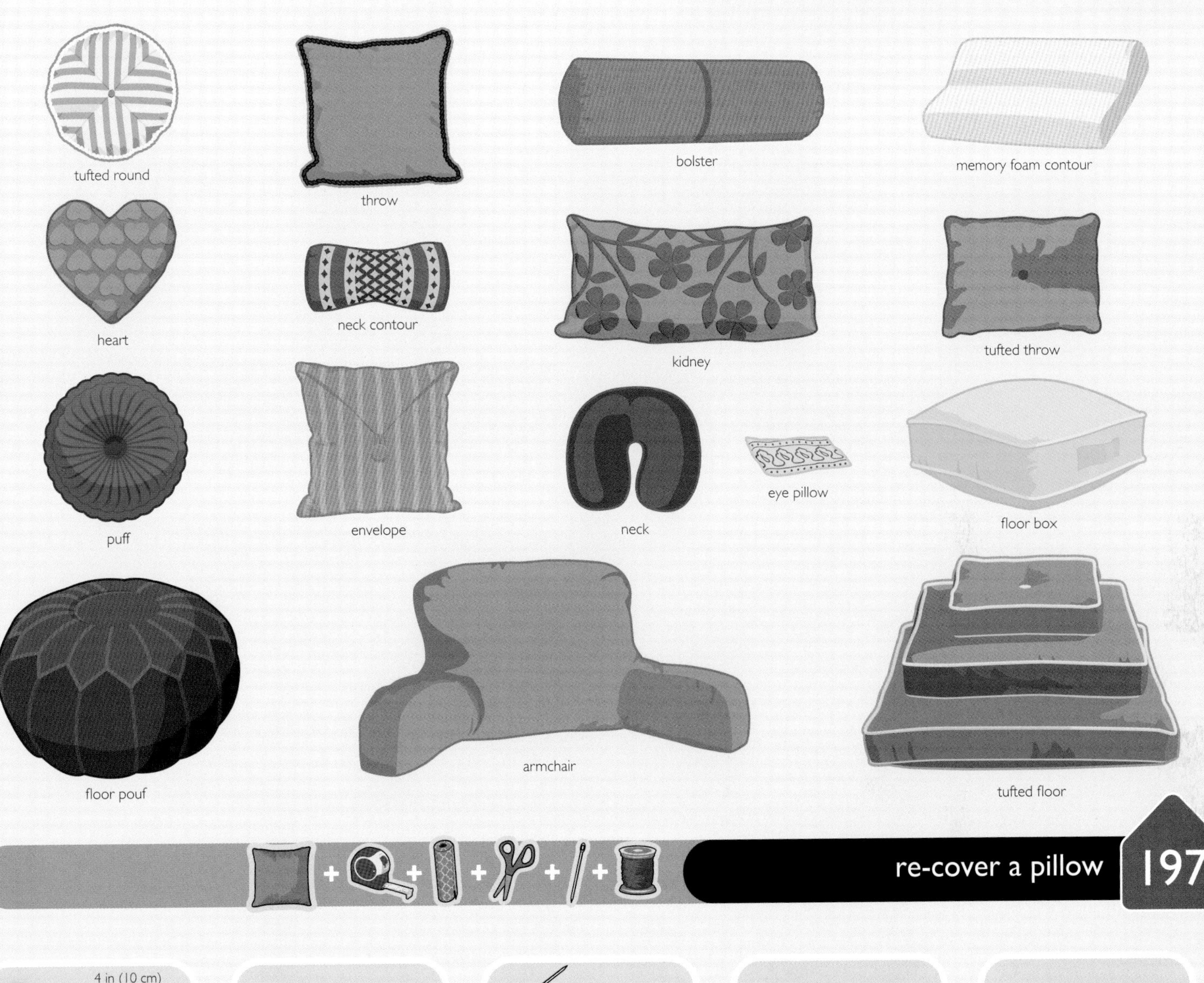

re-cover a pillow 197

Measure pillow; add excess.

Double length; cut.

Hem shorter ends.

Overlap; sew open sides.

Turn inside-out; add pillow.

198 organize a den

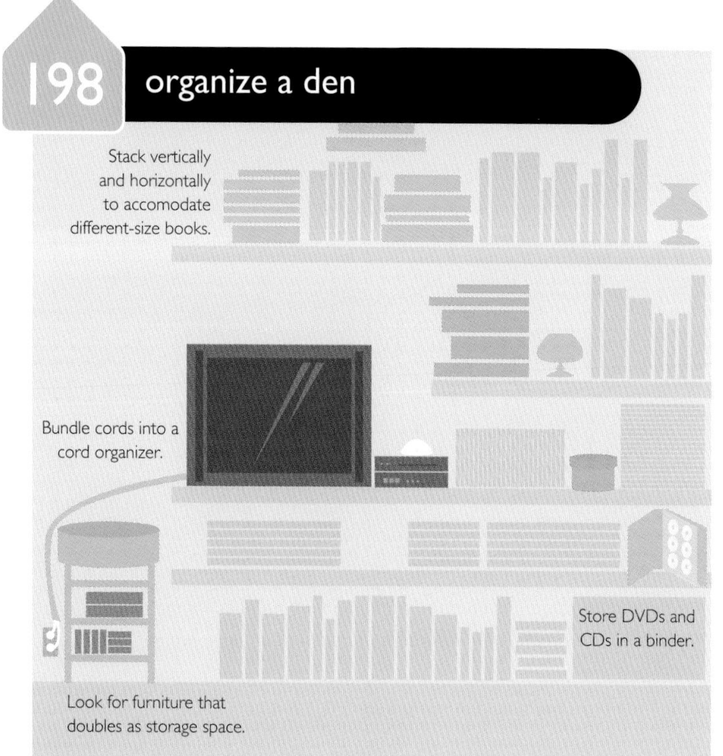

199 tidy a pantry

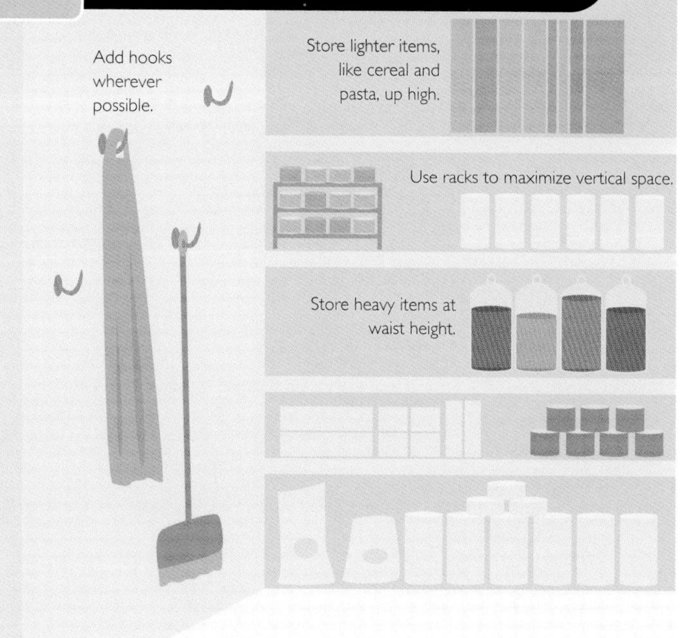

200 straighten a kitchen

201 store clothes efficiently

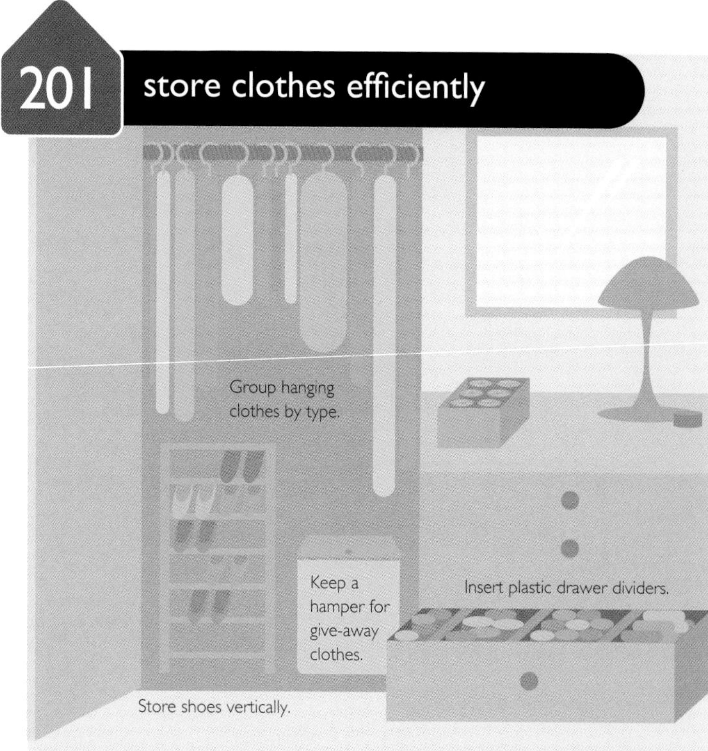

neaten a vanity 202

order a workroom 203

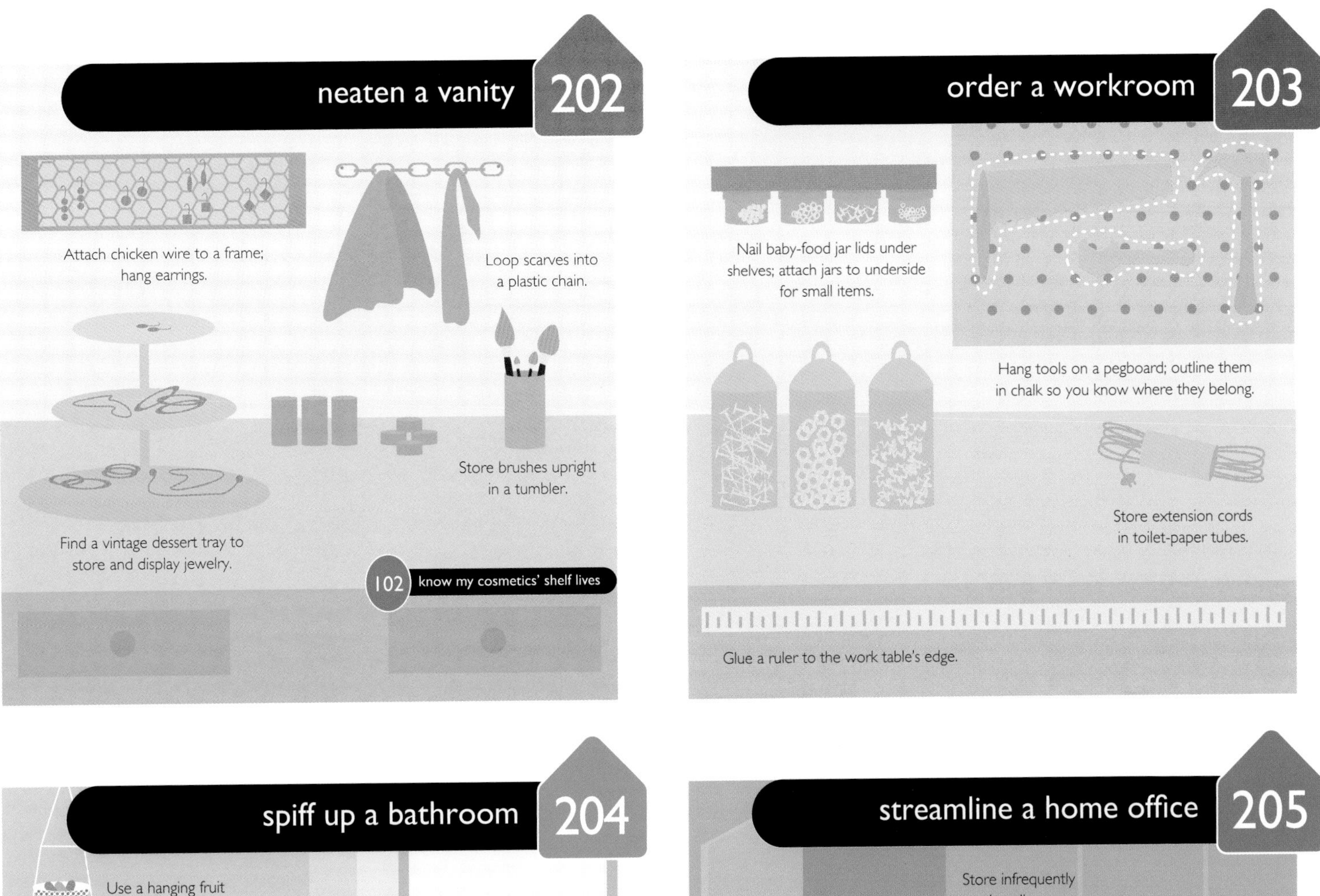

102 know my cosmetics' shelf lives

spiff up a bathroom 204

streamline a home office 205

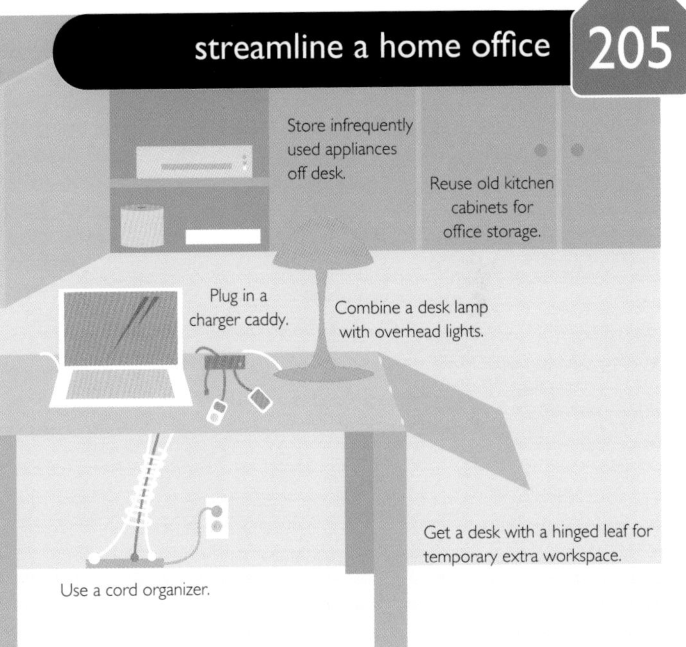

206 shine a wood surface

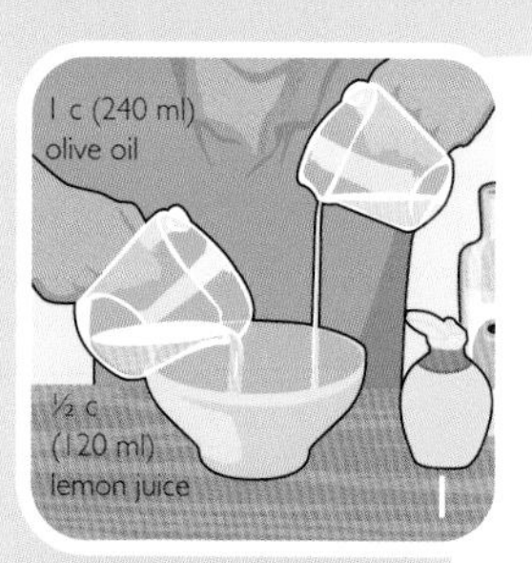

Mix oil and lemon juice.

Dip a soft cloth into mix.

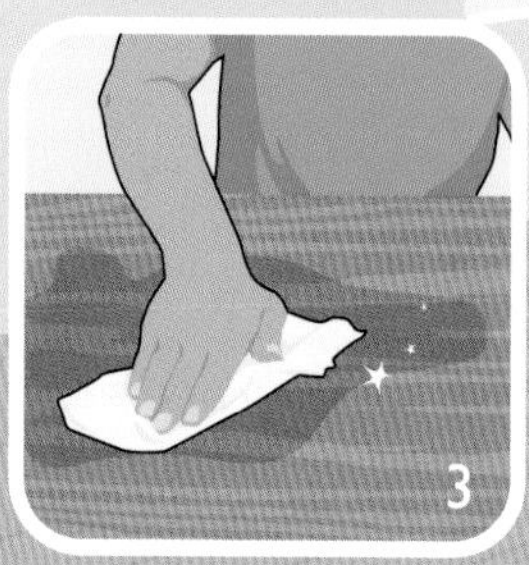

Rub into wood.

Buff with a clean cloth.

207 get rid of water rings

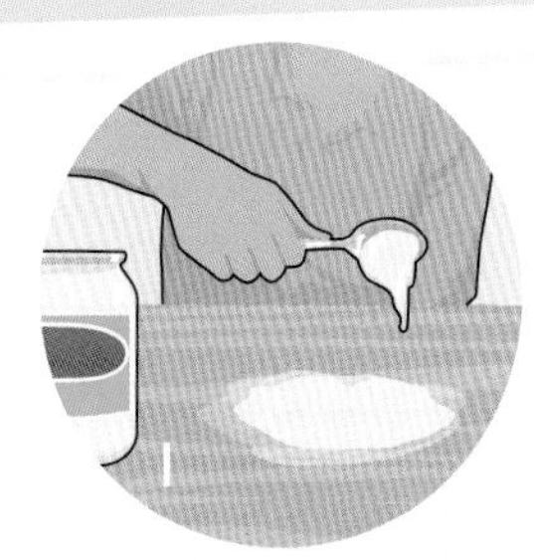

Rub mayonnaise into spots.

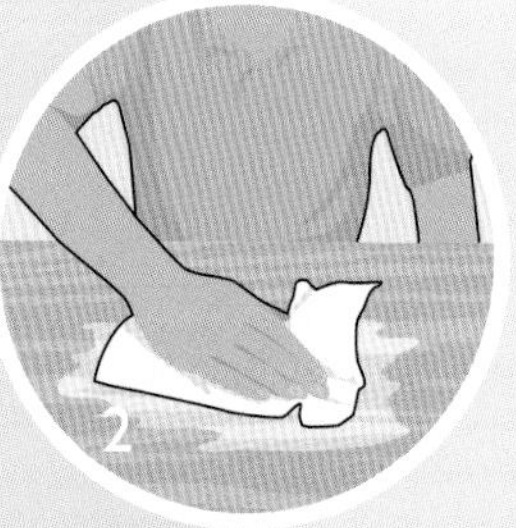

Blot; let sit overnight.

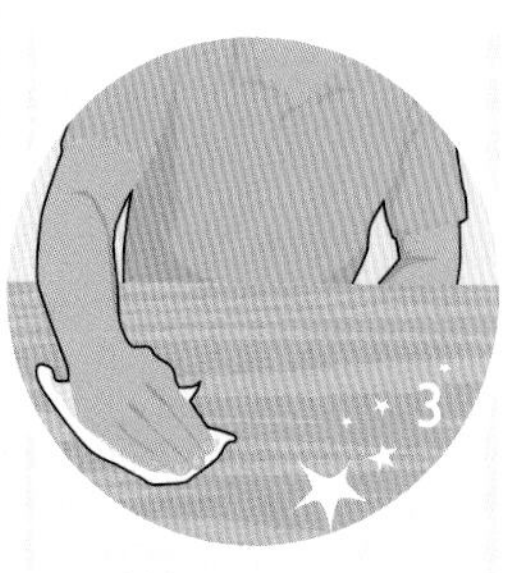

Wipe away excess.

208 buff away floor scuffs

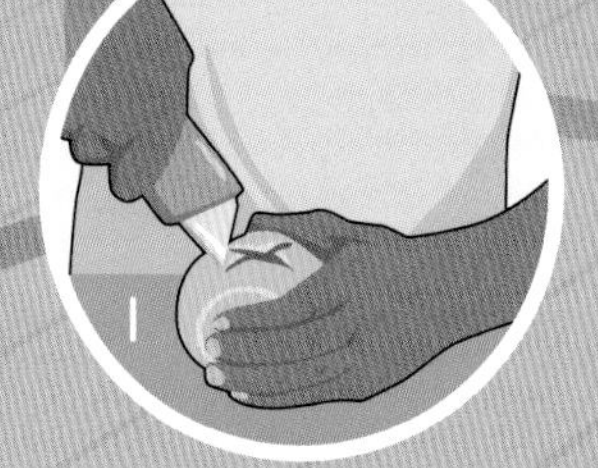

Cut an X in a tennis ball.

Insert broom handle; buff.

209 remove scratches from leather

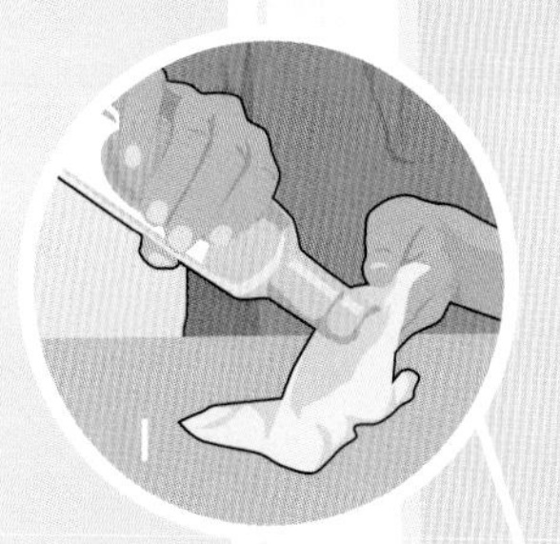

Dab olive oil onto a cloth.

Rub in circular motion.

210 lift wax from a carpet

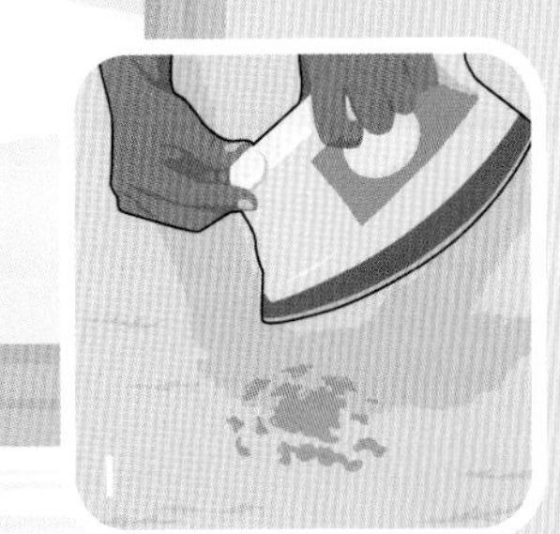

Set iron on lowest heat.

Cover with paper towel.

Gently iron paper.

Repeat until wax is lifted.

211 remove dents from a rug

Place an ice cube in dent.

Let melt; blot any water.

Fluff the rug with a fork.

212 conserve water at home

400 gal (1.5 l) saved a month
Put food coloring in the tank. If it shows up in the bowl, you have a leak.

100 gal (375 l) saved a month
Leave frozen leftovers out overnight to defrost—don't use hot water.

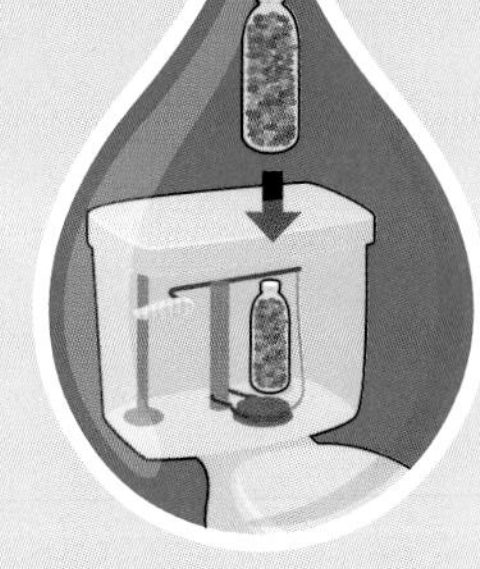

225 gal (850 l) saved a month
Place a bottle filled with pebbles in the tank.

222 compost in my backyard

1,100 gal (4 kl) saved a month
Put mulch, wood chips, or gravel around plants and trees.

135 gal (500 l) saved a month
Turn off water while brushing teeth or shaving.

350 gal (1.25 kl) saved a month
Fill one sink with soapy water and the other with water for rinsing. Don't run the faucet.

650 gal (2.5 l) saved a month
Install a low-flow showerhead.

200 gal (750 l) saved a month
Wash vegetables in a pan of water, instead of under the faucet.

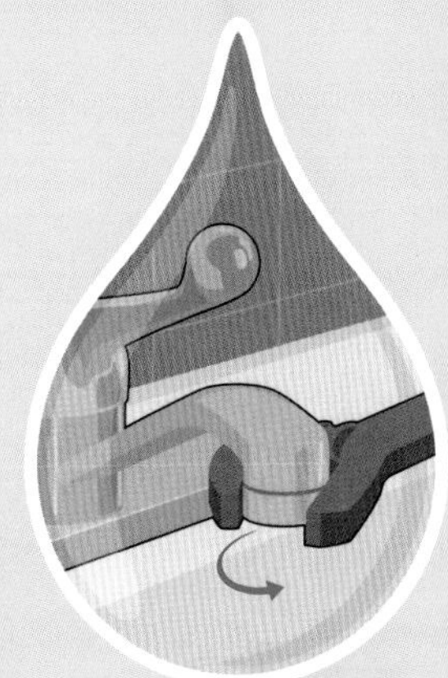

450 gal (1.75 kl) saved a month
Fix any leaky faucets.

1,200 gal (4.5 kl) saved a month
Suds your car with one bucket of soapy water, followed by a quick rinse with the hose. Don't run the hose the entire time.

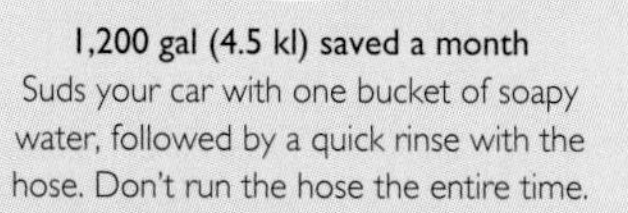

600 gal (2.25 kl) a month
Sweep driveways and walkways. Don't use the hose.

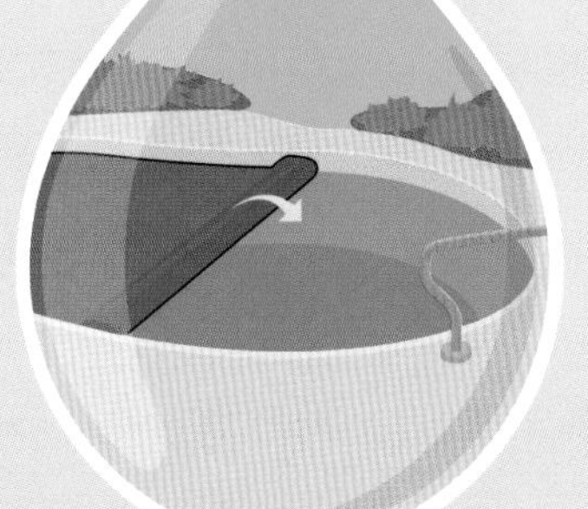

1,000 gal (3.75 kl) a month
Keep pool covered to prevent evaporation.

213 thaw a frozen pipe

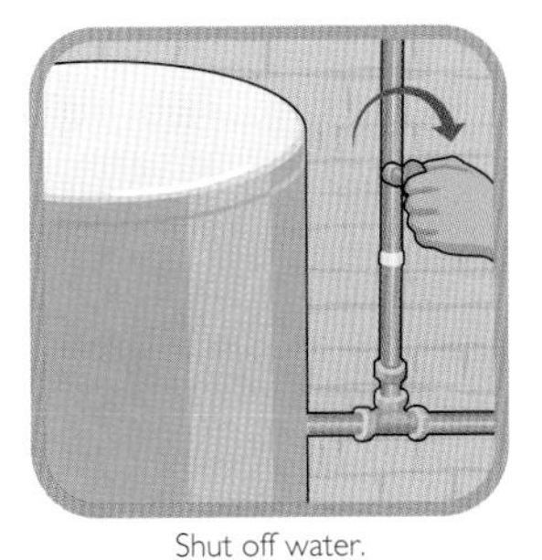
Shut off water.

Make sure frozen pipe is intact.

Thaw pipe with blow-dryer.

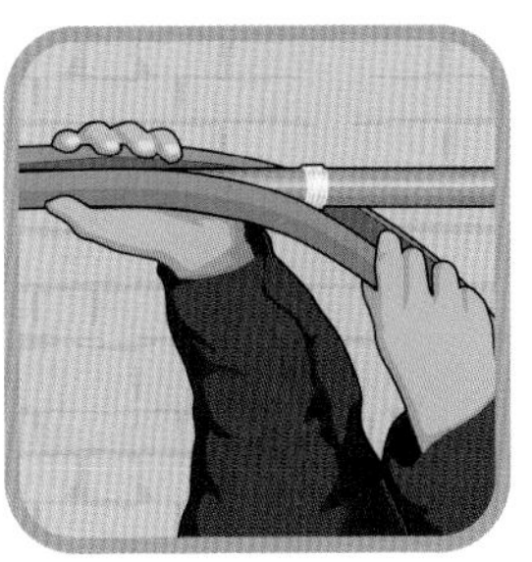
Insulate pipe.

Insulate surrounding walls.

If the pipe has burst, call a plumber right away. Don't thaw it unless you want an icy swimming pool in your basement!

214 replace a faucet

Shut off water supply.

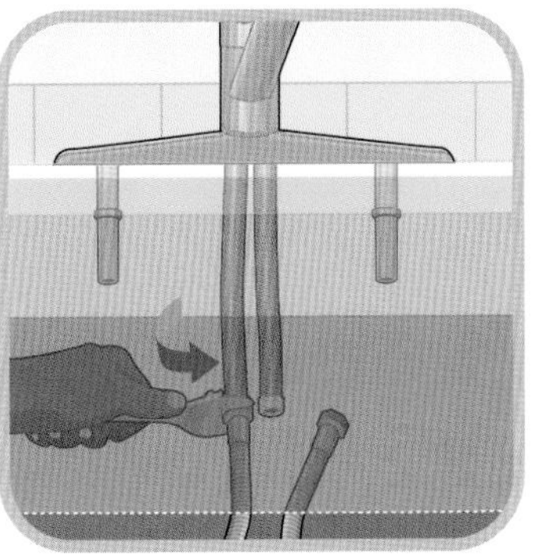
Disconnect faucet from water source.

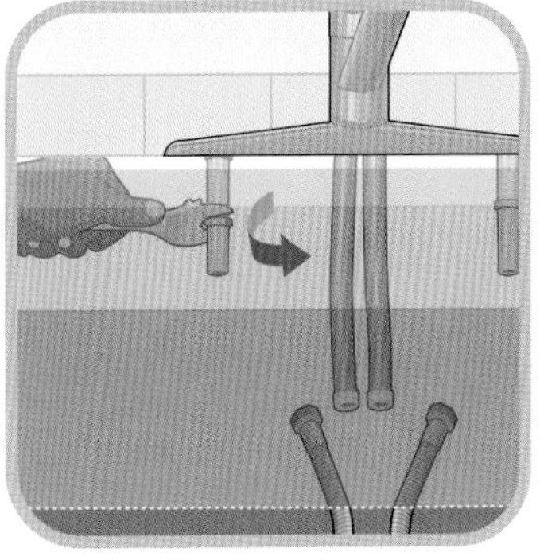
Unscrew washers.

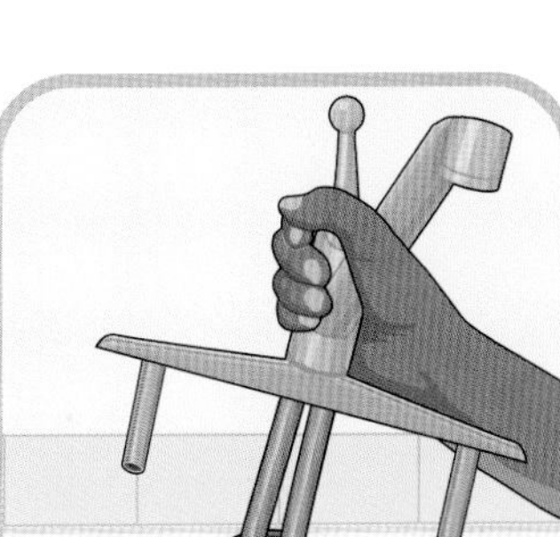
Remove faucet.

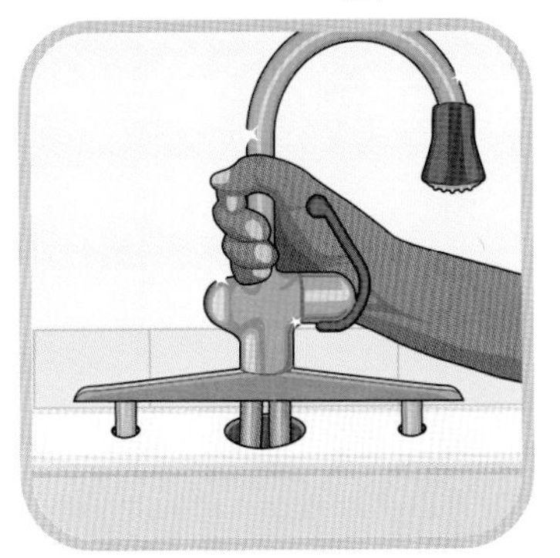
Insert new faucet.

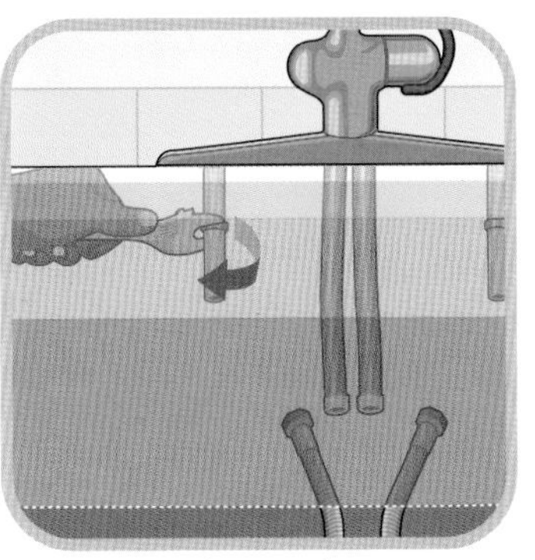
Attach new washers.

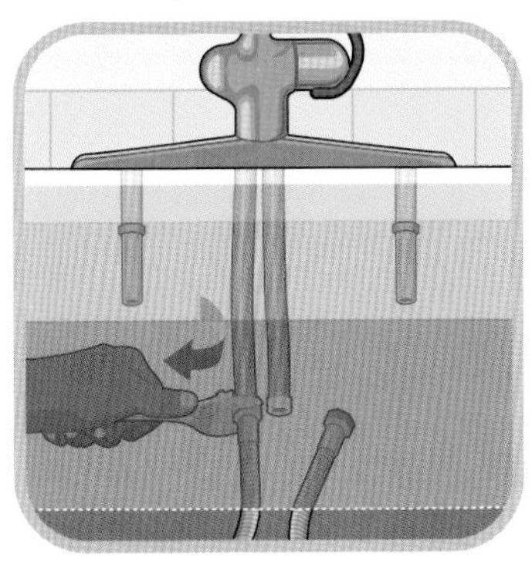
Reconnect water source.

Caulk around faucet base.

grow

215 build a blooming window box

Measure the window.

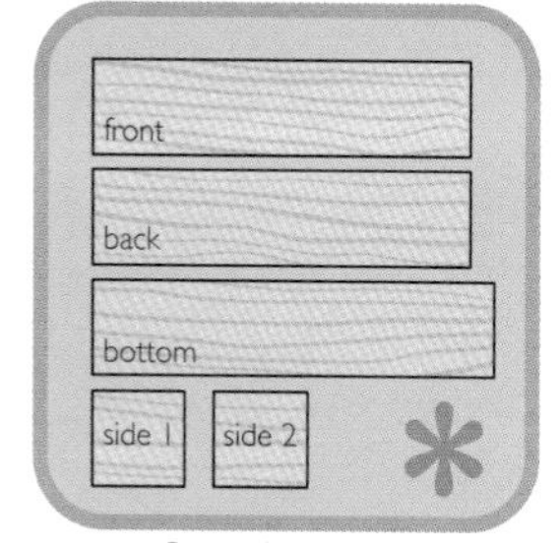

Cut to fit window.

Glue side 1 to front.

Nail in place.

Repeat for all joints.

Glue and nail bottom.

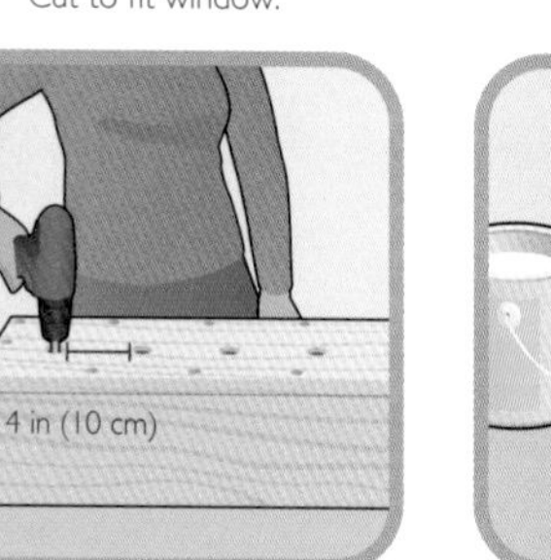

Drill drainage holes.

Prime box; let dry.

Paint; let dry.

Set mesh inside.

To get the correct lengths of the front and back boards, measure the thickness of the board. Multiply that number by two, and then subtract that number from the length of the bottom board. The lengths of sides 1 and 2 should be the same as the width of the bottom board.

Mark window box's height.

Predrill; bolt bracket to wall.

Level second bracket; bolt.

Attach box to brackets.

Plant.

stake tomatoes 217

Tomatoes left to grow along the ground are targets for bugs and rot. Help them rise above the fray by training them on a Florida weave trellis.

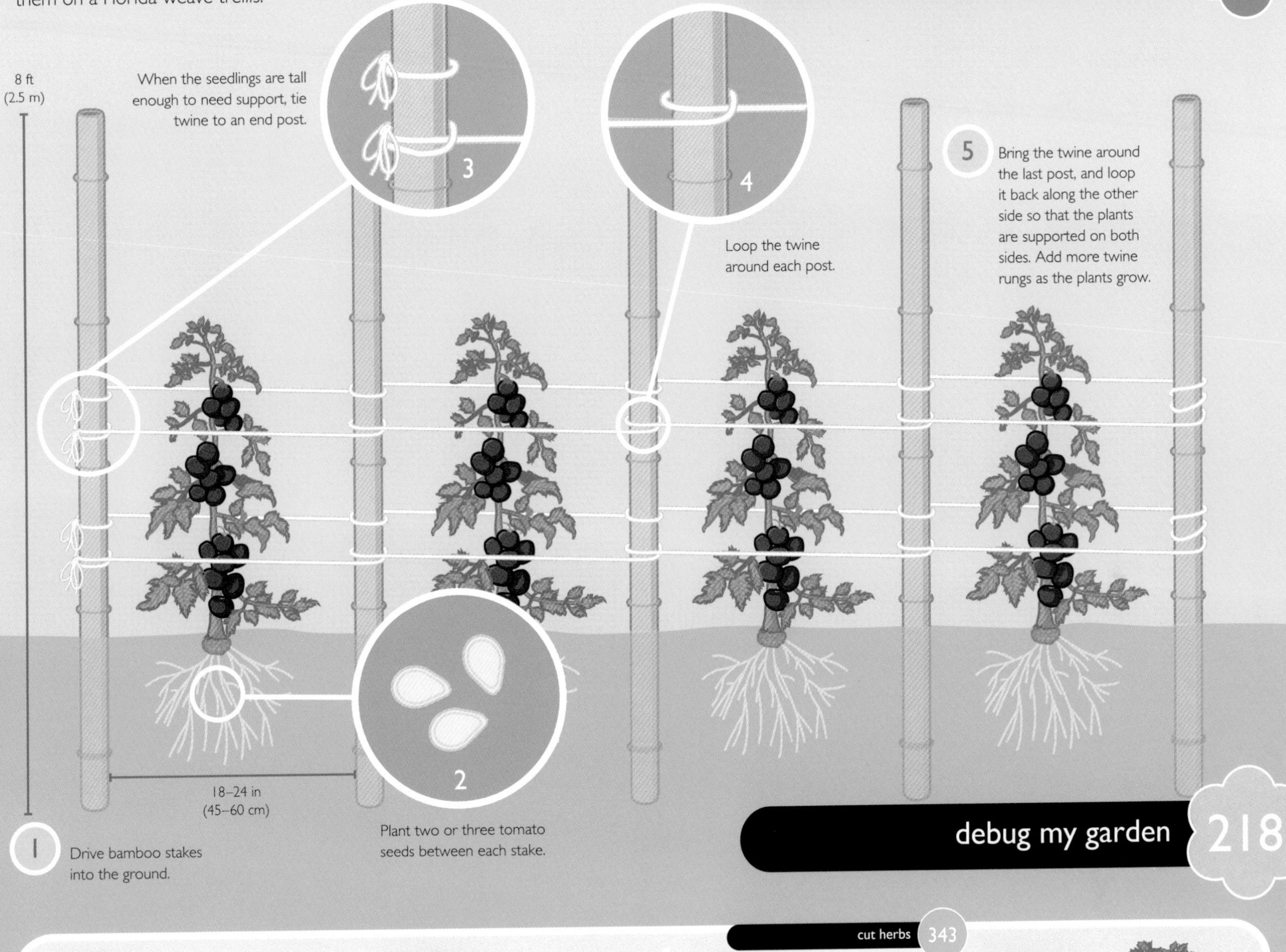

debug my garden 218

cut herbs 343

Spray water mixed with a splash of mild dish soap to kill aphids.

Plant yellow flowers some distance from the tomatoes. They'll draw stinkbugs away from your crop.

A dill plant nearby attracts hornworms and makes them easy to pick off.

Foil cutworms by wrapping the stalk in aluminum foil.

219 sow a seed bomb

Combine seeds and compost.

Mix with clay.

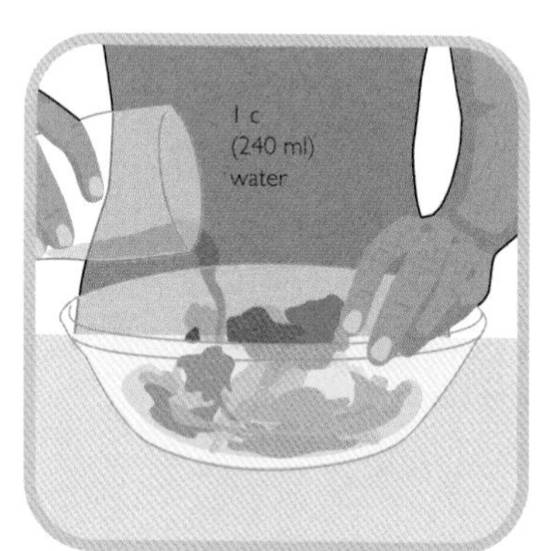

Add water.

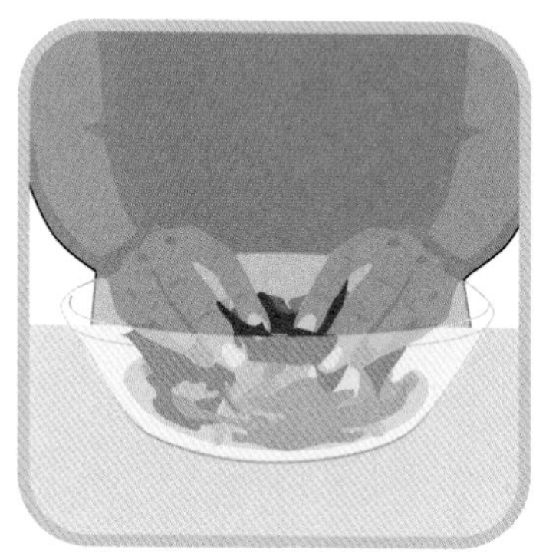

Mix together.

Roll marble-size "bombs."

Dry in the sun.

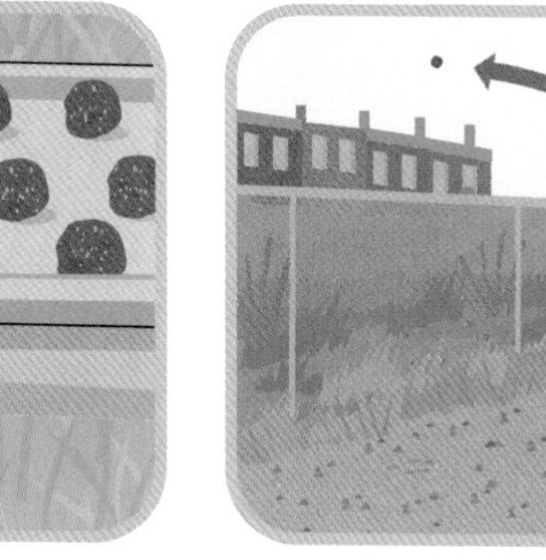

Launch before a rain.

Admire your handiwork.

220 sprout seed paper

Shred paper.

Add equal parts paper and water.

Blend at low speed to make pulp.

58 fold a paper marigold

Stir in seeds.

221 paint moss graffiti

Crumble moss; pour water.

Add water-retention gardening gel.

Add buttermilk.

Pulse blender until gel forms.

Transfer to a bucket.

Paint onto wood or rough concrete.

Mist weekly.

Watch your art grow.

* Water-retention gel, available at gardening stores, helps plants thrive in challenging environments—like the side of a wall.

Roll pulp onto mesh screen.

Use a fan to speed drying time.

Peel off the dry paper; decorate.

Enjoy a plantable greeting card.

222 compost in my backyard

Use a backyard compost pile to keep your organic waste out of a landfill and provide your garden plants with superfood. Layer waste as shown, then water and aerate it. You'll know it has transformed from a pile of garbage into sweet, nutritious compost when you can't discern any bits in it—when it's just dark, rich dirt.

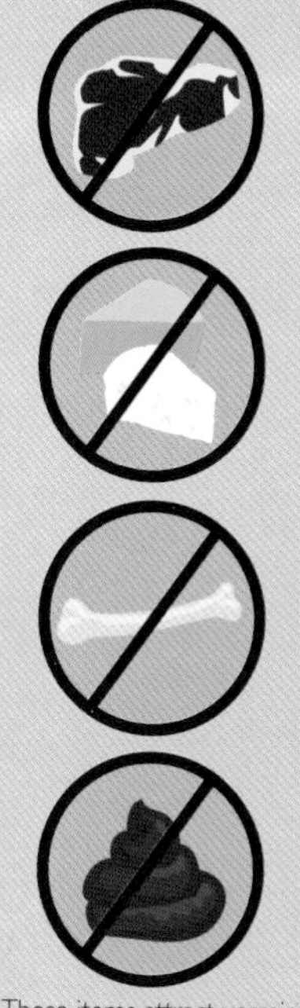

These items attract vermin. Compost them at your own risk!

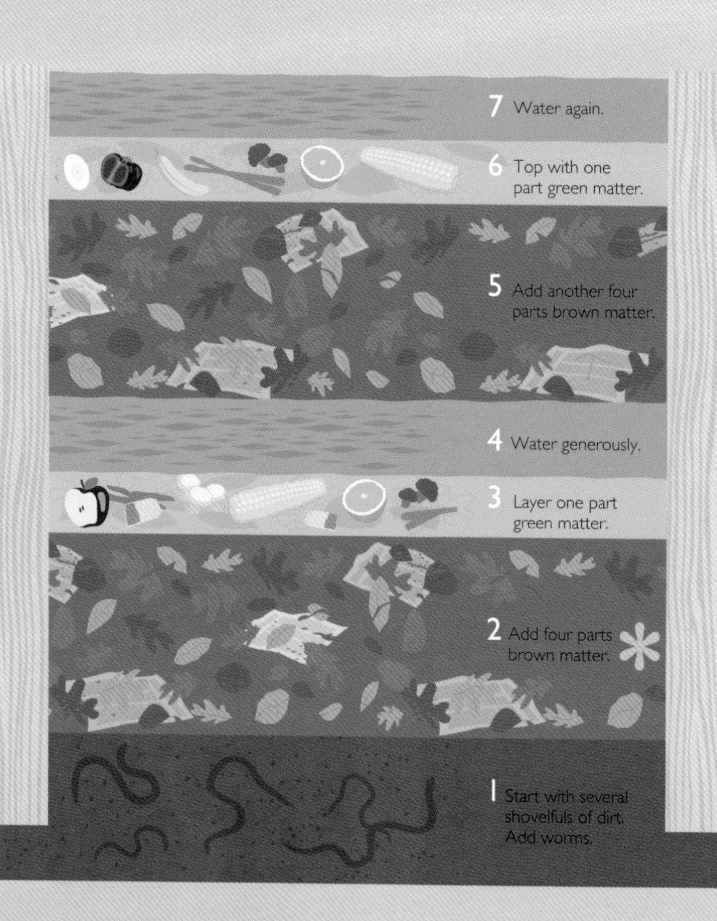

Water the pile weekly.

Turn to aerate as needed.

* The ideal brown matter is carbon-rich: mostly dead leaves, with some newspaper, wood, and sawdust added. Green matter is nitrogen-rich, composed of items like veggies, eggs, green plant leaves, and tea. Tasty!

223 brew compost tea

Put compost into burlap sack.

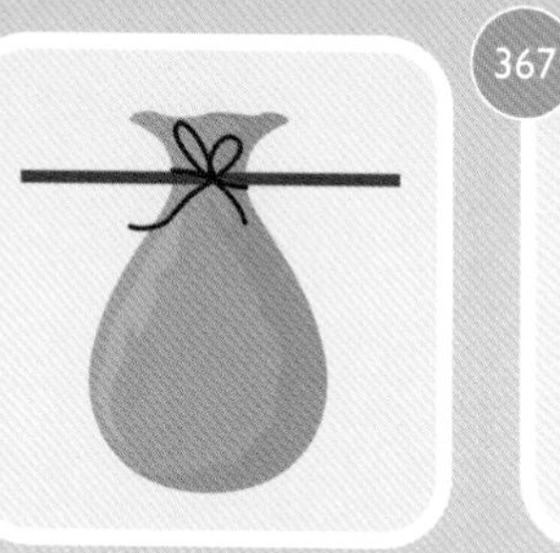

Tie to a stick.

367 brew a restorative tea

Soak in water.

Fill watering can with "tea."

Feed plants.

train bougainvillea 224

Place a stake near the trellis.

Plant three seedlings near stake.

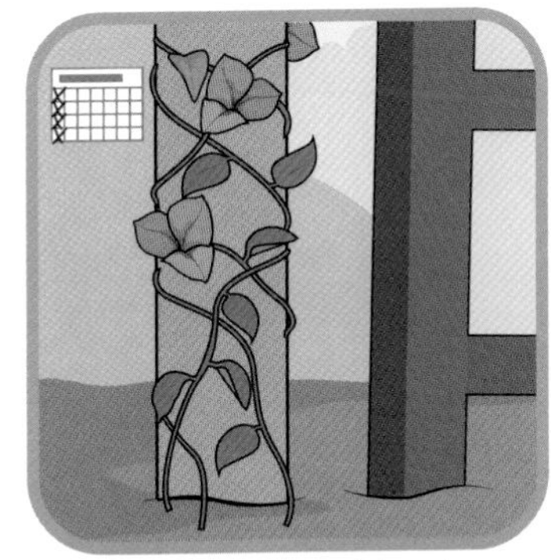
Braid the stems around the stake.

Trim offshoots as plants grow.

Continue braiding and trimming.

Pinch off ends, leaving three leaves.

Secure to trellis with twine.

Prune as desired after it flowers.

* By trimming and pruning the bougainvillea, you encourage the growth of a single vine that you can train up the trellis, rather than many unruly offshoots and buds.

make a bromeliad bloom 225

Drain the water from plant.

Enclose with an apple; wait.

Keep warm in the daytime.

Keep it cool at night.

Await bloom.

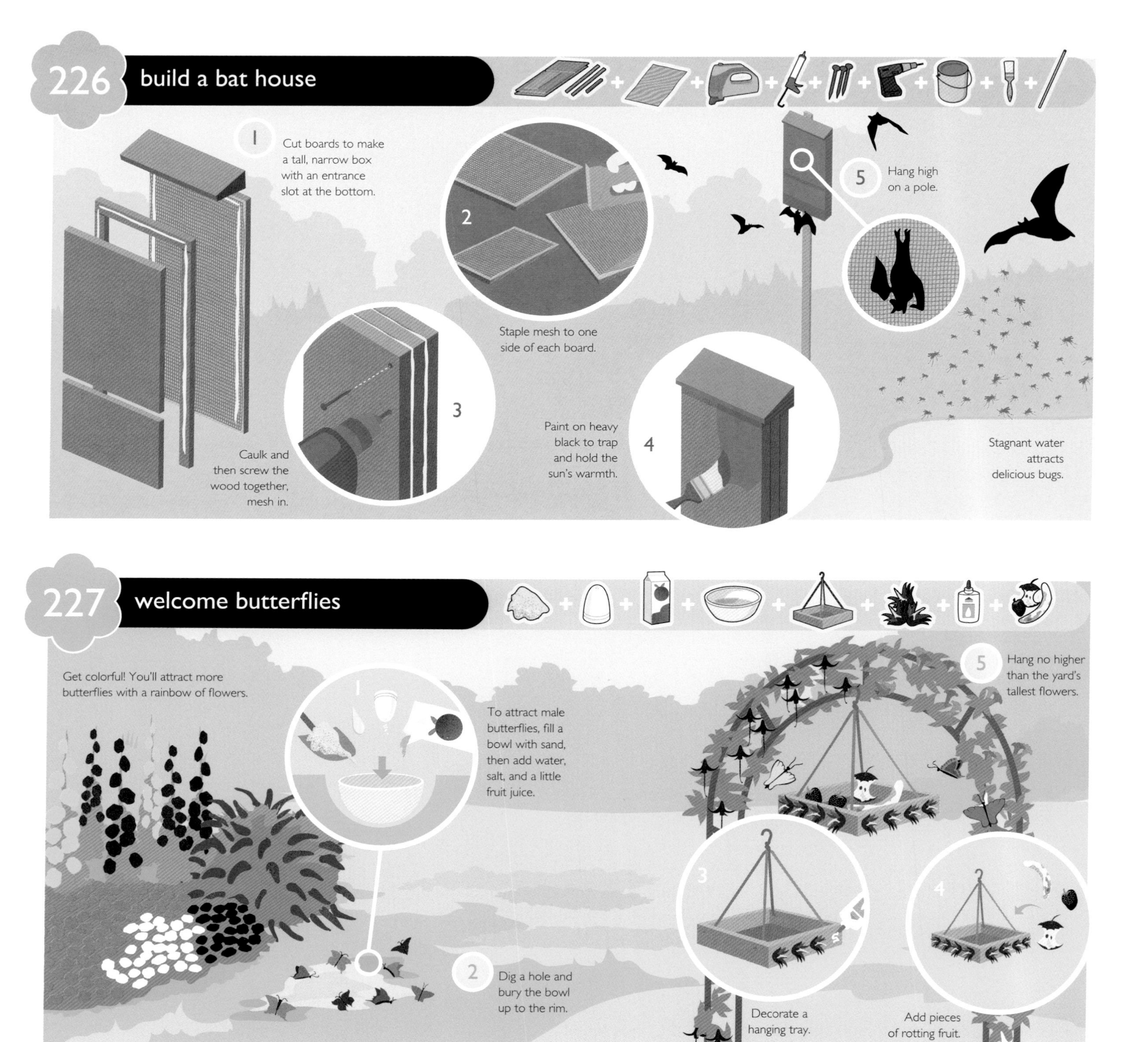
226
build a bat house
1 Cut boards to make a tall, narrow box with an entrance slot at the bottom.
2 Staple mesh to one side of each board.
3 Caulk and then screw the wood together, mesh in.
4 Paint on heavy black to trap and hold the sun's warmth.
5 Hang high on a pole.
Stagnant water attracts delicious bugs.
227
welcome butterflies
Get colorful! You'll attract more butterflies with a rainbow of flowers.
1 To attract male butterflies, fill a bowl with sand, then add water, salt, and a little fruit juice.
2 Dig a hole and bury the bowl up to the rim.
3 Decorate a hanging tray.
4 Add pieces of rotting fruit.
5 Hang no higher than the yard's tallest flowers.

228 make a mason-bee hive

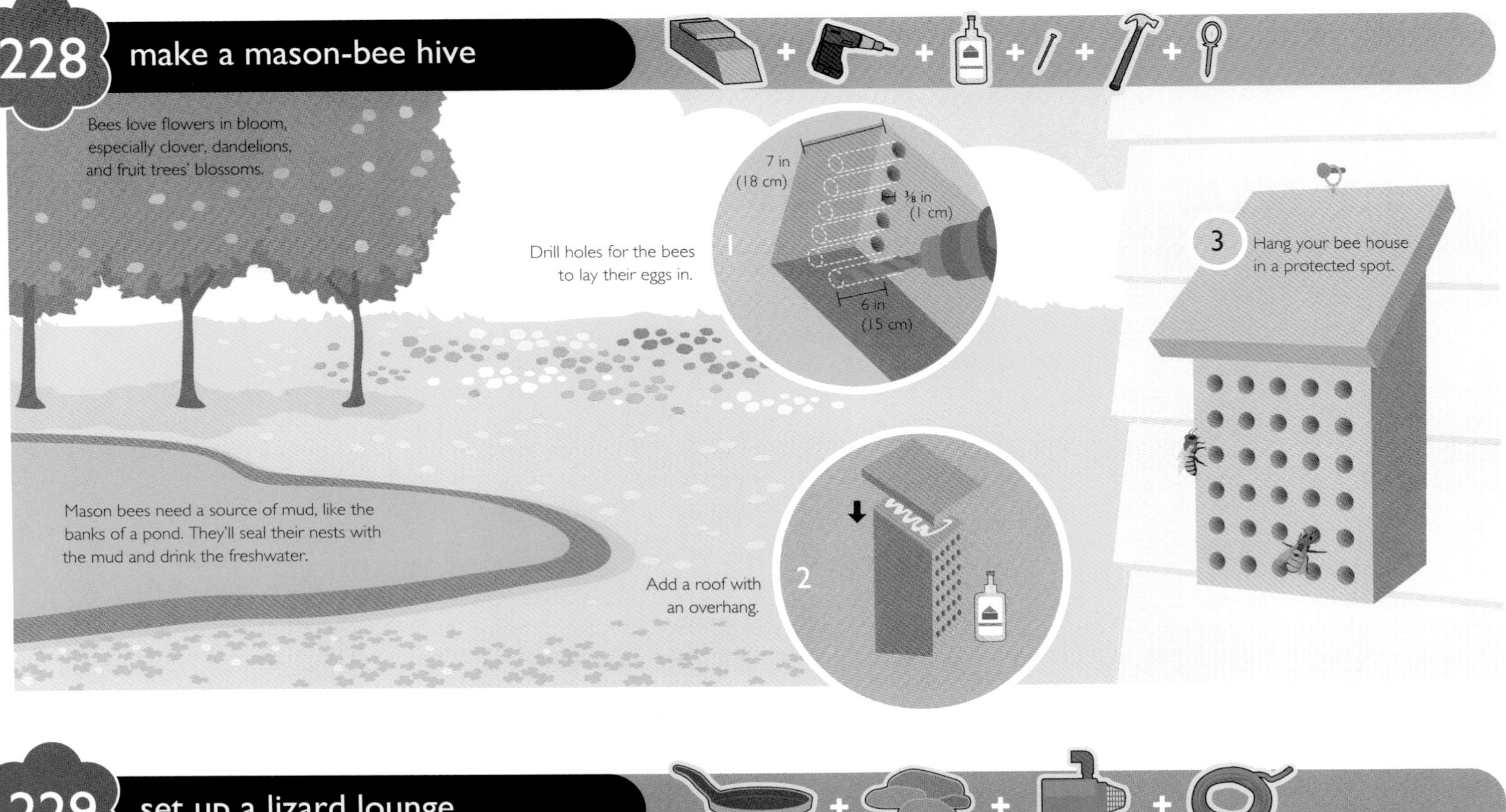

229 set up a lizard lounge

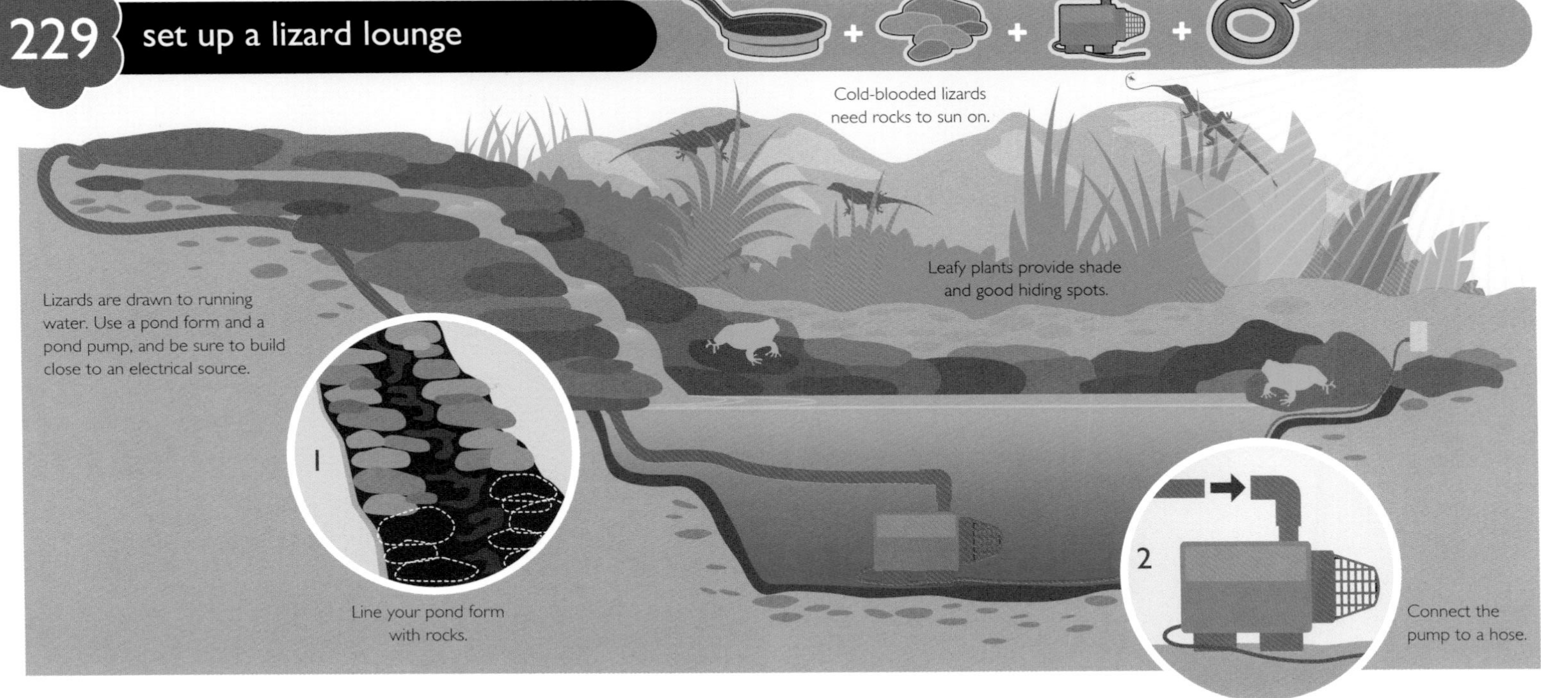

230 detect impending llama spew

Ears snap back.

Ears pin down; nose rises.

Neck straightens.

Lump rises in throat.

Duck!

231 freeze a snack for my octopus

Freeze one half of prawn in ice.

Flip; add more water.

Freeze again.

Drop into tank.

Hours of octo-fun!

232 help a snake with a bad shed

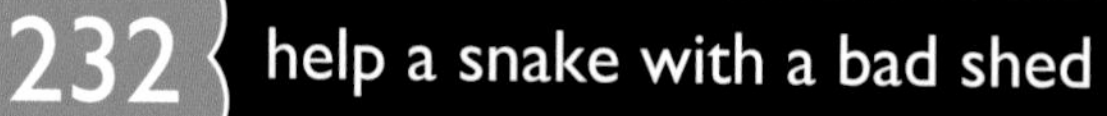

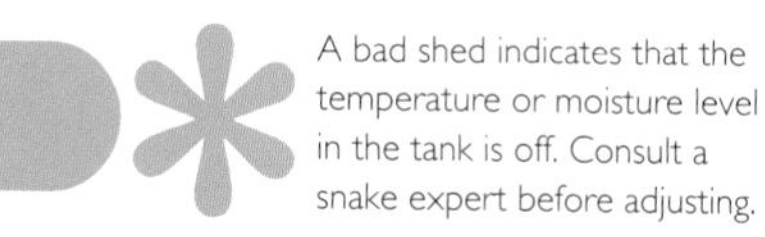
A bad shed indicates that the temperature or moisture level in the tank is off. Consult a snake expert before adjusting.

Identify bad shed.

Place in lukewarm bath.

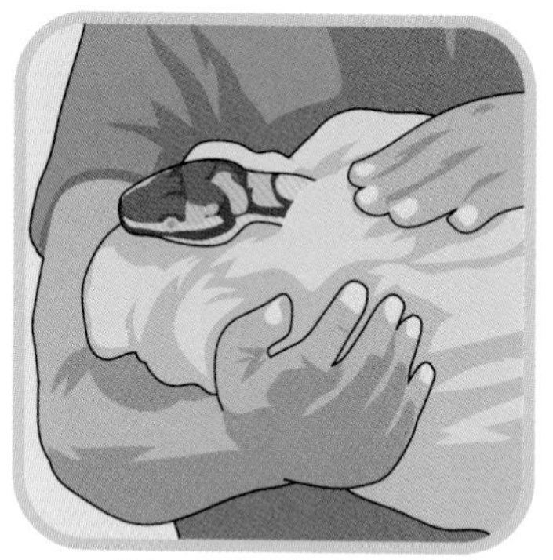
Wrap in damp hand towel.

Let snake crawl out.

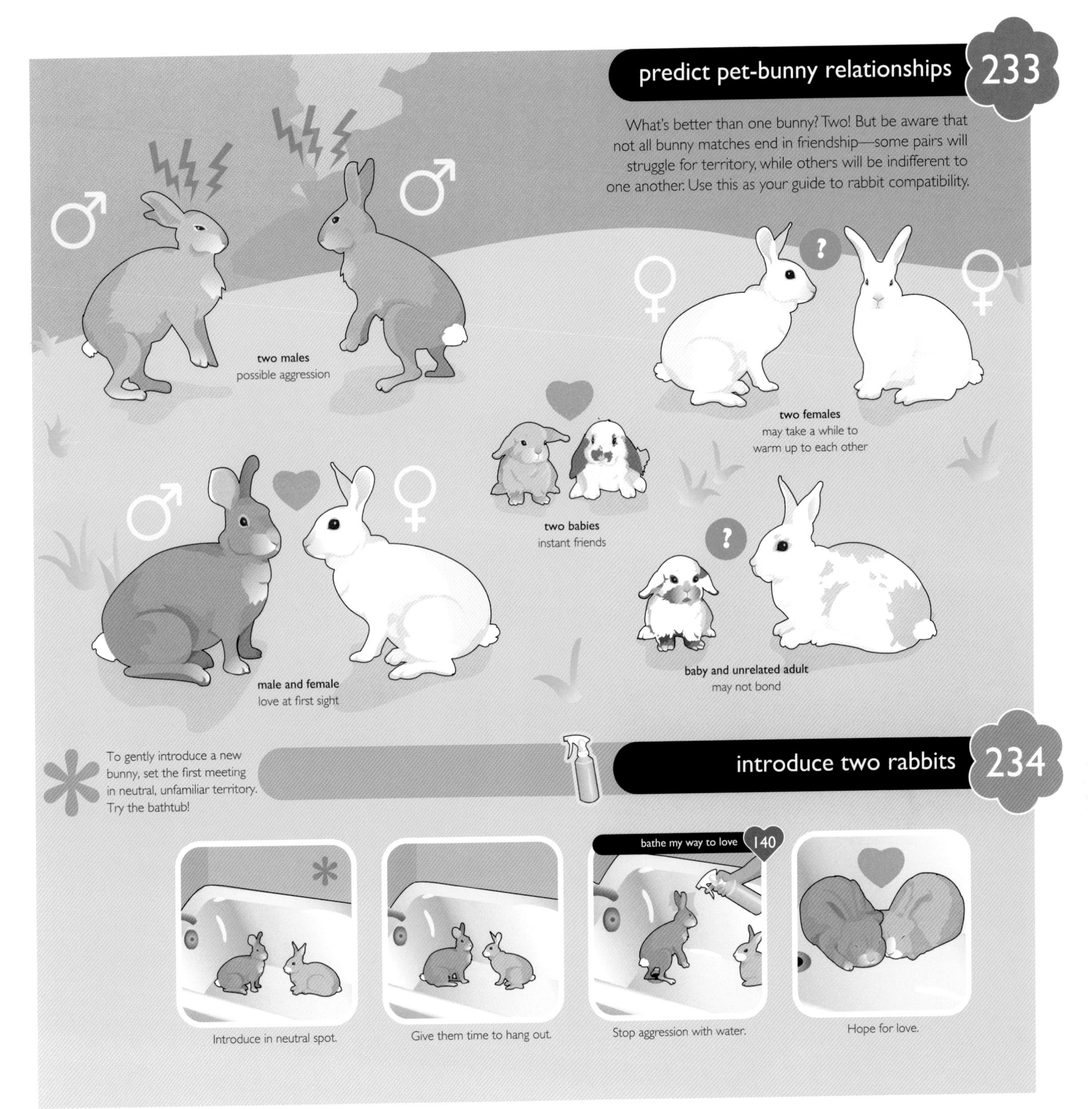
predict pet-bunny relationships
233
What's better than one bunny? Two! But be aware that not all bunny matches end in friendship—some pairs will struggle for territory, while others will be indifferent to one another. Use this as your guide to rabbit compatibility.
two males
possible aggression
two females
may take a while to warm up to each other
two babies
instant friends
male and female
love at first sight
baby and unrelated adult
may not bond
To gently introduce a new bunny, set the first meeting in neutral, unfamiliar territory. Try the bathtub!
introduce two rabbits
234
bathe my way to love
140
Introduce in neutral spot.
Give them time to hang out.
Stop aggression with water.
Hope for love.

235 make catnip-infused oil

Combine in a clean jar; close lid.

Leave on a sunny windowsill.

Strain; pour oil into a clean jar.

Discard catnip.

Add a few drops to scratching post . . .

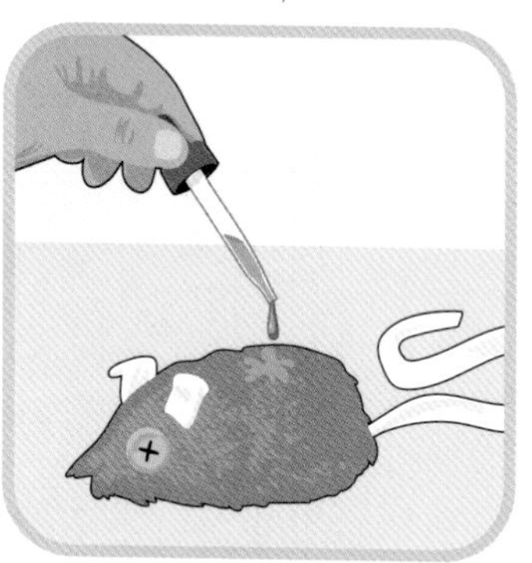

toys . . .

62 blow bubble prints

. . . or nontoxic bubbles.

Store oil in a cool, dry place.

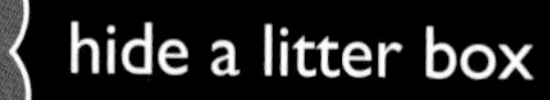

236 hide a litter box

Looking for a discreet yet accessible spot to stash kitty's litter tray without disrupting your decor? Think outside the box with one of these unconventional hiding places.

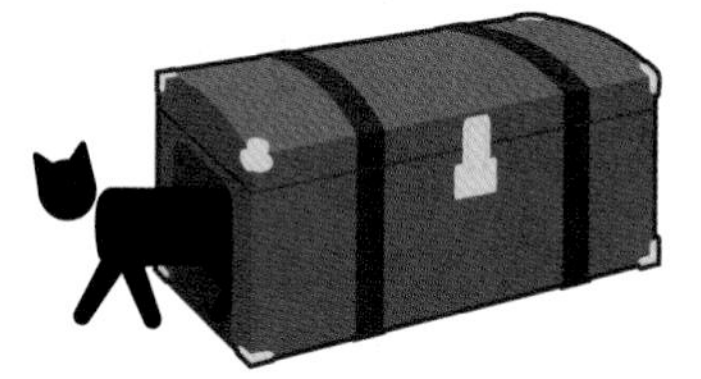

For an antique vibe, cut a hole in the end of an old steamer trunk.

Make an opening in a large planter, then top off with silk flowers for a low-maintenance verdant look.

Art lovers can include kitty's works in their collection—using the base of a plaster pedestal as a vault.

The practical-minded may want to open the side of a cabinet to make a litter box with added storage.

To clean, unfasten the open end, pull out the foam, and throw the cover in the wash.

craft a dog bed

237

Measure your dog's length and width.

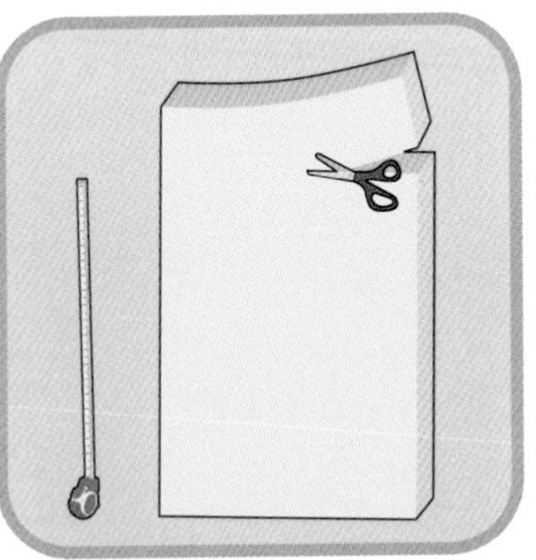

Cut foam to dog's measurements.

sew a rockin' quilt 65

Set on an old quilt.

Fold over; cut to height.

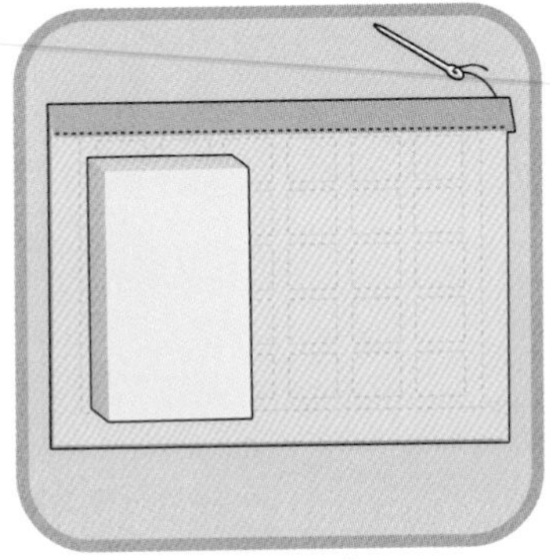

Hem the cut edge.

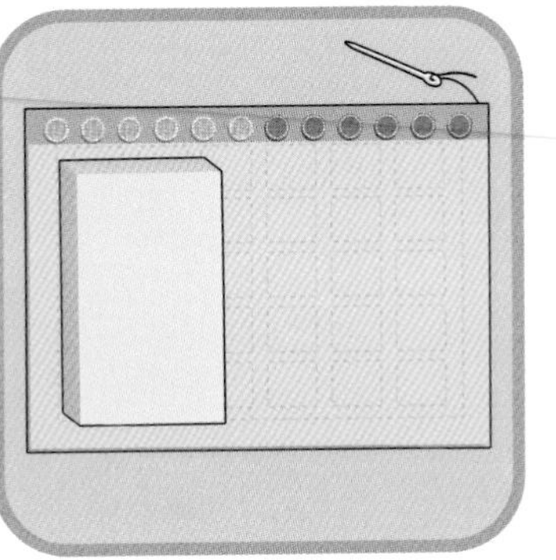

Sew Velcro® dots.

Fasten the dots; sew up open sides.

Sweet doggy dreams!

To make a nutritionally balanced feast for your favorite pup, prepare items from the columns below. Mix together with ½ tsp iodized salt and 2 tbsp vegetable oil. Bone appétit!

feed fido a home-cooked meal

238

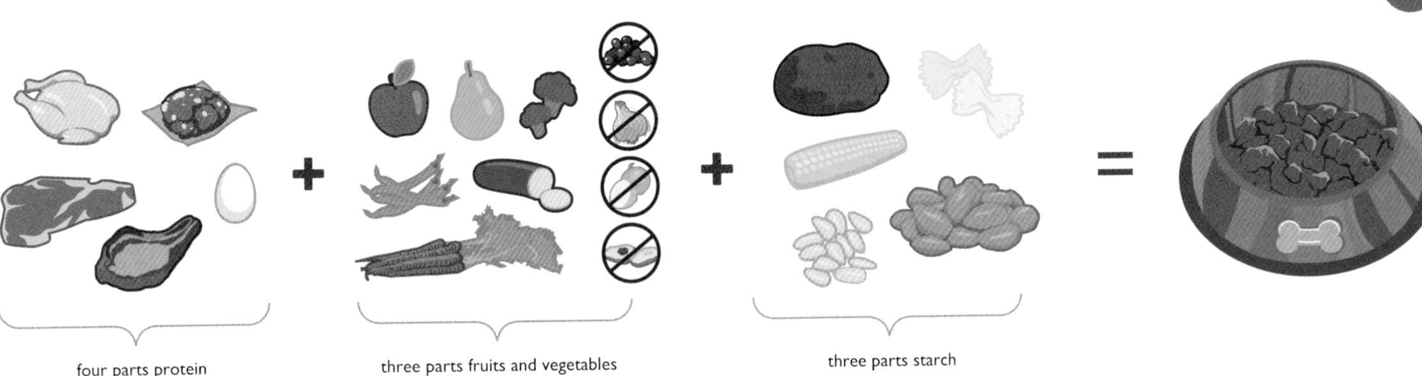

four parts protein
Cook meat in water; remove bones and cut into small chunks.

three parts fruits and vegetables
Boil until tender; purée.

three parts starch
Cook until tender.

239 prepare a nursery

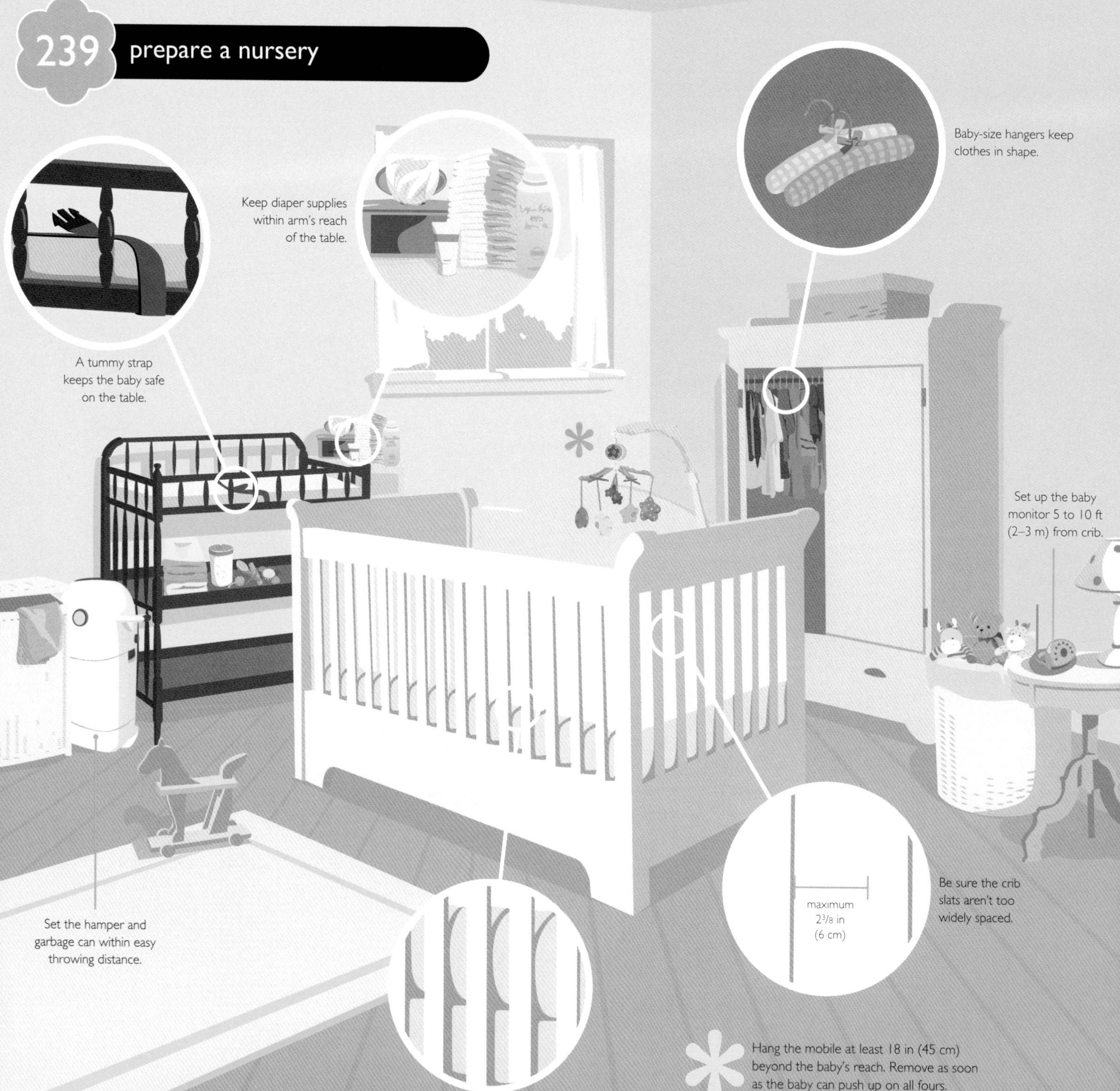

Hang the mobile at least 18 in (45 cm) beyond the baby's reach. Remove as soon as the baby can push up on all fours.

240 pick up a new baby

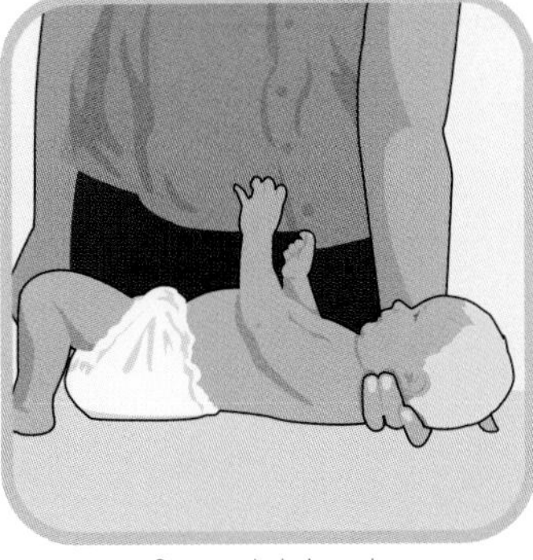

Support baby's neck.

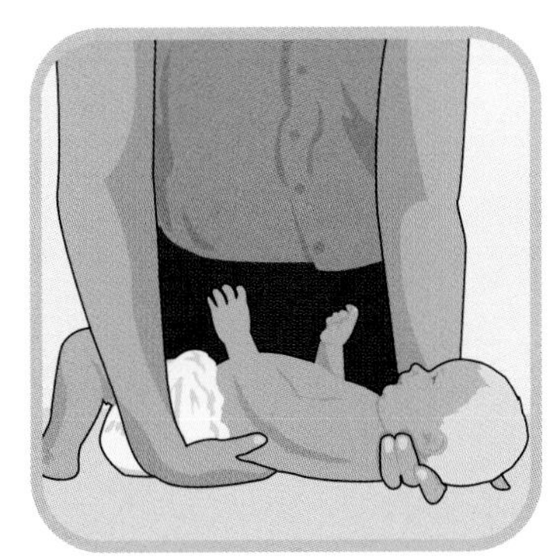

Place other hand under hips.

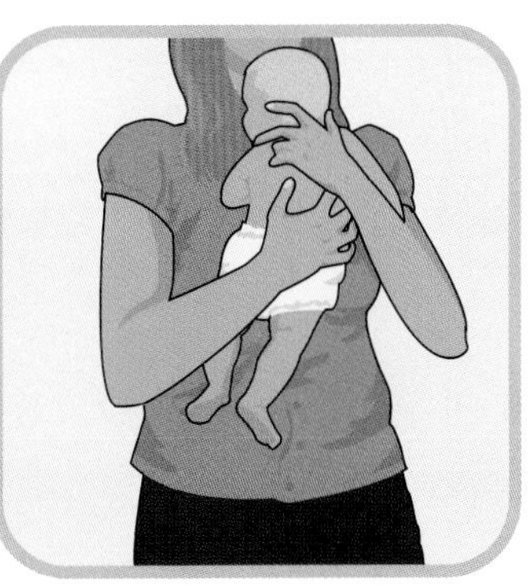

Gently bring to shoulder.

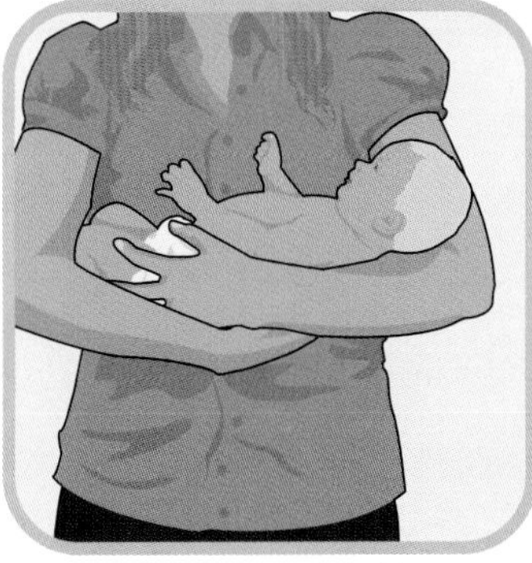

Switch to a cradle hold.

241 make a teething aid

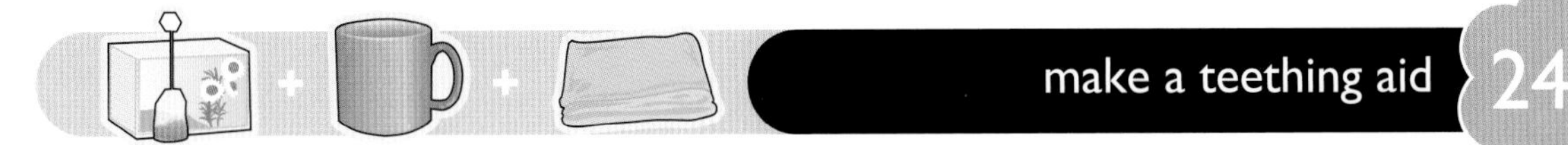

Brew chamomile tea.

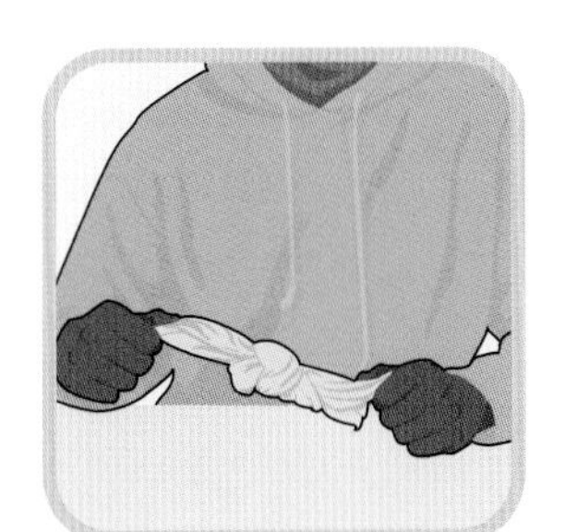

Tie a knot in a washcloth.

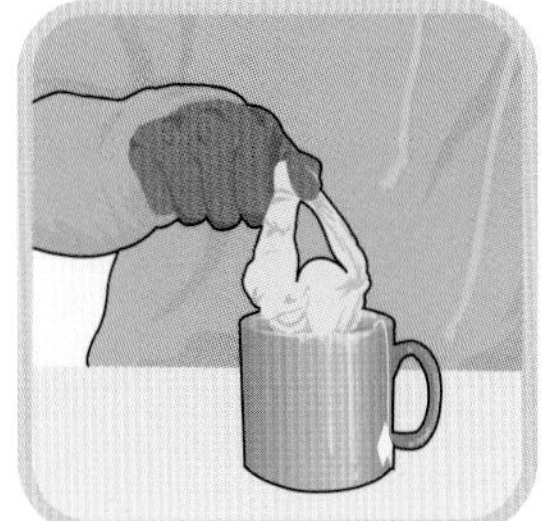

Dip the knot in the tea.

Freeze.

freeze smoothie pops 363

Give to your teething baby.

242 treat a car-sick kid

Stop frequently for walks.

Keep fresh air circulating.

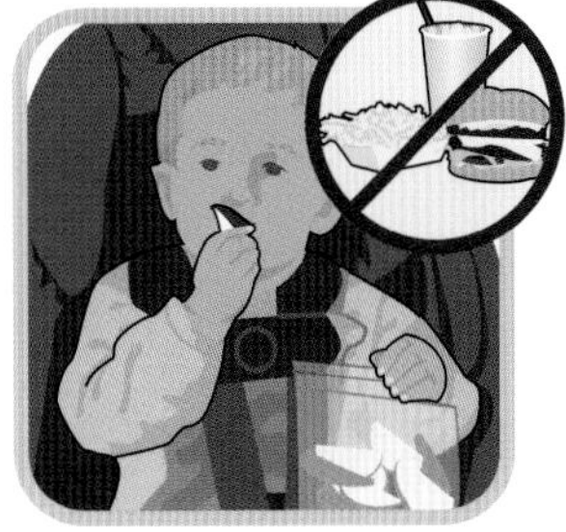

Feed him healthy snacks.

Have him look forward.

Try ginger or mint candies.

243 soothe with a boo-boo bunny

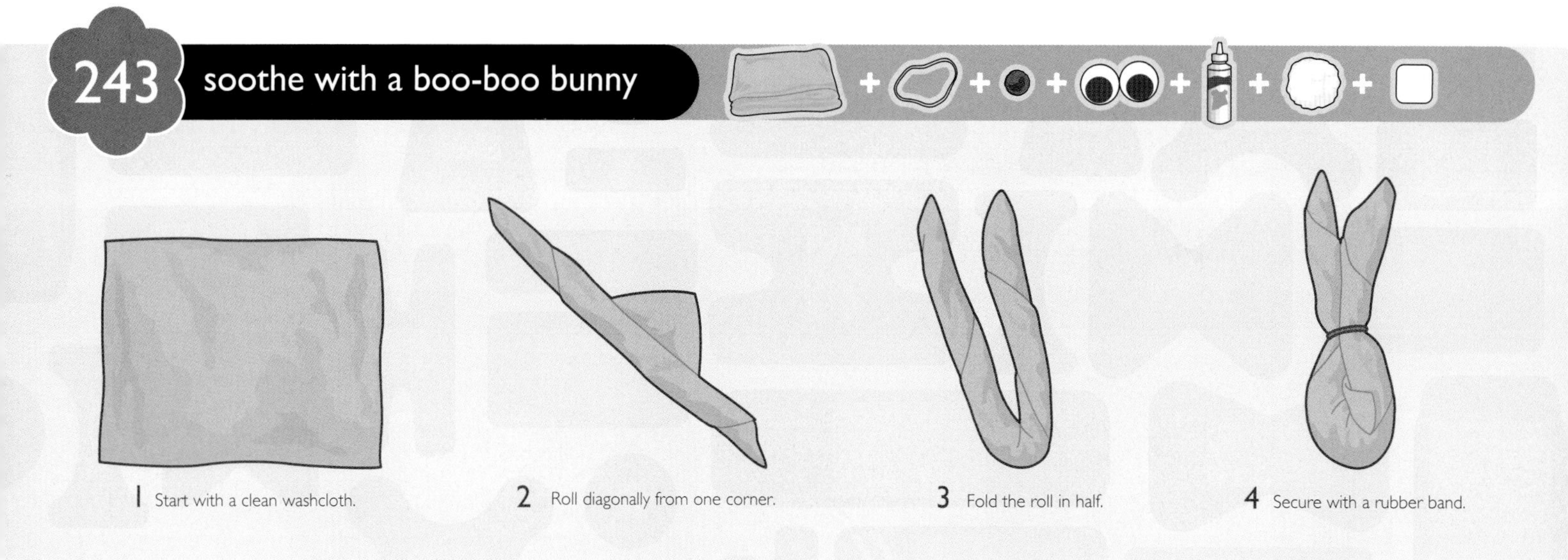

244 play with a baby

When you share these simple, sweet activities with a little one, it's not just fun and games—you're helping baby learn a range of vital developmental skills.

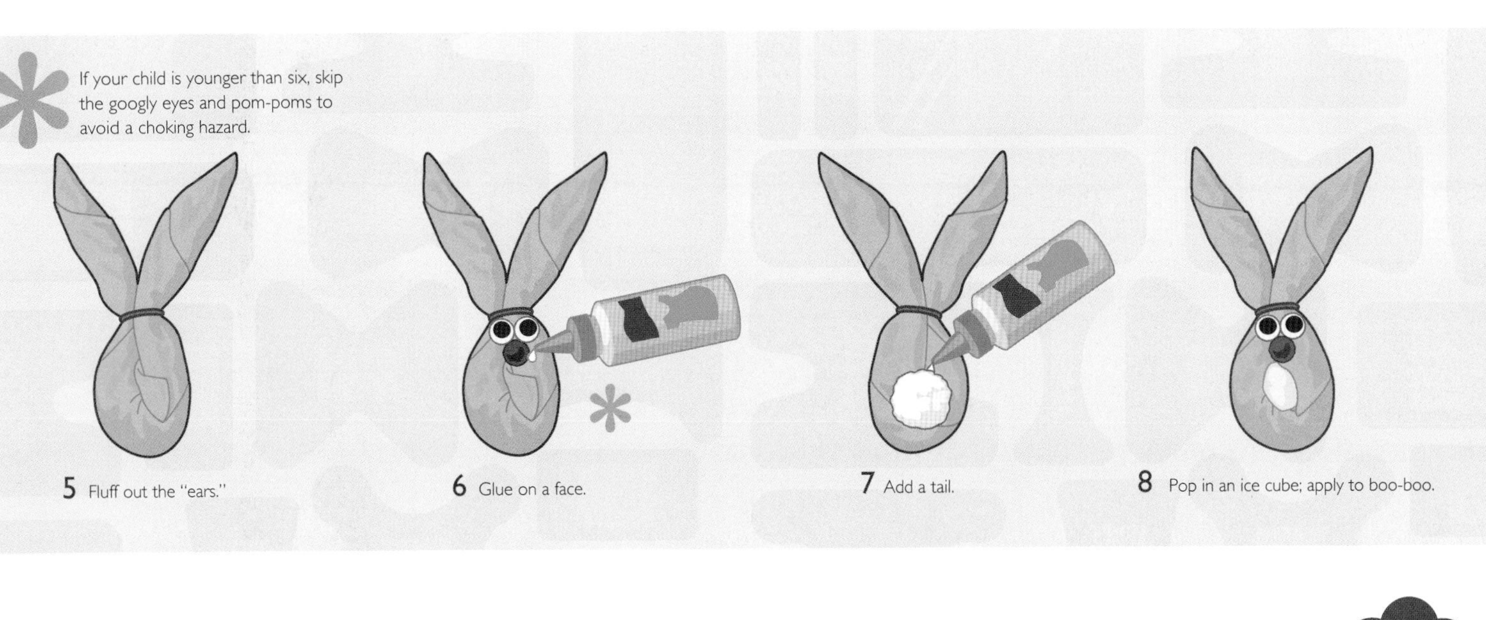

If your child is younger than six, skip the googly eyes and pom-poms to avoid a choking hazard.

5 Fluff out the "ears."

6 Glue on a face.

7 Add a tail.

8 Pop in an ice cube; apply to boo-boo.

play with a toddler 245

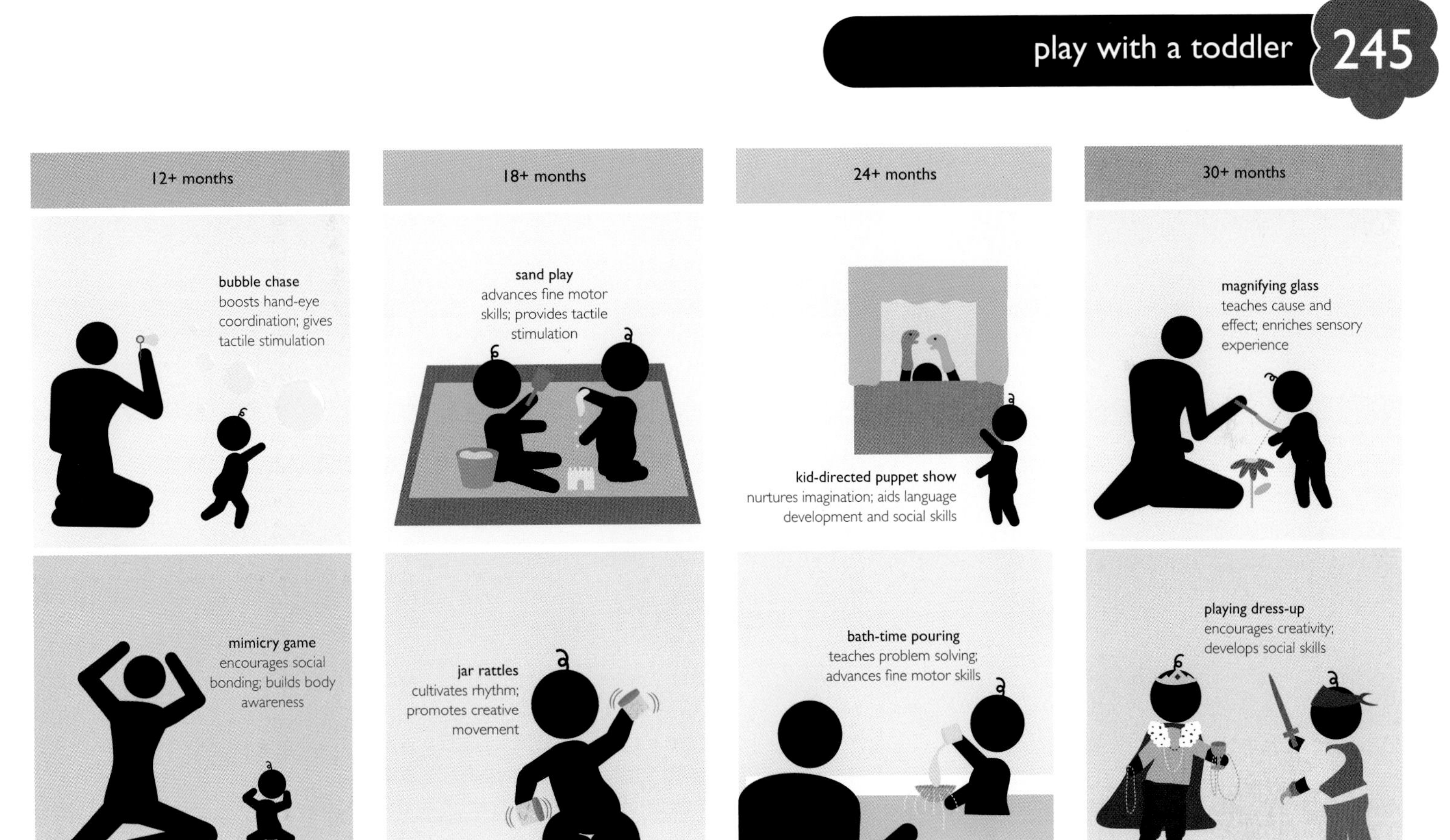

246 fly a kite

Face downwind.

Hand kite to a friend.

Unravel string as friend backs away.

Friend holds kite up high.

Friend tosses the kite.

Walk into the wind with arms up.

Let out string until kite is high enough.

To bring down, slowly reel in string.

247 build a blanket fort

Create perimeter and center support.

Add blankets; weight or fasten edges.

Create an entryway.

Decorate and add vital supplies.

Divide a group of kids into two teams. Have one team choose (or invent) a lost civilization and bury objects that the society would have used. Then have the other team dig up the artifacts and piece together the life of the lost city, just like real archaeologists would.

fake an antiquity 47

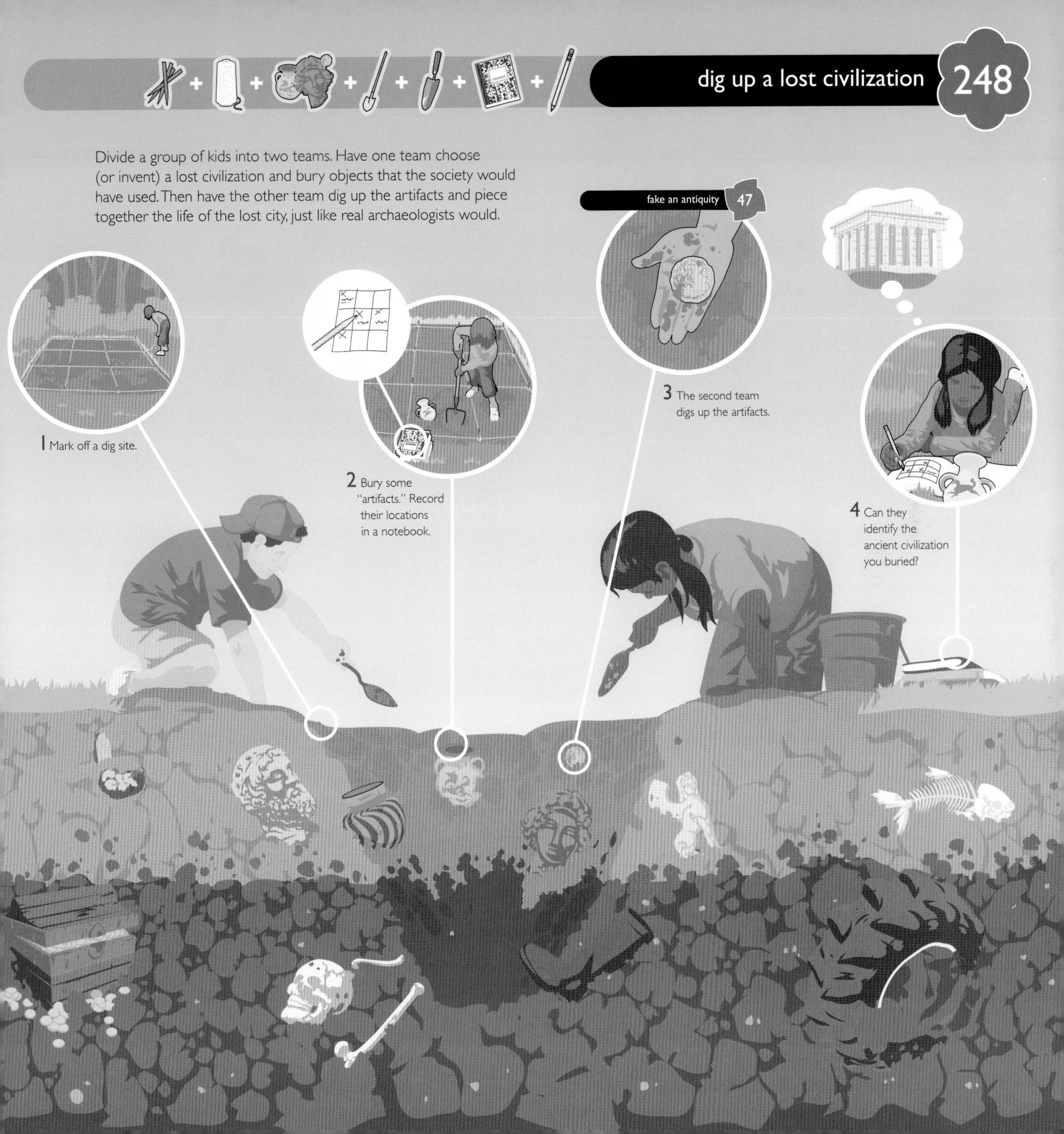

249 navigate adolescence

Being a teenager can be like playing a fast-paced game—one with common trials and traps, as well as tricks that can make the game easier. Whether you're a jock or a musician, a crafter or a geek, every teen can benefit from this coming-of-age tip sheet, and better enjoy the fun maze of adolescence.

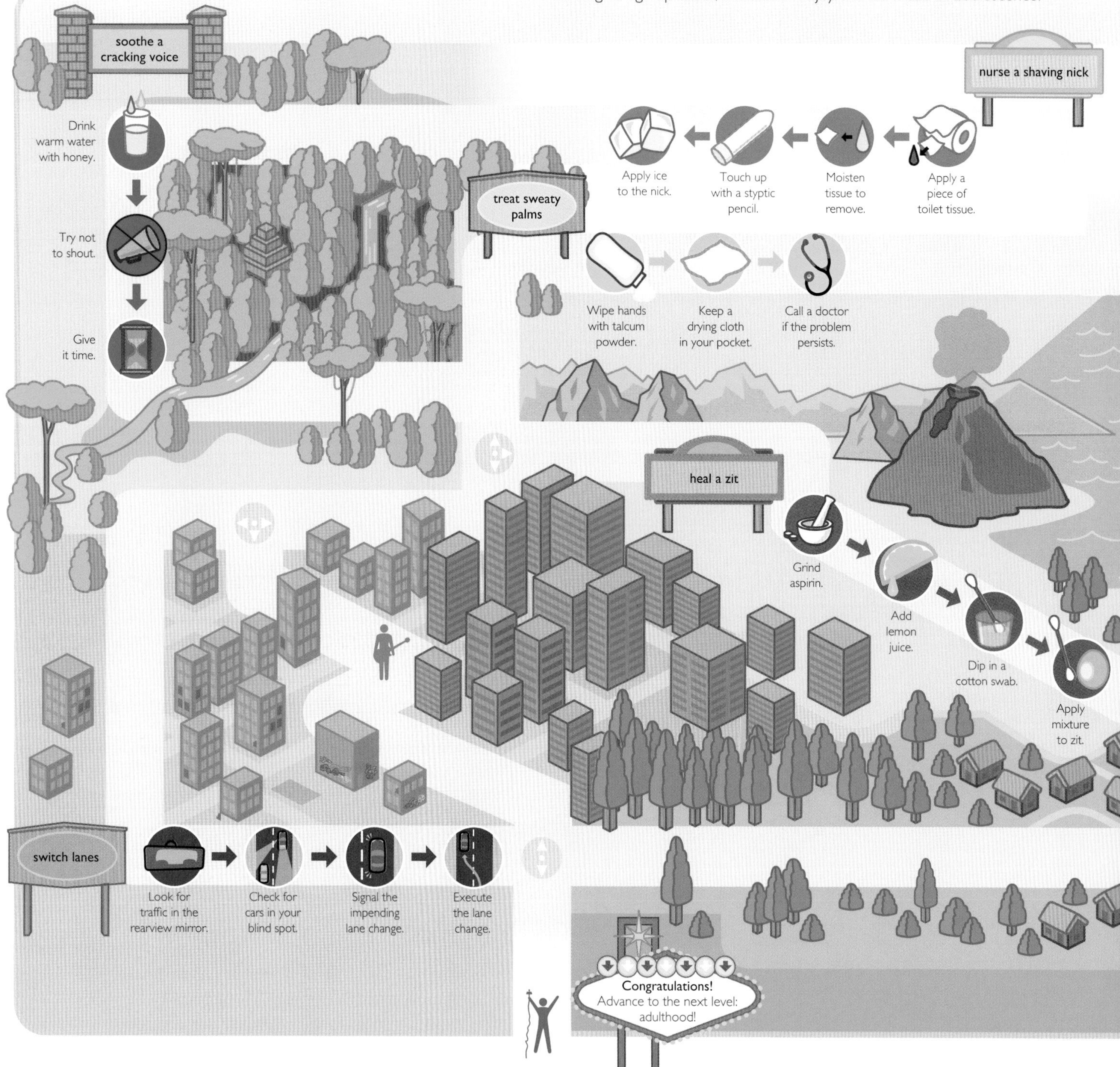

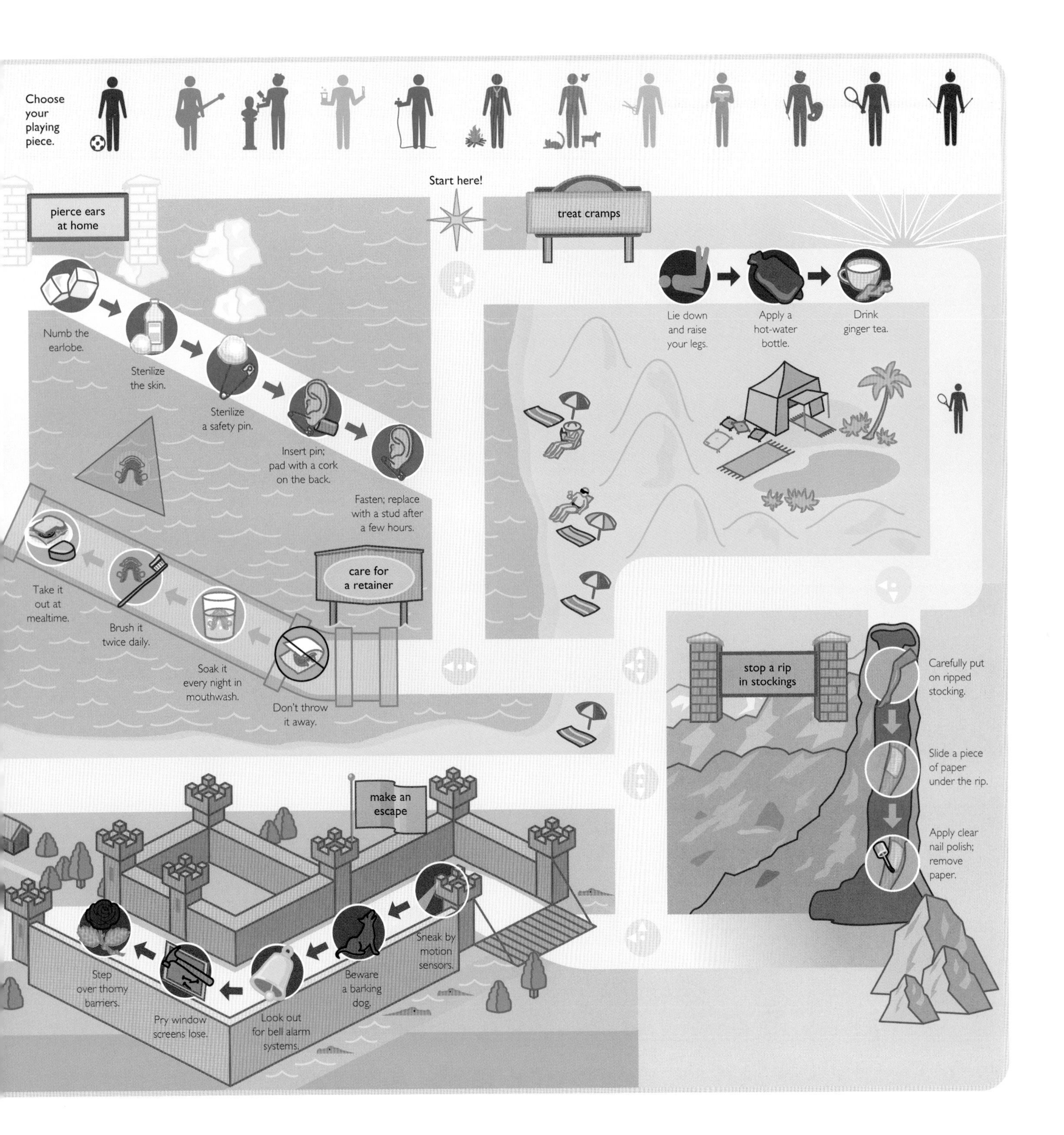

Choose your playing piece.
Start here!
pierce ears at home
Numb the earlobe.
Sterilize the skin.
Sterilize a safety pin.
Insert pin; pad with a cork on the back.
Fasten; replace with a stud after a few hours.
treat cramps
Lie down and raise your legs.
Apply a hot-water bottle.
Drink ginger tea.
care for a retainer
Don't throw it away.
Soak it every night in mouthwash.
Brush it twice daily.
Take it out at mealtime.
stop a rip in stockings
Carefully put on ripped stocking.
Slide a piece of paper under the rip.
Apply clear nail polish; remove paper.
make an escape
Sneak by motion sensors.
Beware a barking dog.
Look out for bell alarm systems.
Pry window screens lose.
Step over thorny barriers.

help

250 prepare for an emergency

Discuss local hazards.

Practice responses.

Pick an emergency contact.

Have a family meeting place.

Add emergency numbers.

Select an out-of-state friend or relative to check in with in case your family becomes separated during an emergency. Make sure everyone has the phone number.

251 burglar-proof my home

Start a neighborhood watch group.

Install an alarm system.

Add outdoor motion-sensor lights.

Keep tree limbs trimmed.

203 order a workroom

Store ladders and trash cans indoors.

Keep garage doors locked.

Secure sliding windows with dowels.

Always look before answering door.

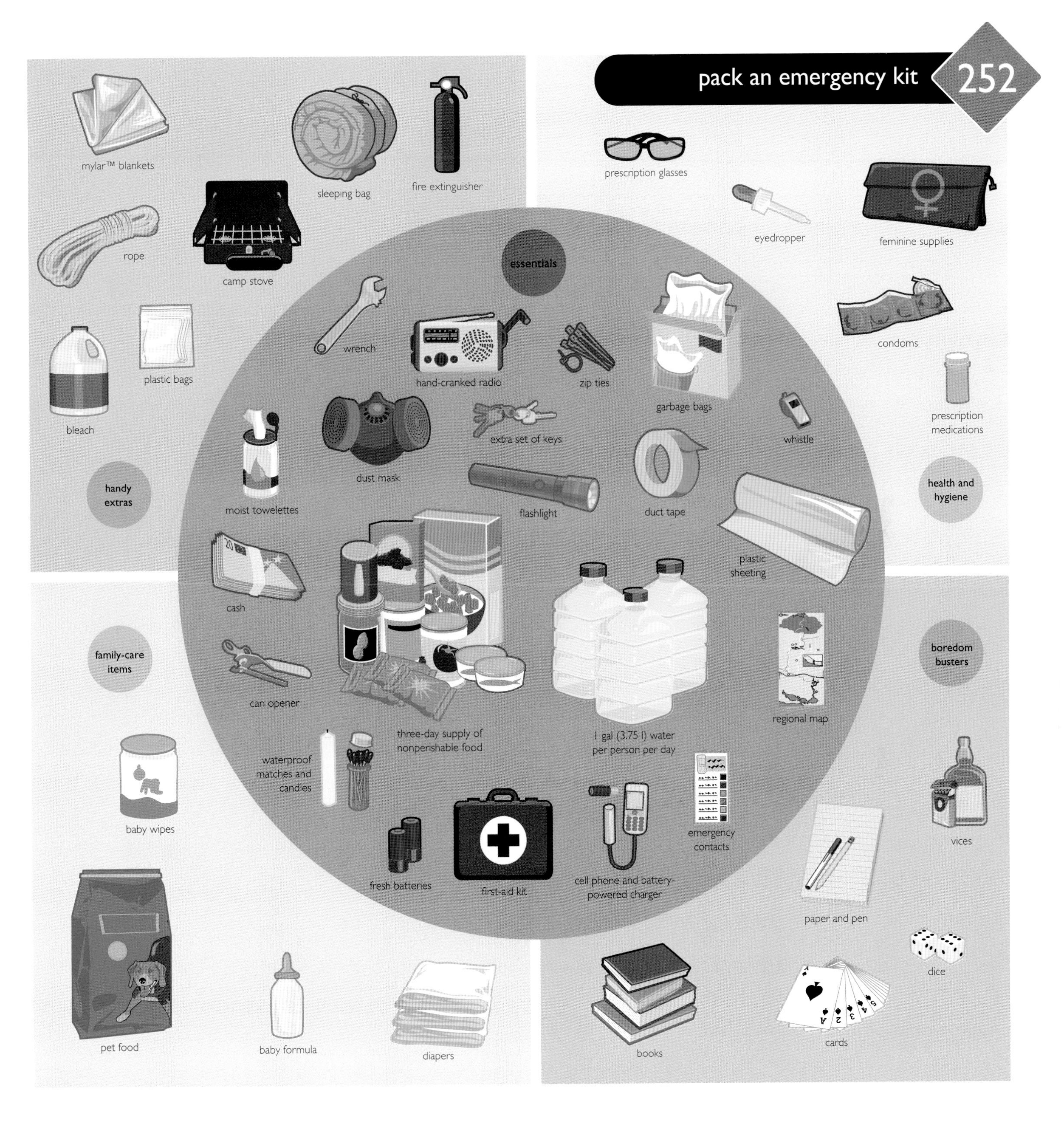
essentials
wrench
hand-cranked radio
zip ties
garbage bags
dust mask
extra set of keys
whistle
moist towelettes
flashlight
duct tape
plastic sheeting
cash
can opener
three-day supply of nonperishable food
1 gal (3.75 l) water per person per day
regional map
waterproof matches and candles
emergency contacts
fresh batteries
first-aid kit
cell phone and battery-powered charger
handy extras
mylar™ blankets
sleeping bag
fire extinguisher
rope
camp stove
plastic bags
bleach
health and hygiene
prescription glasses
eyedropper
feminine supplies
condoms
prescription medications
family-care items
baby wipes
pet food
baby formula
diapers
boredom busters
vices
paper and pen
dice
books
cards

253 firescape my yard

If you live in a dry area, you should look into "firescaping"—that's designing the space and conditions around your home with fire protection in mind. Here are some tips to get you started.

Clear away low-hanging branches, and any deadwood or debris, within 30 ft (9 m) of your home.

Don't plant conifers—they contain highly flammable oils and resins.

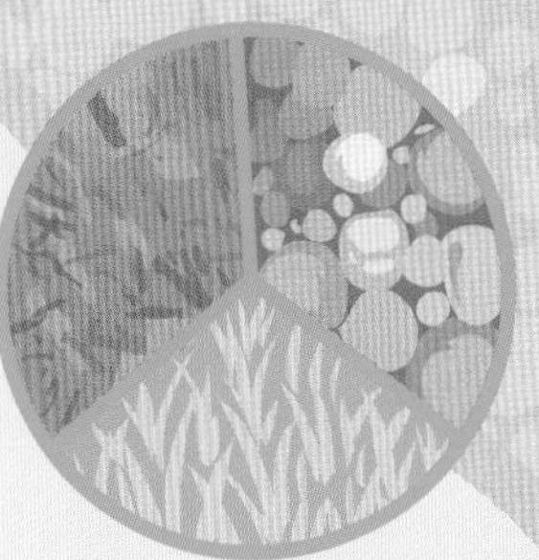

Cover bare spaces with stone, mulch, or high-moisture plants.

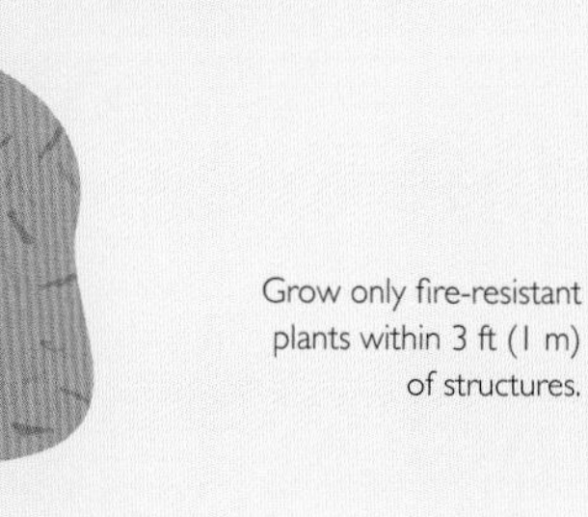

Grow only fire-resistant plants within 3 ft (1 m) of structures.

Store firewood at least 30 ft (9 m) from your home.

Keep plants moist all year with drip irrigation.

Preparation is key to keeping calm during a fire. Practice these basic fire-safety techniques with your family so everyone knows just what to do in an emergency.

Install a spark arrester on your chimney.

Check windows often to make sure none are stuck.

Stay low and cover your mouth with a damp cloth.

Have fire-escape ladders on hand near windows.

pick up a new baby 240

Lower kids down first if you must escape through a window.

Keep hallways clear. Clutter can hinder your escape.

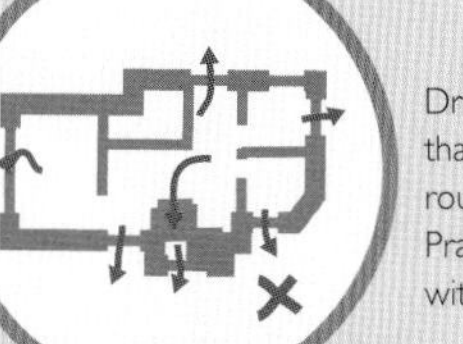

Draw up an exit plan that provides two escape routes from each room. Practice it periodically with your household.

Touch each door before opening. If it feels hot, take a different route.

Test smoke alarms twice a year and replace batteries as needed.

Don't try to put the fire out yourself.

Have a meeting place nearby as the endpoint of your escape plan.

Cover external vents with fine metal screen to keep forest-fire embers from blowing in.

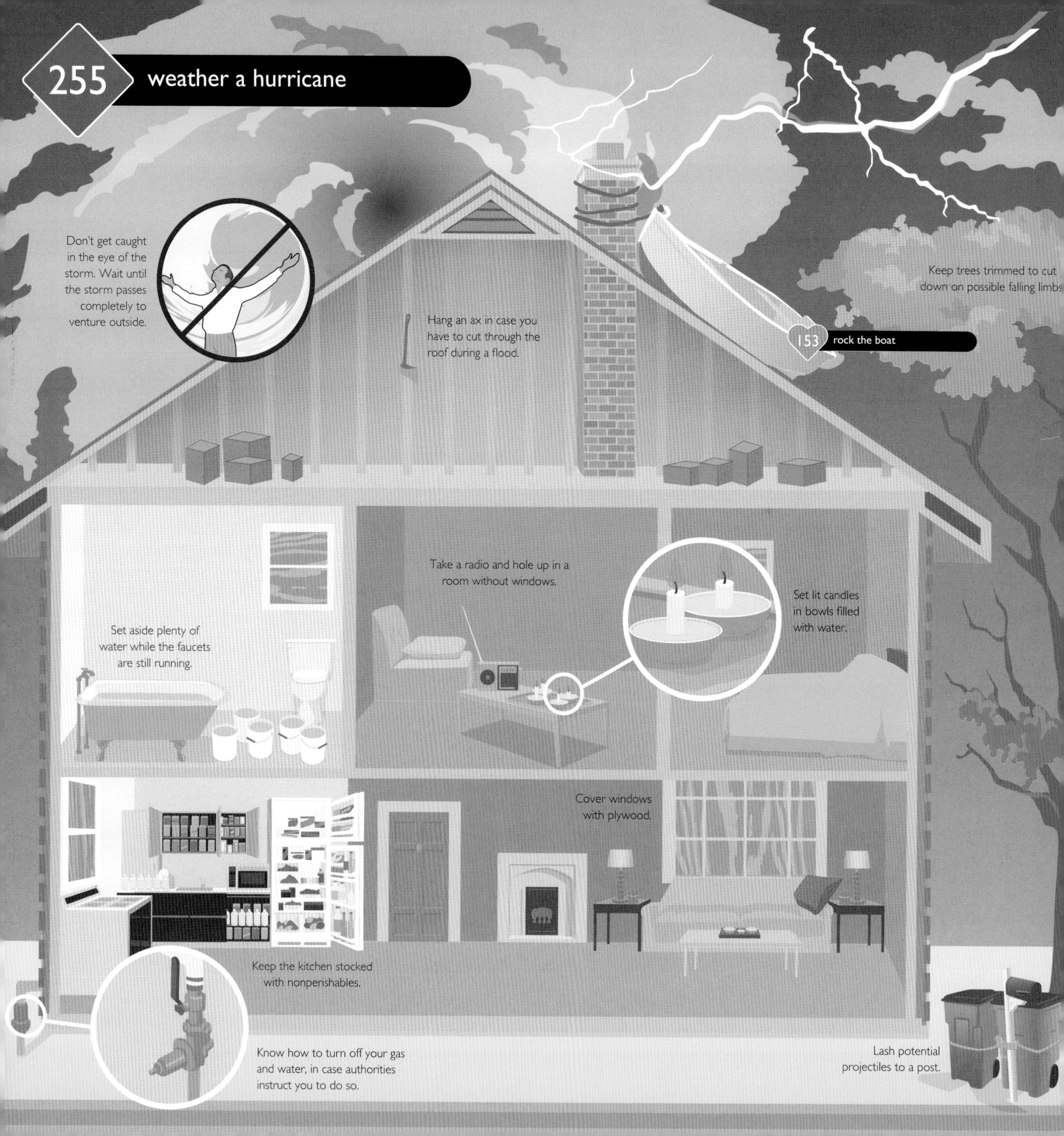
255
weather a hurricane
Don't get caught in the eye of the storm. Wait until the storm passes completely to venture outside.
Hang an ax in case you have to cut through the roof during a flood.
Keep trees trimmed to cut down on possible falling limbs
153 rock the boat
Take a radio and hole up in a room without windows.
Set lit candles in bowls filled with water.
Set aside plenty of water while the faucets are still running.
Cover windows with plywood.
Keep the kitchen stocked with nonperishables.
Know how to turn off your gas and water, in case authorities instruct you to do so.
Lash potential projectiles to a post.

withstand a heat wave 256

Drink lots and lots of water.

Turn out unnecessary lights.

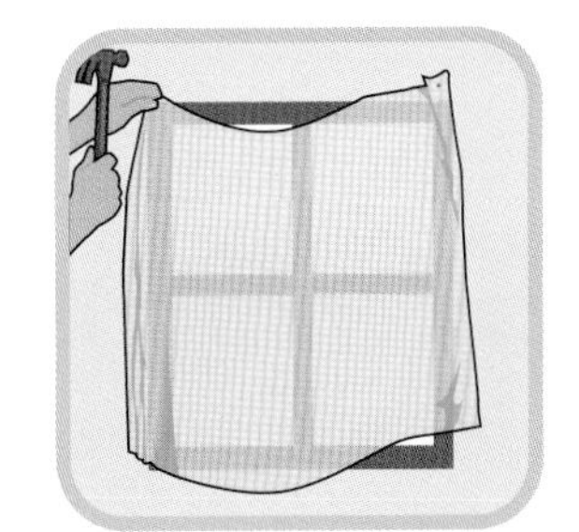
Block windows with sheets.

Circulate air through house.

Visit air-conditioned places.

handle a winter power outage 257

thaw a frozen pipe 213
Bring children and pets indoors.

Insulate drafty doorways.

Sleep together near a fire.

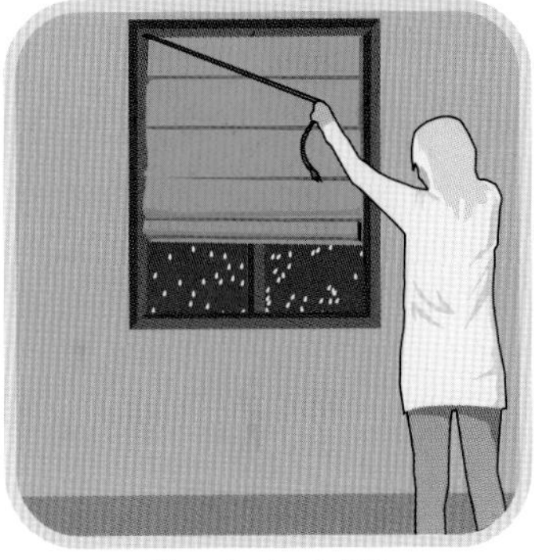
Close shades at night.

Let daylight in for warmth.

Store perishables outside.

Stay hydrated.

Take turns shoveling.

258 survive an avalanche

Wear an avalanche beacon.

Try to jump above break line.

Move perpendicular to flow.

Grab a sturdy tree or rock.

"Swim" on top of snow.

If submerged, cover face.

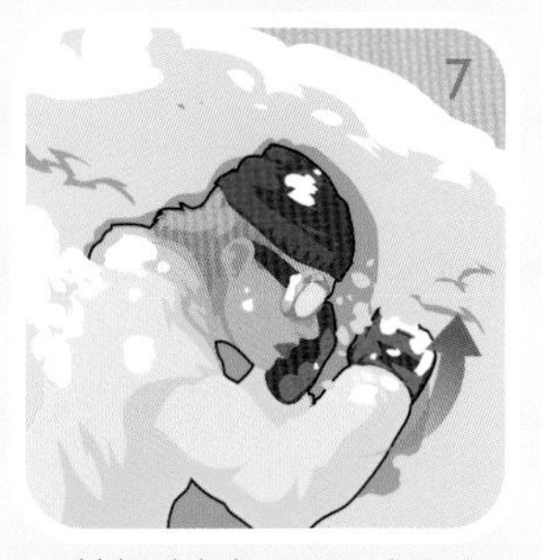

Make air hole as snow slows.

Wait calmly for help.

259 prevent snow blindness

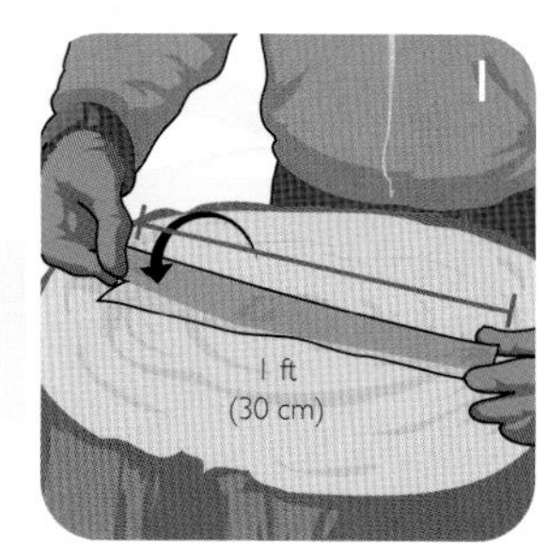

Cut a strip of duct tape; fold.

Make a long slit.

Fasten around head with tape.

Blacken cheeks with soot.

Snow debris and broken trees indicate previous avalanches—be wary of repeat slides.

be avalanche aware 260

Avoid avalanche-prone areas in the forty-eight hours after rough weather or a thaw. If you must go, pack a collapsible shovel, a snow probe, and an avalanche beacon.

Slopes of 30 to 45 degrees are most likely to avalanche, but even slopes of 25 to 60 degrees can slide in certain conditions.

Smooth, grassy slopes without rocks or trees are most dangerous.

A heavy, compacted layer of snow resting on a powdery layer is highly unstable.

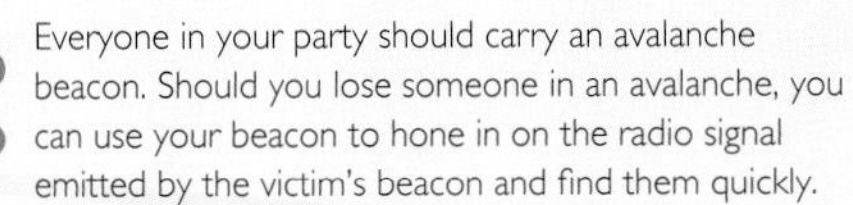

* Everyone in your party should carry an avalanche beacon. Should you lose someone in an avalanche, you can use your beacon to hone in on the radio signal emitted by the victim's beacon and find them quickly.

set up a snow camp 300

find an avalanche victim 261

Go to location of last sighting.

Set beacon to receive mode.

Search downhill.

Beacon narrows search area.

Poke with snow probe.

Dig downhill from victim.

Uncover head first.

Send for help.

262 perform the fireman's carry

While this is an excellent way to transport an unconscious person in most circumstances, don't be fooled by the name: the fireman's carry is not recommended for use during fires.

Pull victim to feet; raise arm.

Kneel; tip victim over your shoulder.

Reach through legs; grab victim's arm.

Stand, lifting with your legs.

263 treat hyperventilation

Address source of anxiety.

Encourage slower breathing.

Bend legs; elevate head.

Cover nose, mouth; breathe.

If not recovered, call for aid.

264 help after an accident

Respond calmly.

Call for help.

Elevate victim's legs.

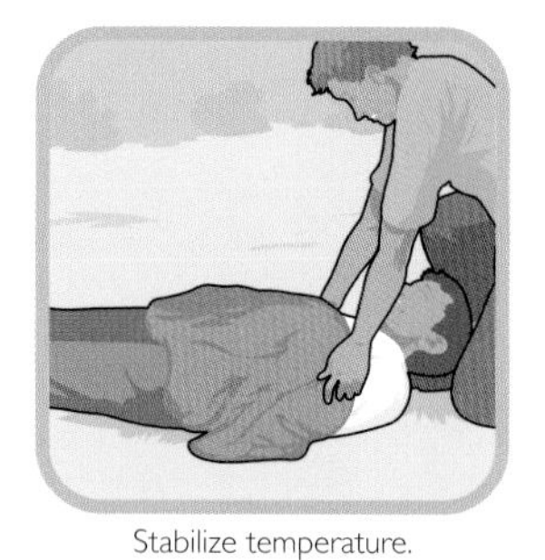

Stabilize temperature.

Help is on the way.

Speak soothingly.

escape a riptide 265

A riptide is a very strong current that swiftly draws water (and swimmers) at the surface out into open water. Riptides are rarely over 100 ft (30 m) wide, so swimming parallel to the shore is the best way to escape one.

Avoid water that's been churned up.

If caught in riptide, remain calm.

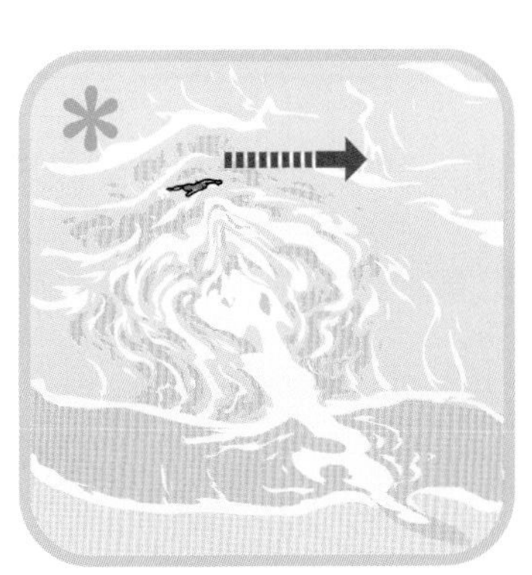
Swim parallel to beach.

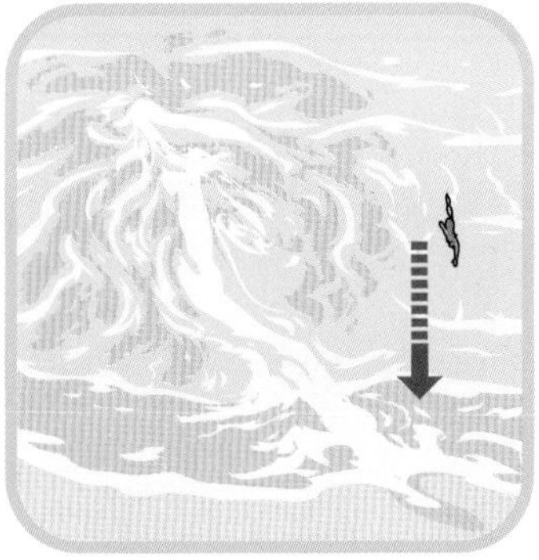
Once out of current, swim to shore.

rescue a swimmer 266

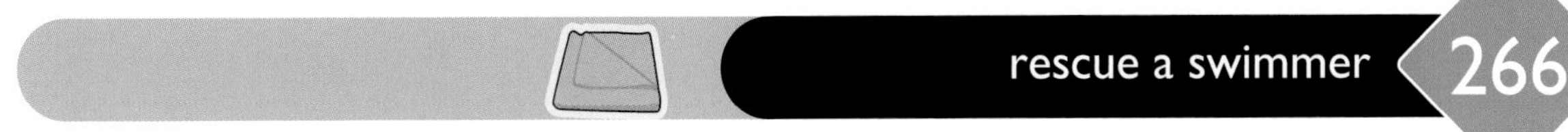

Grab clothing or a towel.

Swim out.

Stop short; throw towel.

Let victim get a grip.

Swim in, pulling victim.

ease heat exhaustion 267

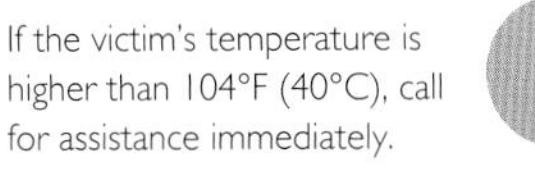
If the victim's temperature is higher than 104°F (40°C), call for assistance immediately.

Check victim's temperature.

Get victim into the shade.

Remove layers.

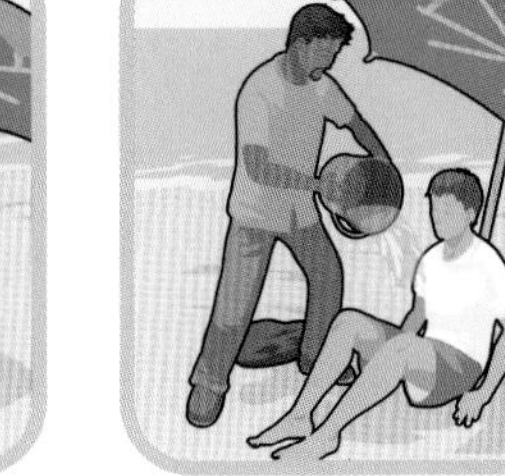
Bathe with cool water.

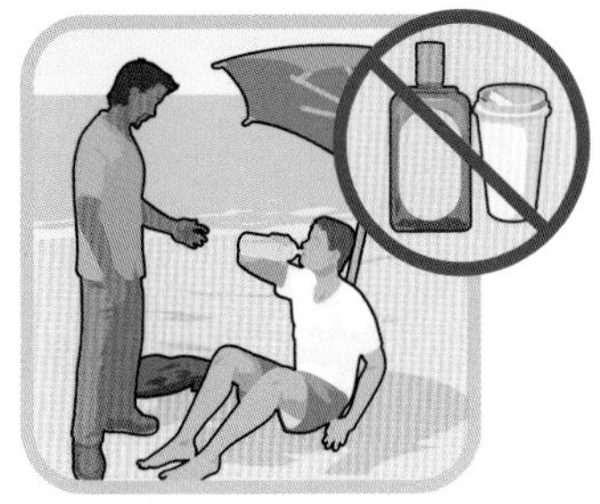
Give victim plenty to drink.

268 outlast an angry gorilla

Crouch to show submission.

Groom gorilla if possible.

If not, face away.

Avoid eye contact.

380 power slam like a luchador

If attacked, curl into a ball.

269 endure a jellyfish attack

Rinse with salt water.

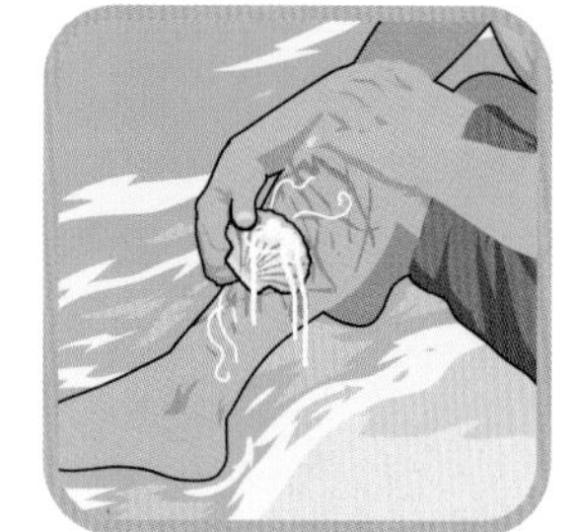
Scrape away tentacles.

Take ibuprofen.

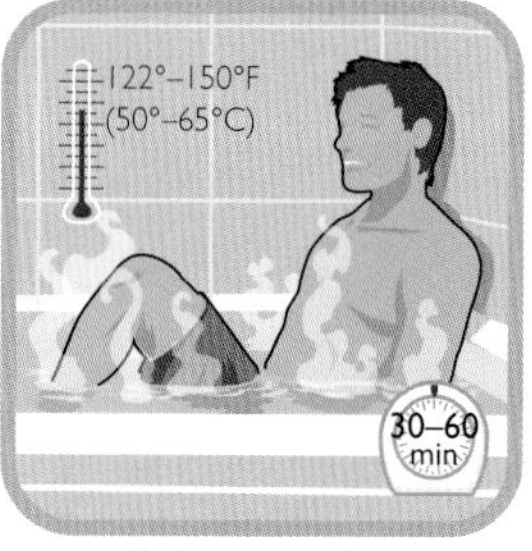

Soak in hot water.

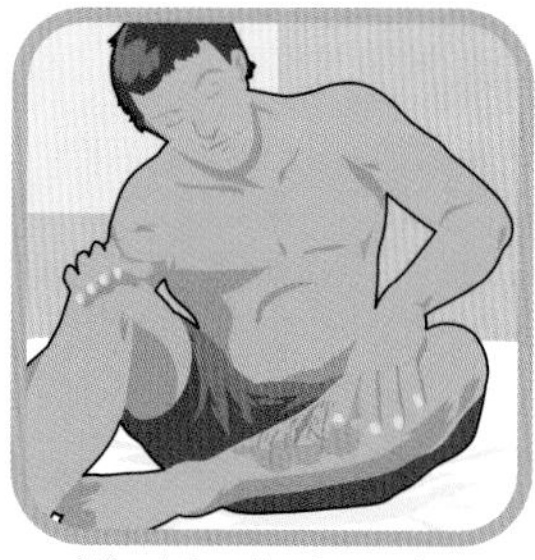
Watch for allergic reaction.

270 free myself from an anaconda

Be on guard in shallow water.

If attacked, take a deep breath; hold it.

Bite anaconda's tail as hard as you can.

Grab a rock; strike head repeatedly.

271 be bear aware

Pitch your camp in a quiet, open area where you'll be visible to wildlife—and vice versa.

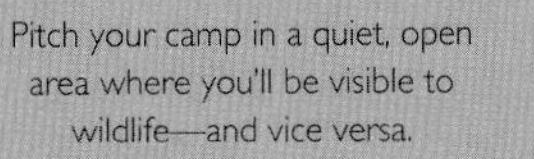

Suspend food out of reach.

Do your cooking at least 100 ft (30 m) downwind from the campsite.

Raise a ruckus and try to make yourself look bigger if a bear approaches.

If threatened, don't climb up a tree (black bears and some grizzlies can follow you up). Stand still if charged—the bear might be bluffing.

Don't leave any food in or near your tent.

272 know a black bear from a grizzly

black bear

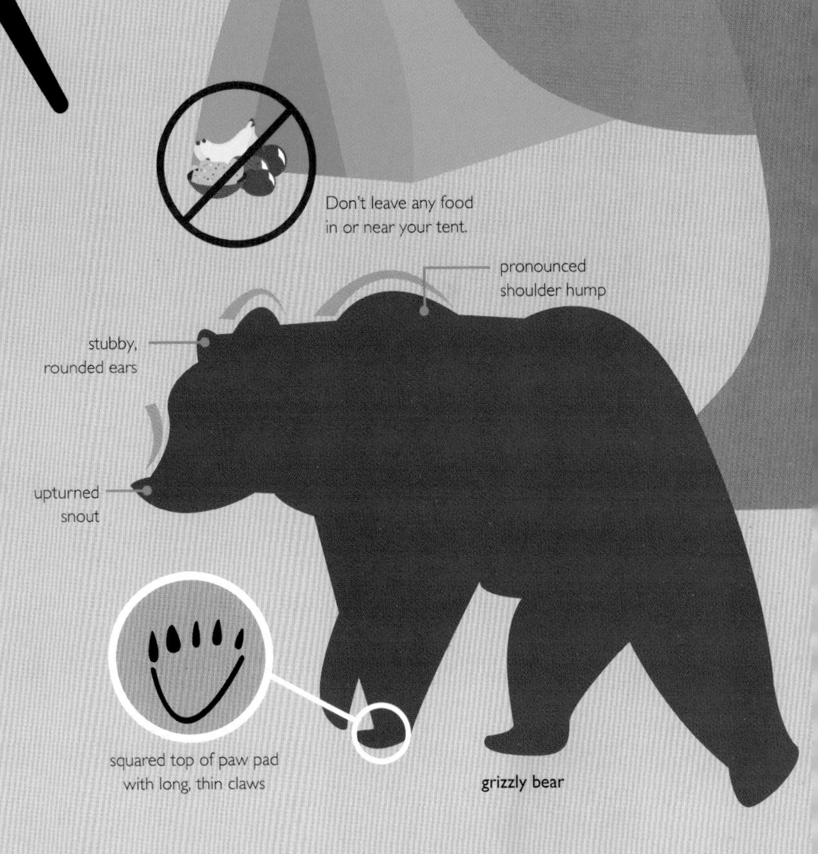

grizzly bear

273 eat a scorpion

If you're stranded in the desert with nothing else to nibble, scorpions can provide the protein you need (and a satisfying crunch). Don't forget all these other edible bugs, too!

Find a tasty scorpion.

Stun it with liquor.

Stab to hold in place.

218 debug my garden

Remove stinger.

Skewer.

Roast until brown, crispy.

Eat in one bite.

collect water from a branch 274

conserve water at home 212

Weight a plastic bag.

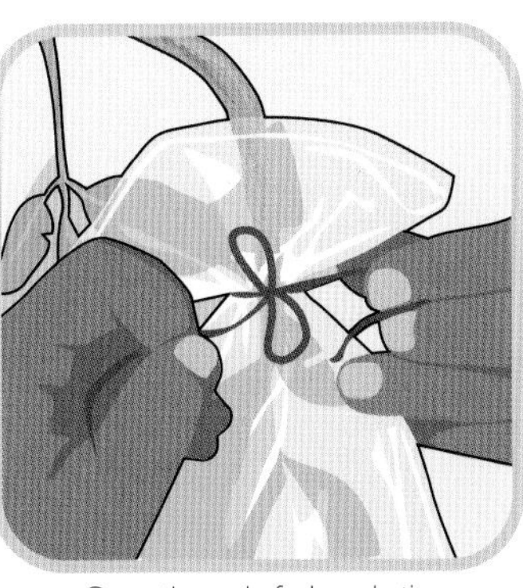

Cover the end of a branch; tie.

Wait.

Collect water; boil before drinking.

stay warm in a fire bed 275

Set out a layer of rocks.

Build a large fire on top.

Wait for fire to die down.

Spread coals over rocks.

Cover coals with a layer of dirt.

Lay logs over dirt.

Cover with brush and needles.

Sleep tight!

276 break an attacker's grip

Draw back free arm.

Aim for attacker's hand.

Strike center.

Run.

277 stay safe in a parking garage

Avoid parking between vans.

Find an open, well-lit area.

Park near the exit.

Use elevators instead of stairs.

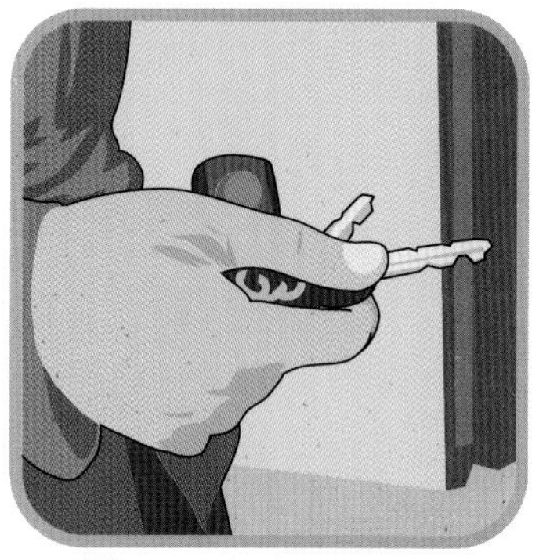
When returning, have keys ready.

Look inside before getting in.

Lock doors immediately.

Leave garage immediately.

ride out a nightclub stampede 278

Look for alternate exits.

Slip between people diagonally.

Continue on without lost shoes.

Shelter behind fixed objects.

avoid a bar fight 279

Avoid conflict if possible.

If not, try to appear insane.

Do something unexpected.

Make yourself look larger.

Scream at top volume.

say "cheers" in any language 288

Throw a beverage.

Menace with chair.

Run away.

go

280 pick a destination

Whether you're an adrenaline junkie or a honeymooner, a family of four or a rugged individualist, there's a world of great destination ideas for your next vacation.

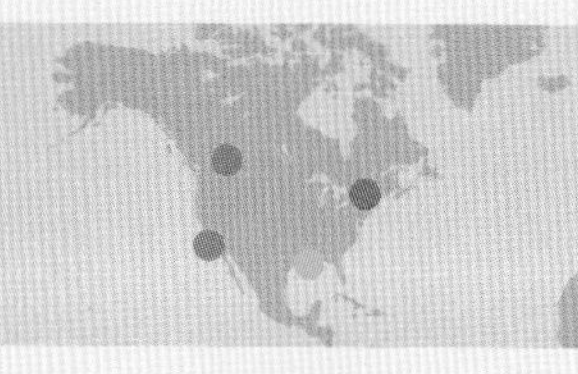
north america

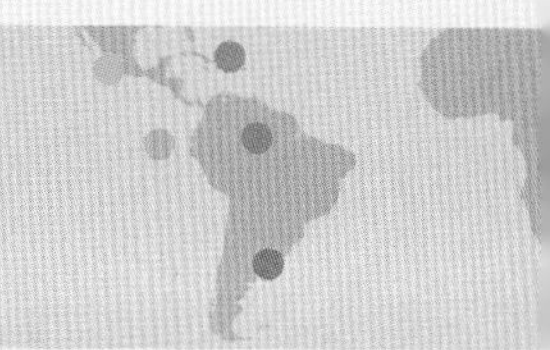
central and south america

54 disappear from a cruise ship

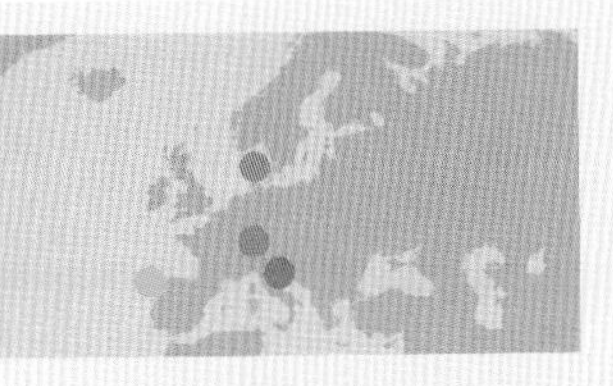

europe

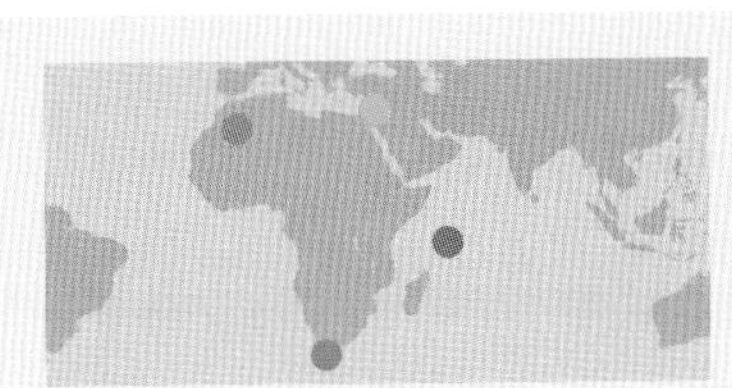

africa and the middle east

asia

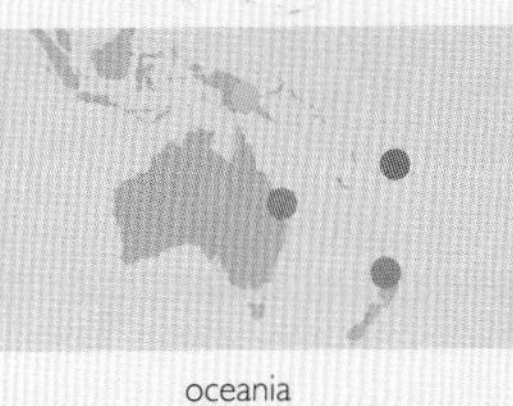

oceania

venice, italy

seychelles islands

pulau pangkor, malaysia

vatulele island, fiji

billund, denmark

merzouga, morocco

chiang mai, thailand

gold coast, australia

southwest coast, iceland

serengeti, kenya

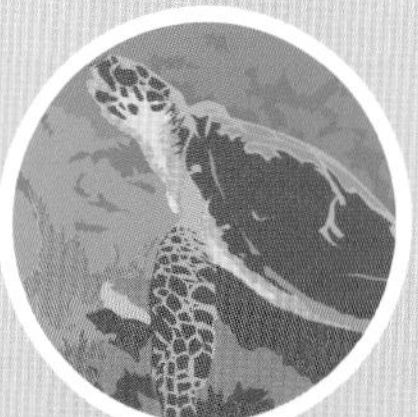

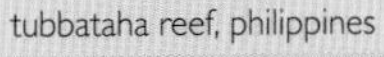

tubbataha reef, philippines

purnululu national park, australia

haute route, switzerland

gansbaai, south africa

himalayan foothills, myanmar

waitomo caves, new zealand

camino de santiago, spain

petra, jordan

almaty, kazakhstan

goroka, papua new guinea

281 prepare my home before a trip

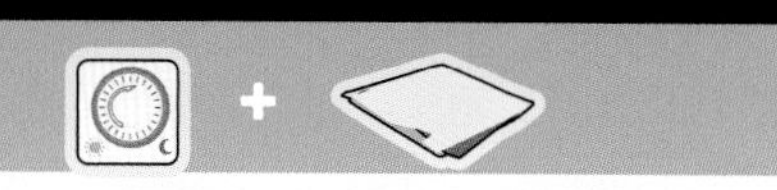

Have mail held at post office.

Set timers on lights.

Purge fridge of perishables.

251 burglar-proof my home

Lock the windows.

Give travel info to a neighbor.

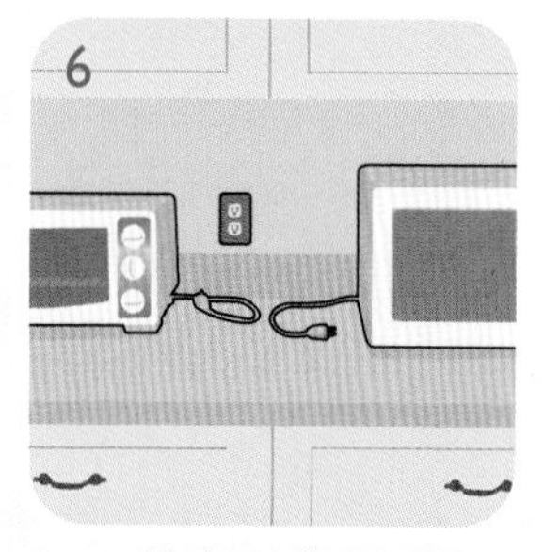

Unplug appliances.

Stop any leaks or drips.

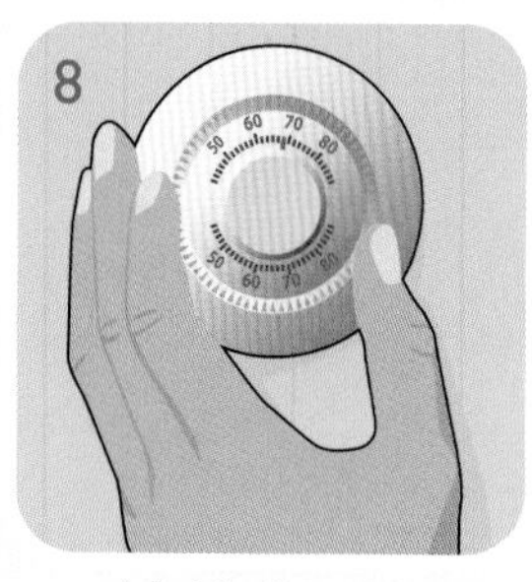

Adjust the thermostat.

282 fly with a toddler

Use approved car seat.

Bring a stroller bag.

Pack more than you'll need.

Give a bottle at takeoff.

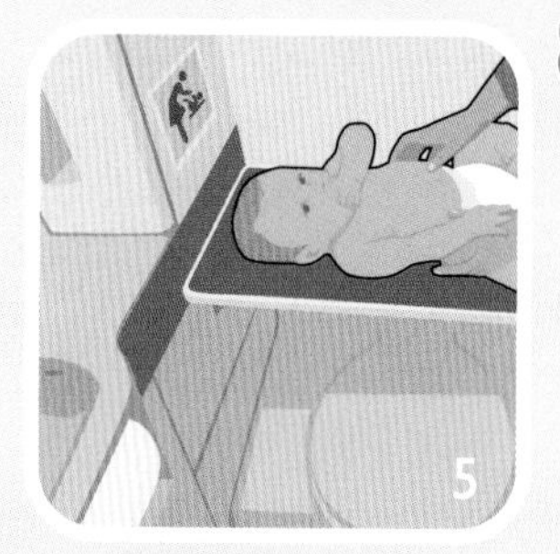

Change in the bathroom.

242 treat a car-sick kid

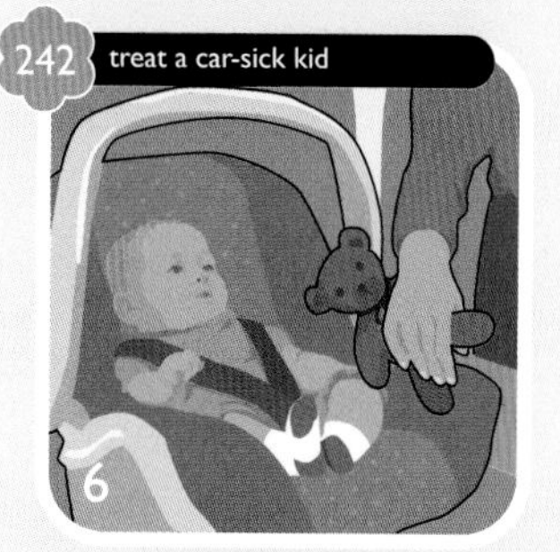

Distract with new toys.

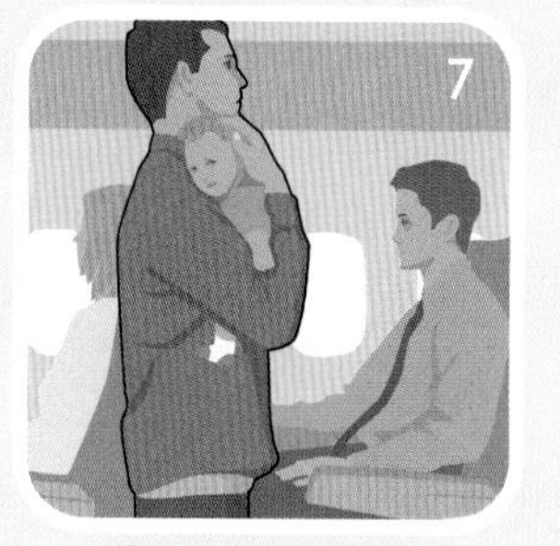

If upset, walk the aisles.

Give bottle during descent.

take a big dog on a plane 283

Clear him for travel at the vet.

Clip nails.

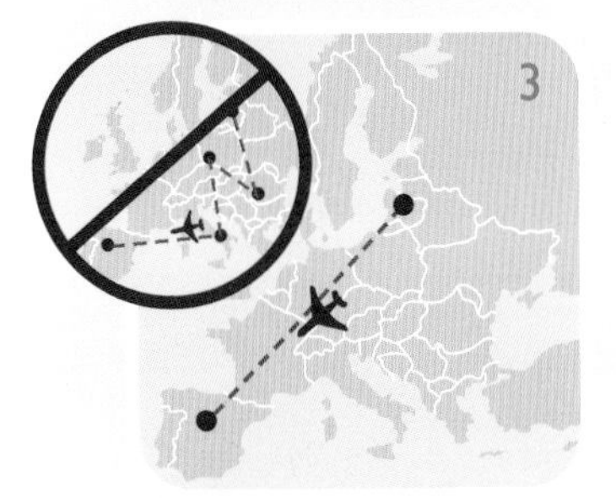

Schedule a direct flight.

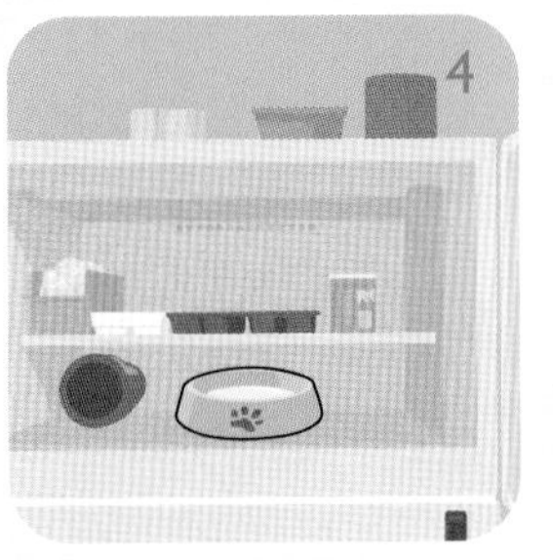

Freeze a water bowl.

Feed well before flight.

Check with ice bowl inside.

Ask steward to check on him.

Feed on arrival.

get myself ready for a trip 284

	six months	Get visas and passport in order.
	six months	Get recommended vaccinations.
	three weeks	Fill prescriptions to last through trip. Make copies of prescriptions (including eyeglasses and contact lenses) to bring along.
	three weeks	Research and purchase electrical converters for each country on your itinerary.
	two weeks	Inform your credit-card companies of your itinerary.
	two weeks	Purchase maps and phrase book; practice useful phrases.
	one week	Tear relevant pages from your guidebooks and staple into packets.
	one week	Get cash, currency, and traveler's checks in order.
	one week	Ask friends or relatives to be your emergency contacts; write down all their contact info.
	one week	Photocopy your passport, tickets, reservation confirmations, and front and back of credit cards.
	three days	Put copies of your documents and contact info in your suitcase, exchange them with your traveling companion, and leave a copy with someone at home.
	two days	Label all luggage with contact info on a tag outside and a card inside.

285 ask without words

Have you ever found yourself in a foreign city desperately needing a camel but with no idea how to ask for one? Of course you have. With this handy chart, those days are behind you. Simply find a hotel employee or friendly local and point to what you need.

hotel amenities

entertainment

dining

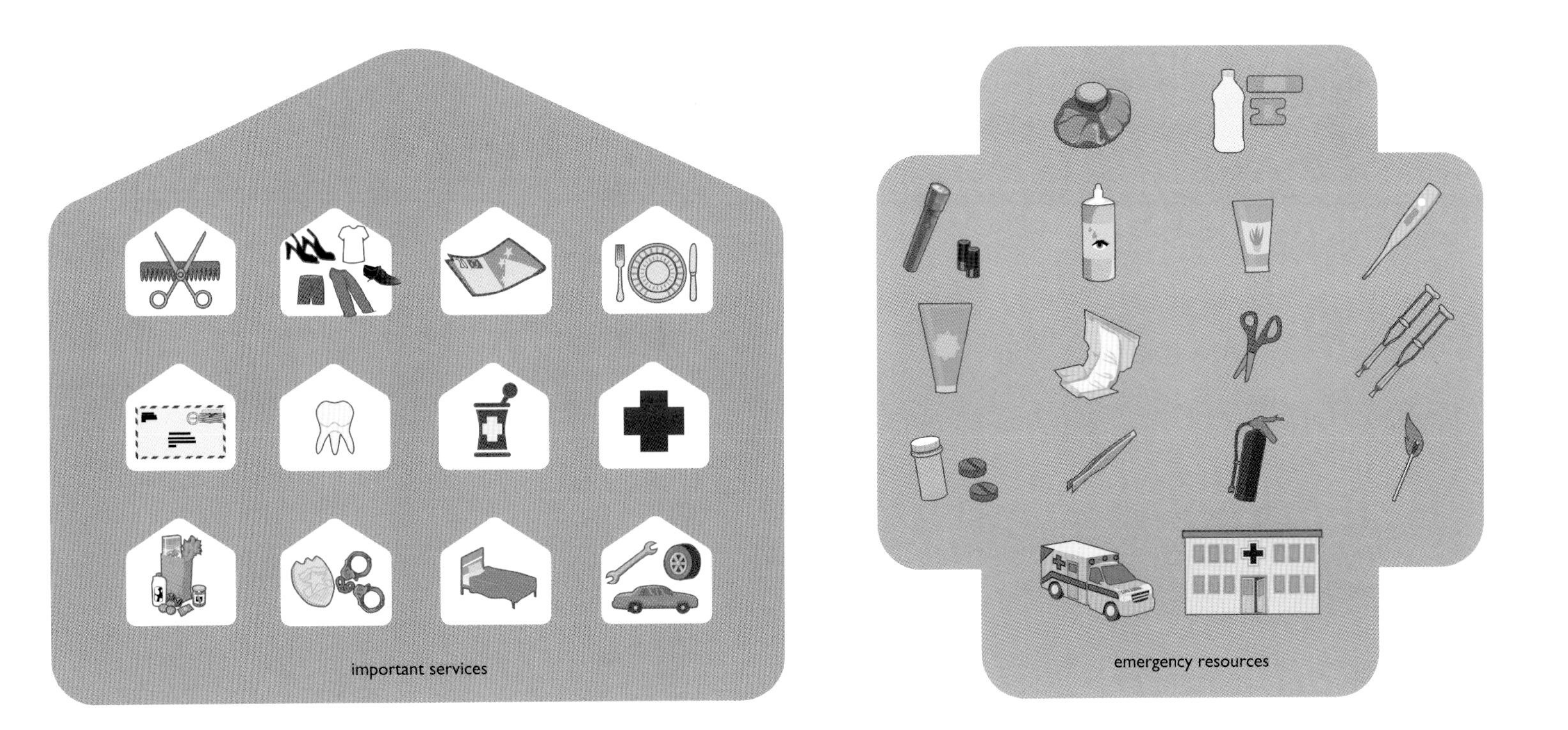
important services
emergency resources

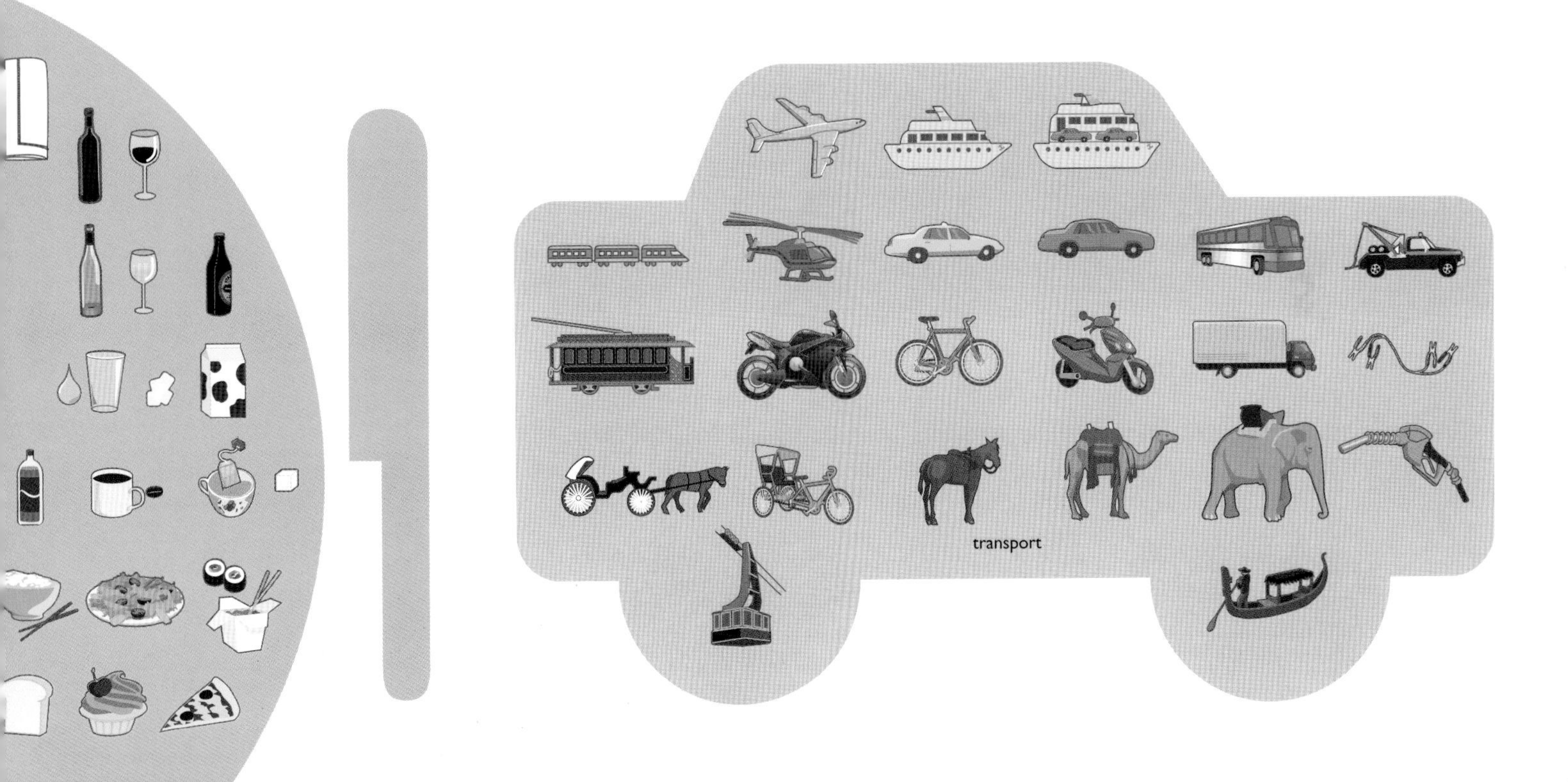
transport
copy me!

286 greet around the world

287 avoid offensive gestures

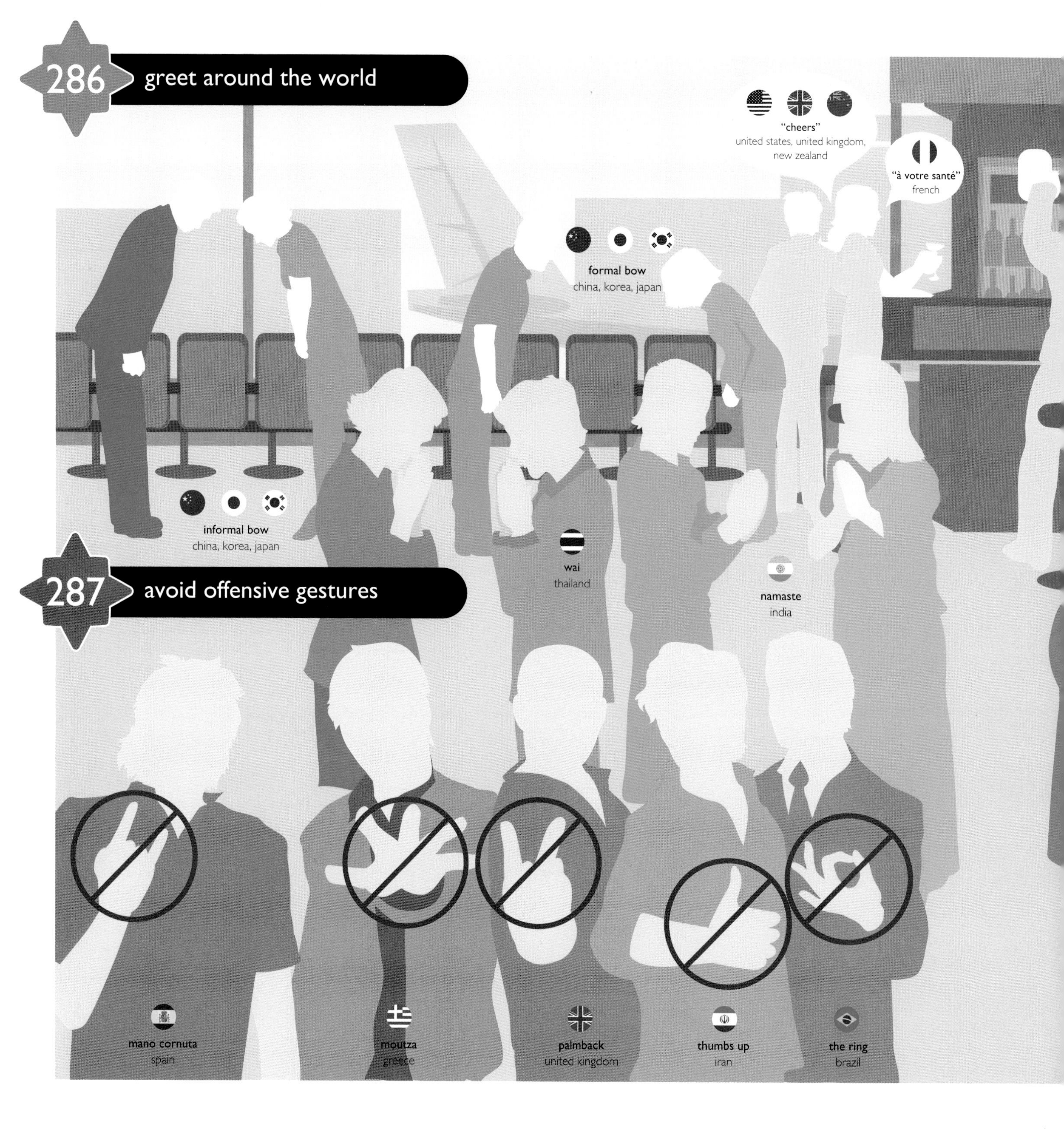

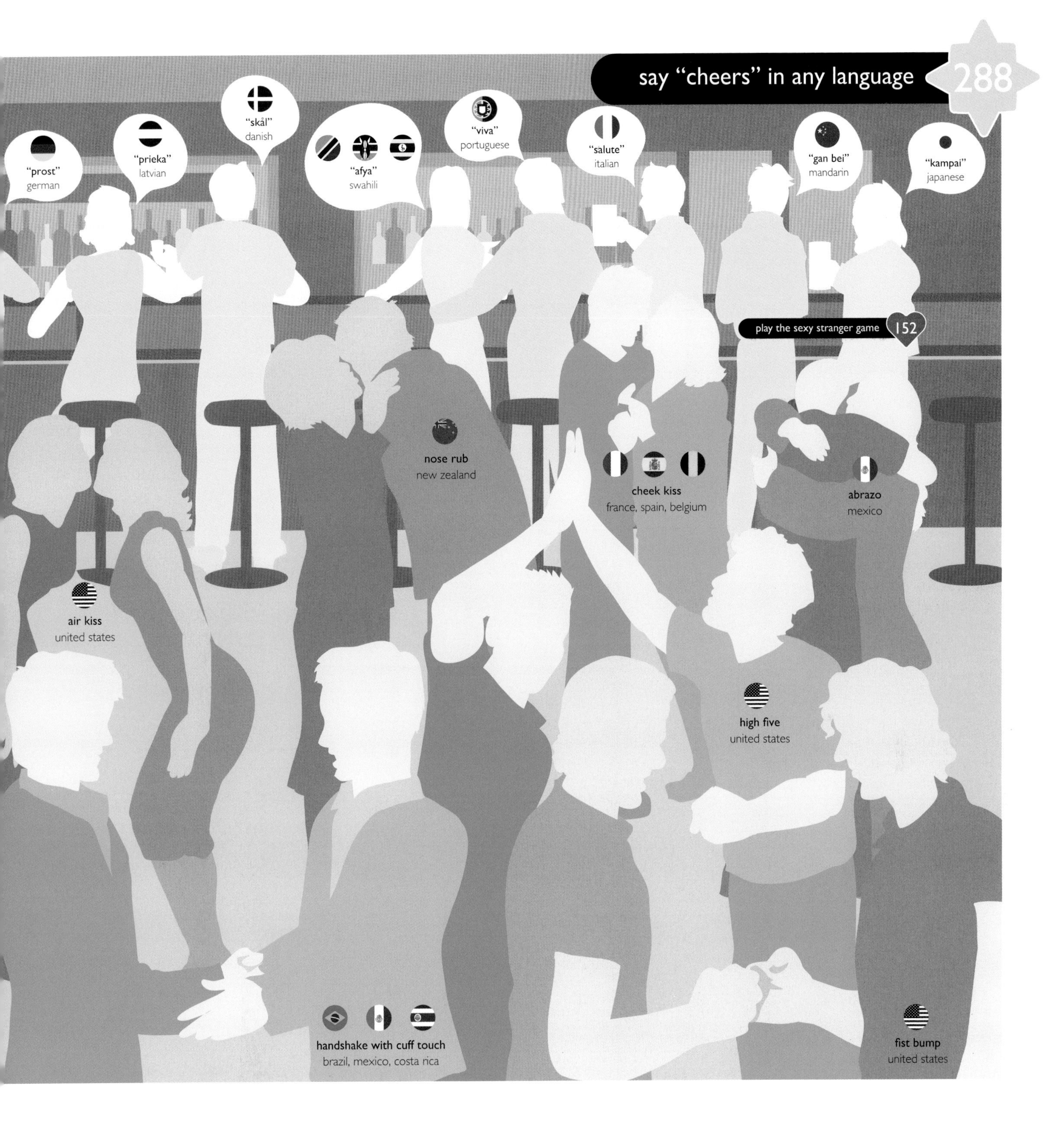
"prost"
german
"prieka"
latvian
"skål"
danish
"afya"
swahili
"viva"
portuguese
"salute"
italian
"gan bei"
mandarin
"kampai"
japanese
play the sexy stranger game 152
nose rub
new zealand
cheek kiss
france, spain, belgium
abrazo
mexico
air kiss
united states
high five
united states
handshake with cuff touch
brazil, mexico, costa rica
fist bump
united states

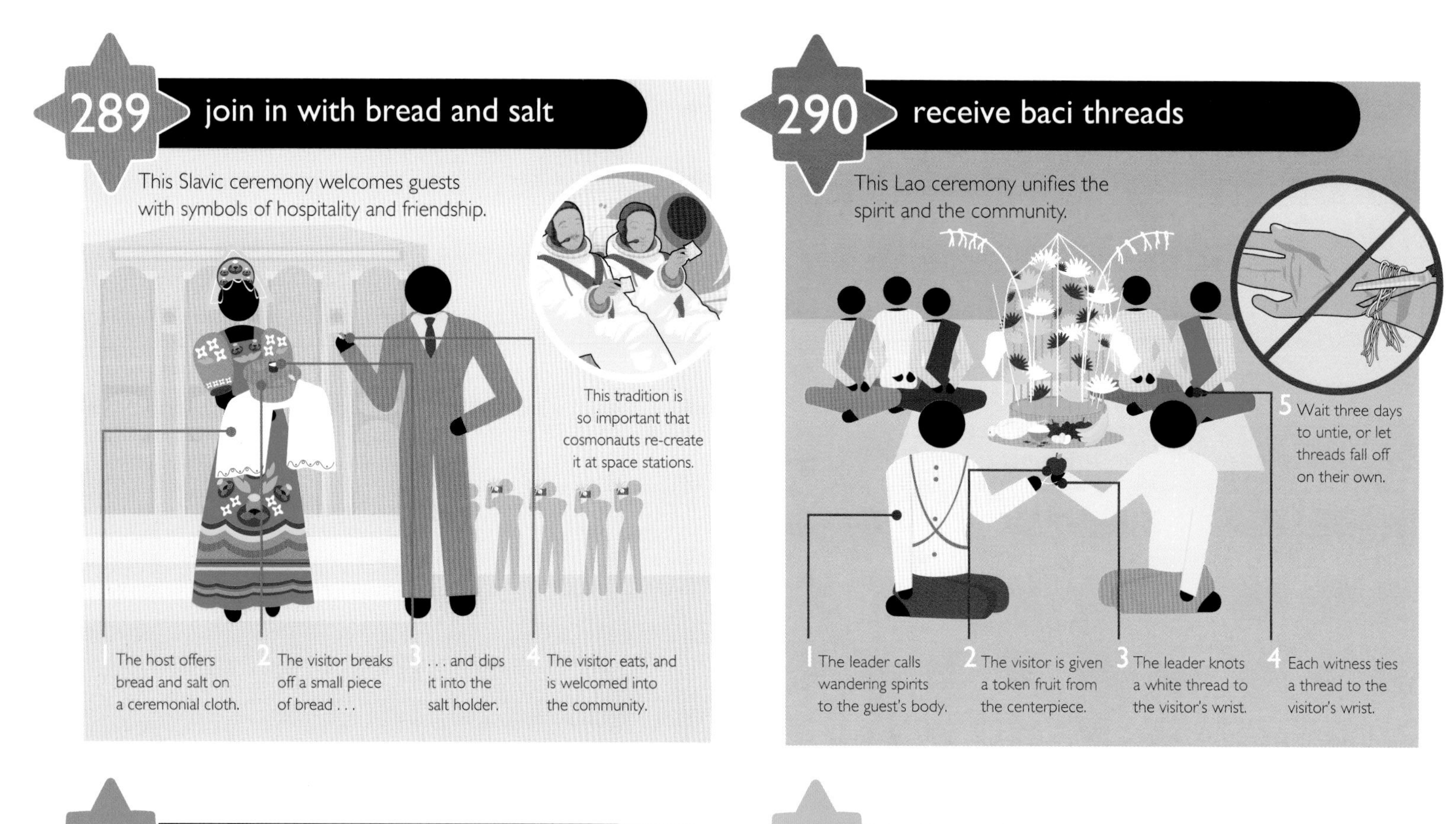

289 join in with bread and salt

This Slavic ceremony welcomes guests with symbols of hospitality and friendship.

This tradition is so important that cosmonauts re-create it at space stations.

1. The host offers bread and salt on a ceremonial cloth.
2. The visitor breaks off a small piece of bread . . .
3. . . . and dips it into the salt holder.
4. The visitor eats, and is welcomed into the community.

290 receive baci threads

This Lao ceremony unifies the spirit and the community.

1. The leader calls wandering spirits to the guest's body.
2. The visitor is given a token fruit from the centerpiece.
3. The leader knots a white thread to the visitor's wrist.
4. Each witness ties a thread to the visitor's wrist.
5. Wait three days to untie, or let threads fall off on their own.

291 drink kava at a sevusevu

Kava is shared after a Tahitian chief accepts an outsider's request to visit his island.

1. Visitor presents kava root and asks permission to visit.
2. The chief drinks first, followed by important witnesses.
3. The visitor shouts the local phrase of thanks and claps once.
4. The visitor accepts the drink with both hands.
5. The visitor hands the shell back, claps three times, and repeats thanks.

292 accept a powhiri challenge

After a symbolic challenge by their Maori hosts, guests may approach the long house.

1. A warrior challenges the visitor with a threatening spear dance.
2. A warrior lays a token branch before the visitor.
3. The visitor picks it up to show his peaceful intent.
4. The host women perform a chant.
5. The visitor may cross the sacred space outside the long house.

Chicha is an indigeneous Andean beer, often enjoyed at festivals and community events.

drink chicha like a local 293

Accept bowl of chicha.

Pour some on ground.

Drink rest of bowl.

Turn upside-down, shake.

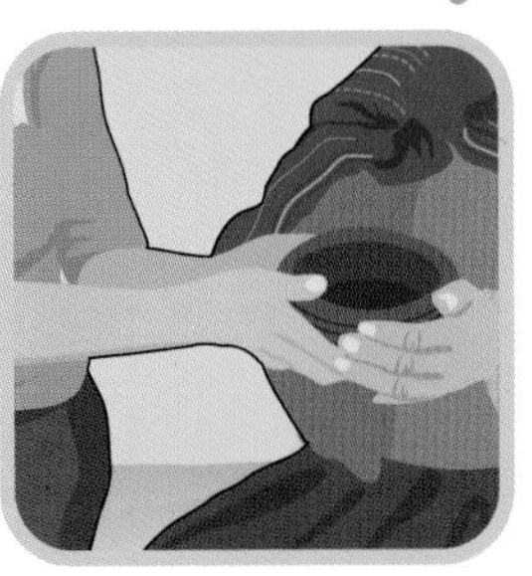

Pass bowl along.

Attaya is a Senegalese tea-serving custom that welcomes visitors and celebrates friendship.

perform an attaya ceremony 294

Boil water on charcoal burner.

Add to boiling water.

Let steep.

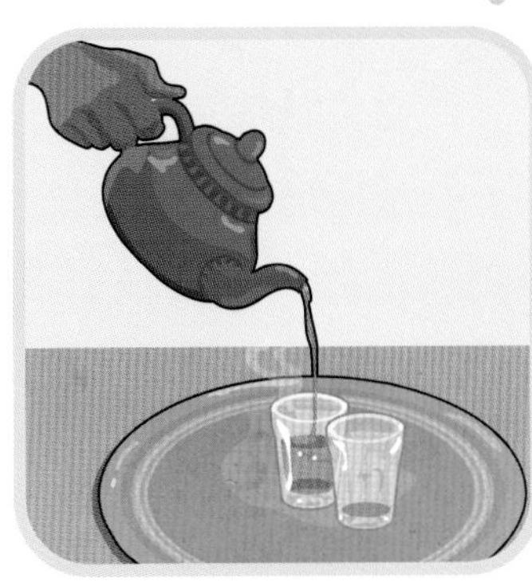

Pour into glass.

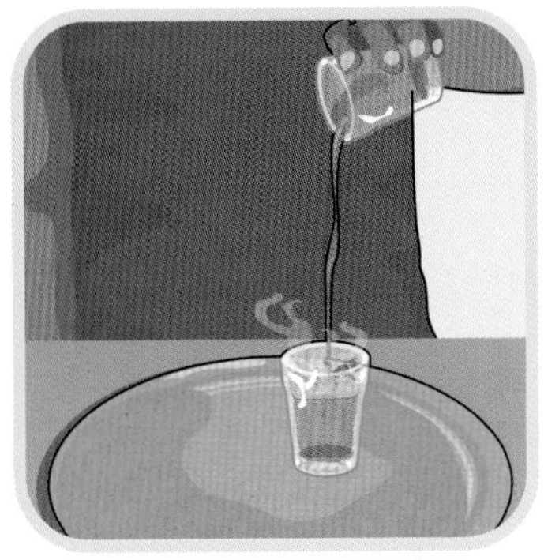

Transfer to new glass, making foam.

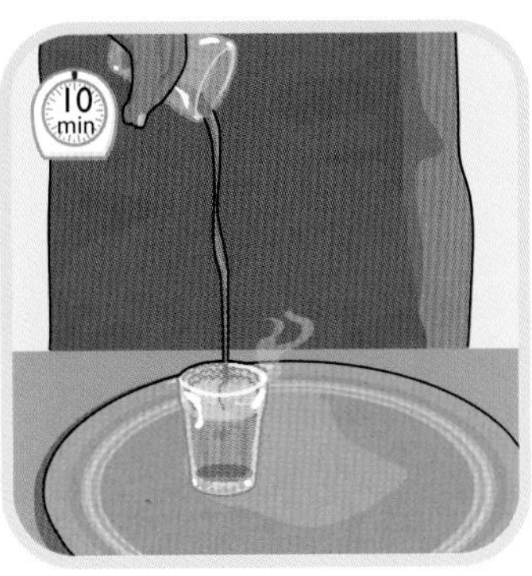

Pour back and forth, building foam.

Reheat on burner.

brew compost tea 223

Serve on a tray.

295 travel by skijoring

Feed dogs prior to excursion.

Harness dogs; attach towline.

Put on your harness and skis.

Lean forward and yell.

Command dogs through turns.

238 feed fido a home-cooked meal

Reward dogs at the trip's end.

296 try surf kayaking

Board in shallow water.

Wait for a lull; paddle out.

Push through the breaks.

Turn; paddle with waves.

Use paddle as a rudder.

come about in a sailboat 297

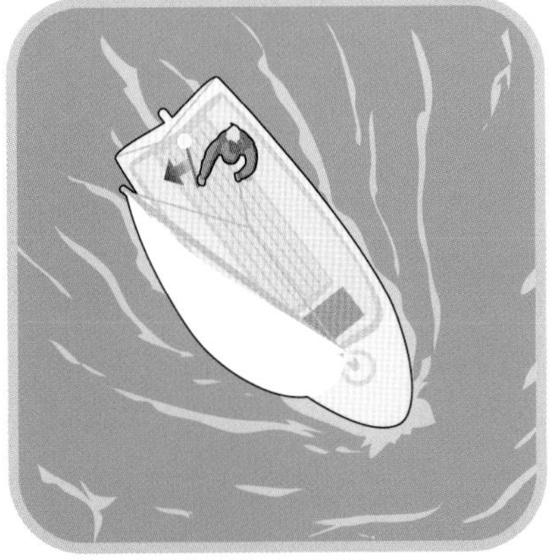
Push the tiller toward the sail.

Crouch down.

Release the main sheet.

Move to the opposite side.

Pull in the main sheet.

Enjoy the ride.

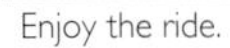

fix a slipped chain while riding 298

Note chain off front ring.

Pedal steadily.

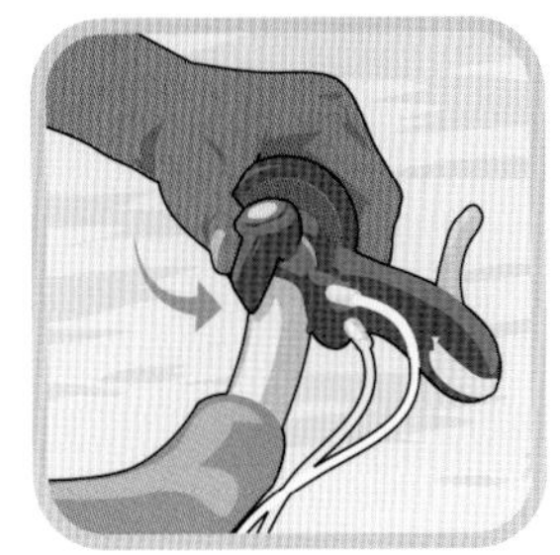
Slowly switch to higher gear.

Chain will catch middle ring.

Bike on.

299
pitch a tent on the beach
Use damp sand to scrub dishes.
Camp well above the line of debris marking high tide.
Don't bring glass bottles.
A tarp lean-to provides shade.
Hang a camp shower to rinse off sand and salt water.
Keep towels and a brush handy at the tent's opening.
Weight tent lines and bury in sand instead of using spikes.
Fully extinguish fires before burying coals.
300
set up a snow camp
418
do a telemark kick turn
Pack down a flat spot with skis, then pitch tent on top.
Tie tent lines to rocks and bury in the snow.
Dig a pit at tent opening for putting on and removing boots.
Don't use camp stove inside tent.
Don't pitch tent in avalanche-prone areas.
Run melted snow through a coffee filter before drinking.

sleep in the rainforest
301
Don't feed the wildlife!
Hold a flame to leeches to remove; don't pull.
Leave perfume and deodorant at home.
Camp well away from pesky insect nests.
Sleep on a raised platform with a rain tarp and mosquito netting.
Pitch tent at least 25 ft (7.5 m) above any water source.
Seal food in plastic containers and store outside of tent.
Shake out sleeping bags before getting in.
camp in the desert
302
Camp on high ground, but avoid dangerous exposed ridges or peaks.
Keep your distance from wildlife.
Use separate coolers for food and drinks.
Pitch tent facing away from the east to avoid a sunrise awakening.
Secure tent lines to rebar, then top with plastic bottles to protect feet and ankles!
Shake out your shoes before putting them on.
Bring (and drink!) 1 gal (3.75 l) of water per person per day.

303 hop a train

Train-hopping is best left to real-life hoboes. If you do try it, be sure to pick the right car for the safest ride, and never hop onto a moving car. Keep an eye out for informative hobo signs left by your fellow travelers!

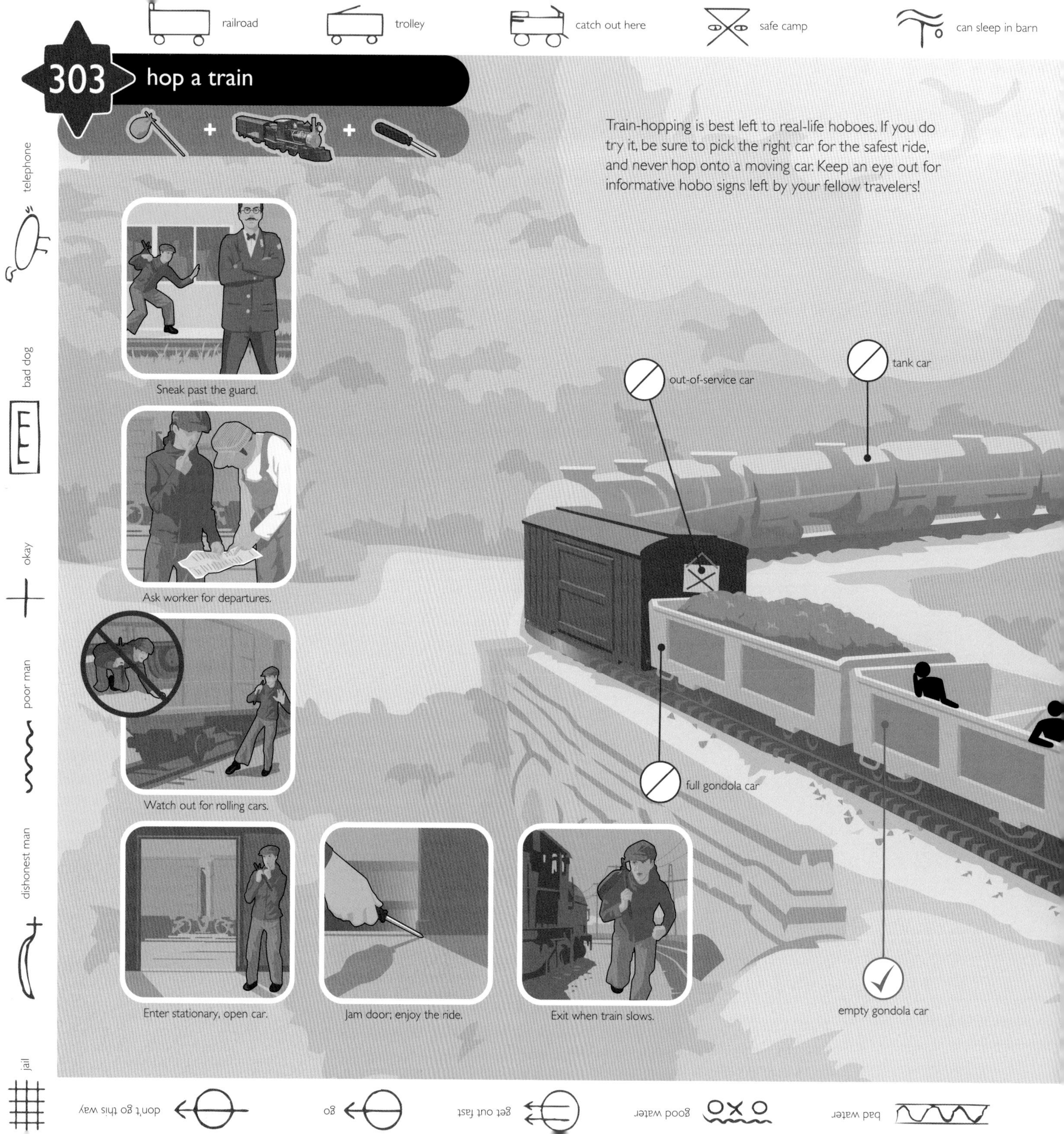

Sneak past the guard.

Ask worker for departures.

Watch out for rolling cars.

Enter stationary, open car.

Jam door; enjoy the ride.

Exit when train slows.

304 pack a backpack for roughing it

Whether you're riding the rails or tramping the trails, you'll need a good backpack that contains only the essentials. Start with these basics, then adapt to your own itinerary.

weather a hurricane 255

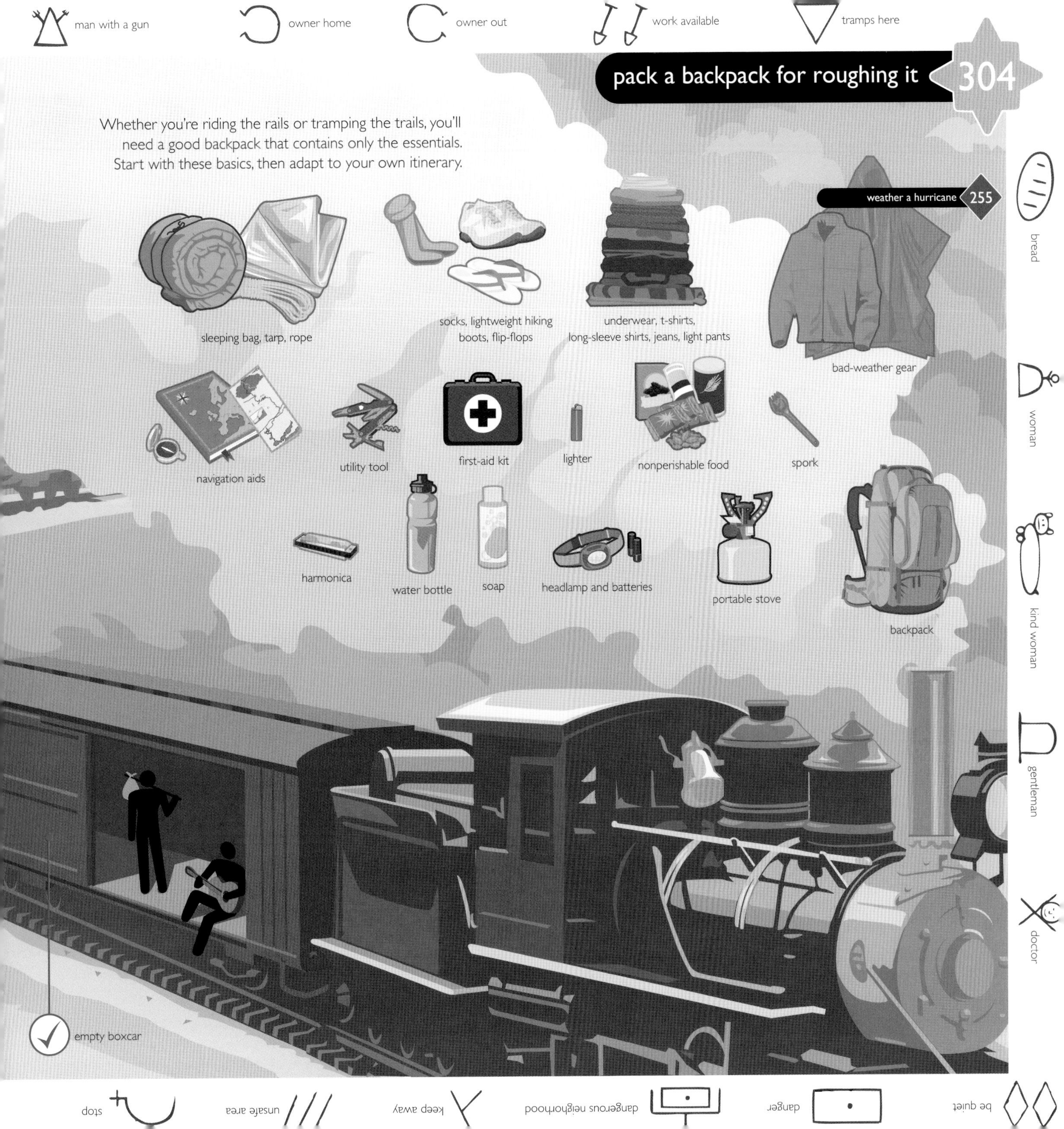

celebrate

305 prep for chinese new year's eve

At the end of each year, the Kitchen God travels to heaven to report on every family's behavior. Start the Chinese New Year preparations by honoring (and bribing!) him.

Offer sweet or sticky foods at his altar.

Burn his picture.

Once the Kitchen God is taken care of, prepare your home (and yourself) for the New Year's celebrations.

Get your hair cut before the New Year.

Hang the symbol for luck upside-down.

Put away sharp objects.

Sweep the house clean.

Finally, it's New Year's Eve. Red packets, firecrackers, and special foods are the universal and beloved features of the big night.

Firecrackers frighten away evil spirits.

Celebrate at home with family and a feast.

Red envelopes of cash are given to children and unmarried friends.

Greet friends with this well-wishing gesture.

Feed the lion a red packet for bonus good fortune.

A monk figure accompanies the lion.

celebrate chinese new year's day 306

During Chinese New Year, neighbors and merrymakers take to the streets to wish each other a prosperous year. The sights and sounds include lucky lion dancers, booming drums, and tons of firecrackers.

A noisy percussion band frightens away bad spirits.

Don't sweep any firecracker paper away during the fourteen days of celebration.

feed a chinese lion dancer 307

Hang lettuce and a red envelope of money over your door.

The lion will come along and "eat" it.

The lion keeps the envelope and spits out the lettuce.

4

Grab a piece of lettuce for good luck in the new year.

308 wish on a daruma doll

Whittle face into craft ball.

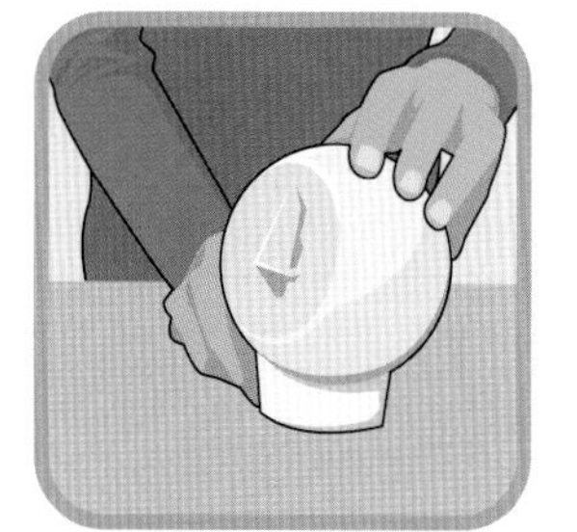
Add paper to make a stand.

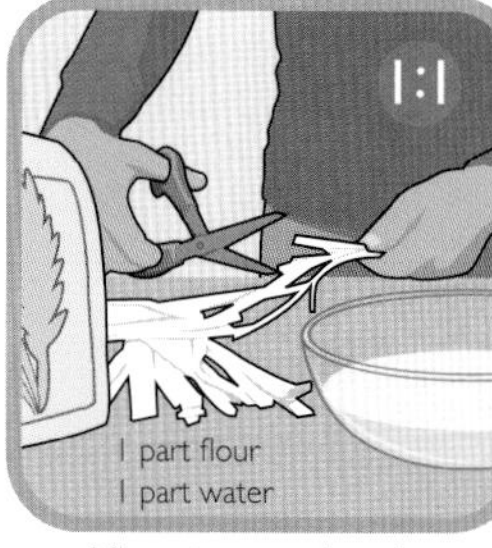

Mix; cut paper; dip strips.

Layer papier-mâché; let dry.

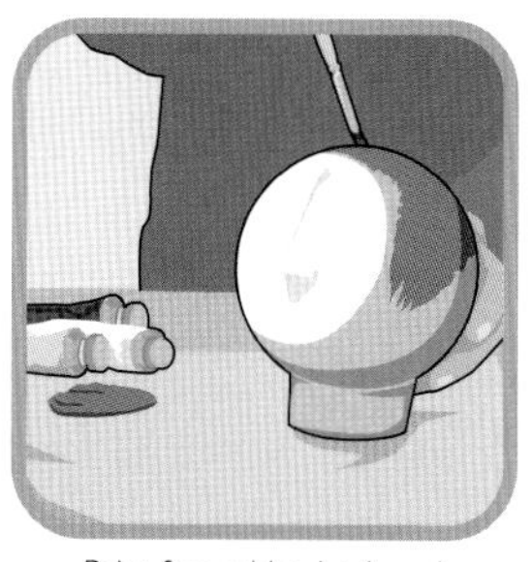
Paint face white, body red.

Paint facial features.

Decorate the body.

Fill left pupil; set a goal.

Display doll in high place.

Fulfill goal; draw right pupil.

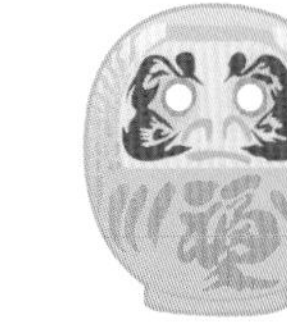
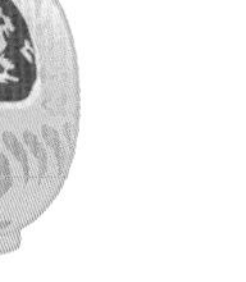

red luck and good fortune

white love and harmony

purple health and longevity

yellow security and protection

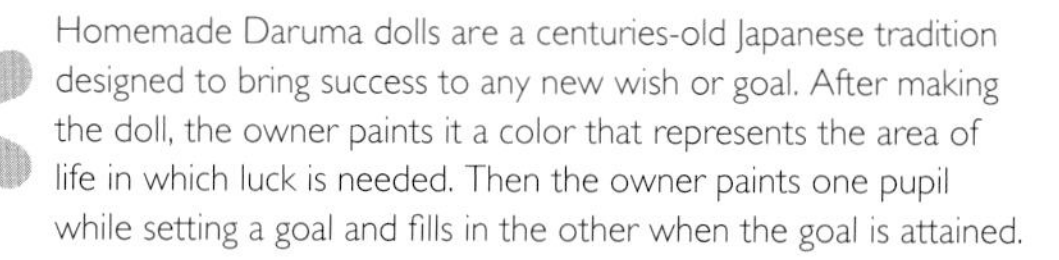

Homemade Daruma dolls are a centuries-old Japanese tradition designed to bring success to any new wish or goal. After making the doll, the owner paints it a color that represents the area of life in which luck is needed. Then the owner paints one pupil while setting a goal and fills in the other when the goal is attained.

309 kick off a mongolian new year

Feast with family.

Greet the sun.

Take ceremonial first steps.

Perform ceremonial greeting.

Feast with extended family.

celebrate a swazi new year 310

Gather foam from ocean.

Make sacred acacia grove.

King eats first pumpkin.

Everyone eats pumpkin.

Offer a few belongings.

tell a new year's fortune 311

Place weight on spoon.

Melt weight over flame.

Pour lead into cold water.

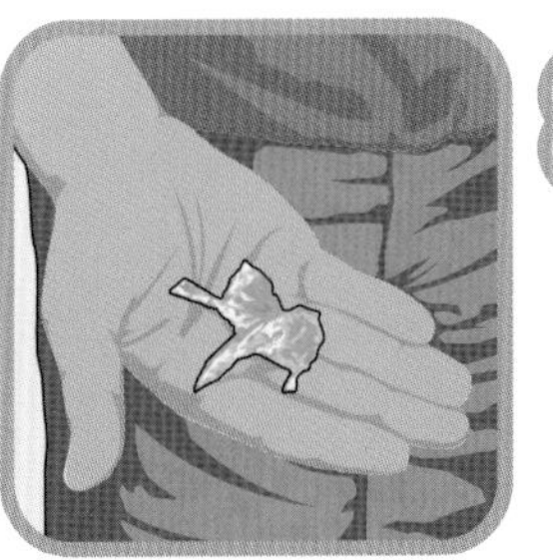

Remove; analyze shape.

Predict "shape" of new year.

light up a spanish new year 312

Pick out old clothes.

hem a pair of pants 122

Stuff and pin to make a dummy.

Bring it into the street at midnight.

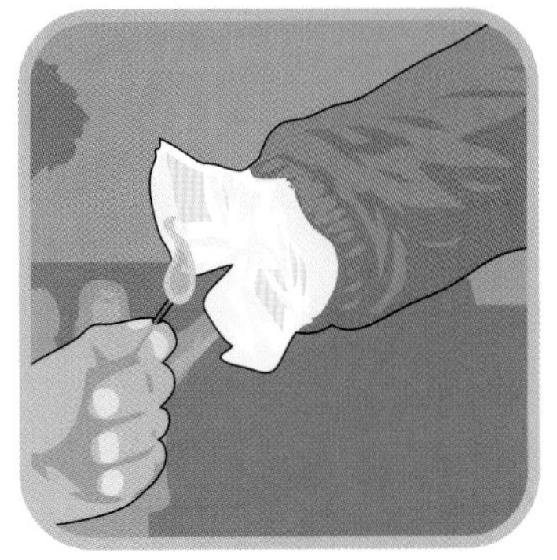

Burn to mark end of old year's self.

313 bake hamentaschen for purim

Beat together.

Mix in flour.

Roll dough into log; wrap.

Freeze overnight.

Let thaw.

Slice.

Add jam to center.

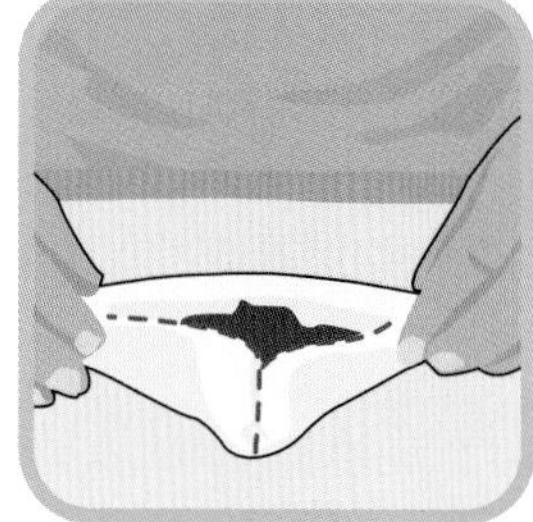

Fold triangles; pinch corners.

Place on lined baking sheet.

Bake until golden.

Purim is a rowdy Jewish holiday celebrated with costumes, plays, noisemakers, and special treats. Hamantaschen symbolize the holiday's villain, Haman.

314 send a purim basket

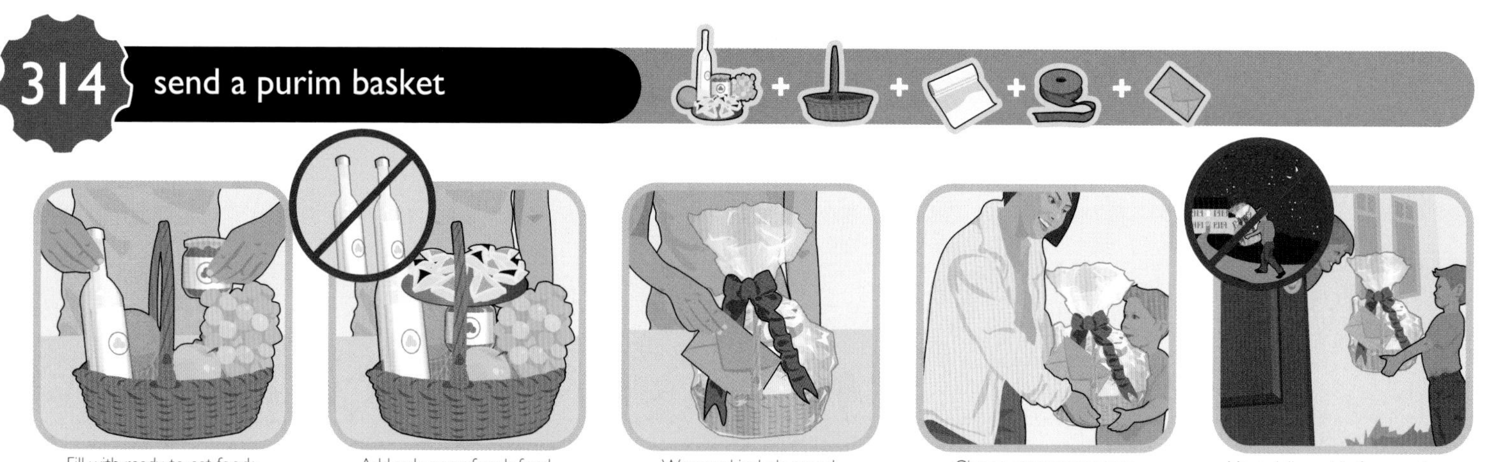

Fill with ready-to-eat foods.

Add only one of each food.

Wrap and include a card.

Give to young messenger.

Have delivered before sundown.

decorate a gingerbread house 315

Welcome Christmas in an old-fashioned way with a festive gingerbread house. Use plenty of icing to attach these whimsical decorations—you'll make a house that's as sweet to look at as it is to taste.

make icing "glue" 316

Separate eggs; save whites.

Combine with whites.

Beat until stiff.

Add water if it hardens.

86 mosaic a stepping stone

317 make a st. lucia crown

Select foam ring that fits.

Insert candle cups.

Add greens and ribbon.

Add candles; light.

Lead a St. Lucia procession.

Before donning your St. Lucia's wreath, be sure to cover your head with a damp cloth. It will protect against candle flames and wax drips.

318 light santa fe-style luminarias

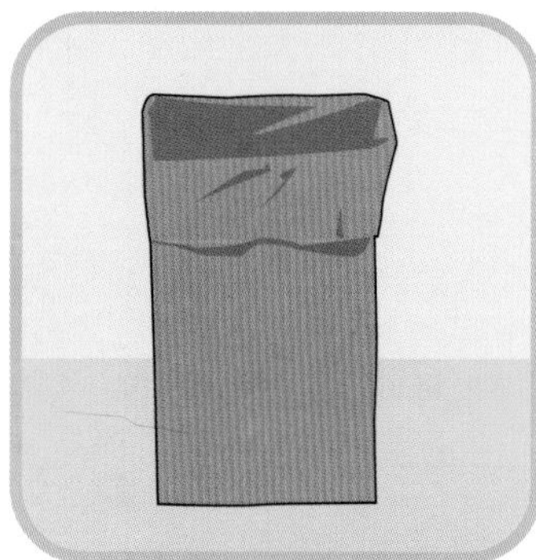
Fold down the top quarter.

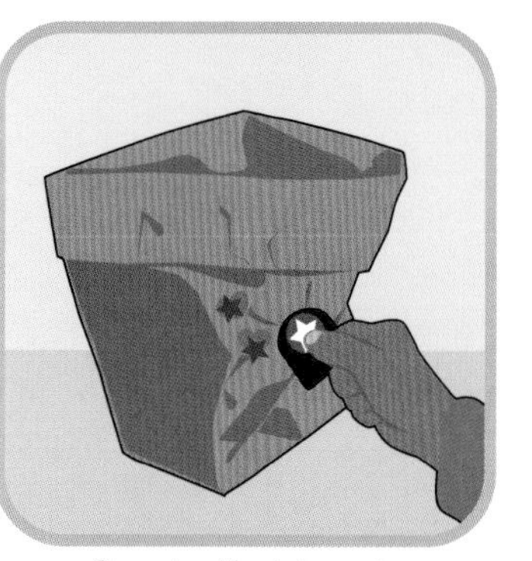
Decorate with a hole punch.

Cover the bottom with sand.

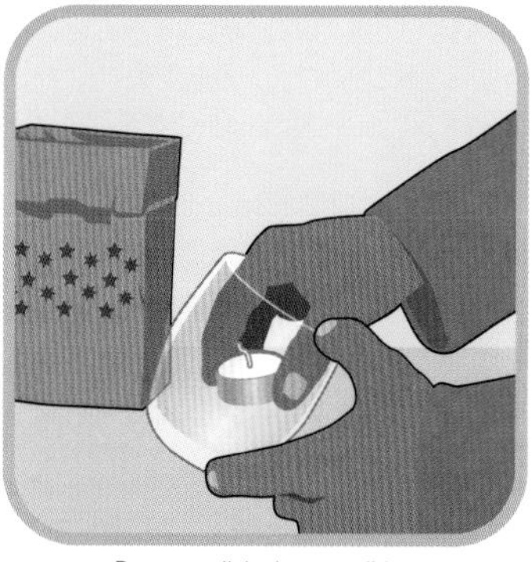
Put a tea light in a small jar.

Nestle the jar into the sand.

Line up in rows outdoors.

Light with a long match.

Bask in the holiday glow.

release a taiwanese sky lantern
319
1
2
3
4
5
Attach fuel cell to lantern.
Unfold lantern.
Swing to fill with air.
Light fuel cell.
Release, making a wish.
float a thai krathong lantern
320
1
2
Fold banana leaves.
Pin leaves to banana-tree slice.
3
4
Arrange flowers on top.
Add incense and candle.
5
Hold the finished krathong to your head; make a wish.
6
Light the candle and incense and release it into the river. Your bad luck will float away with it!

321
string a lei
×50
Gather plumeria flowers.
Pinch off the stems.
4 ft (1.25 m) dental floss
Thread floss onto lei needle.
Attach clip near end of floss.
Push needle through center.
Slide flower down floss.
5 in (12 cm)
Continue, leaving end free.
224 train bougainvillea
Remove clip; knot ends.
322
say it with hula
Hula is more than a dance—it's an ancient and beautiful language. Here are some of the graceful arm motions that hula dancers use to tell a story.

323 cook in a hawaiian imu pit

An imu pit is the center of the luau, a traditional Hawaiian feast. Give the pit three hours to heat up, and another six to fully cook the kalua pig.

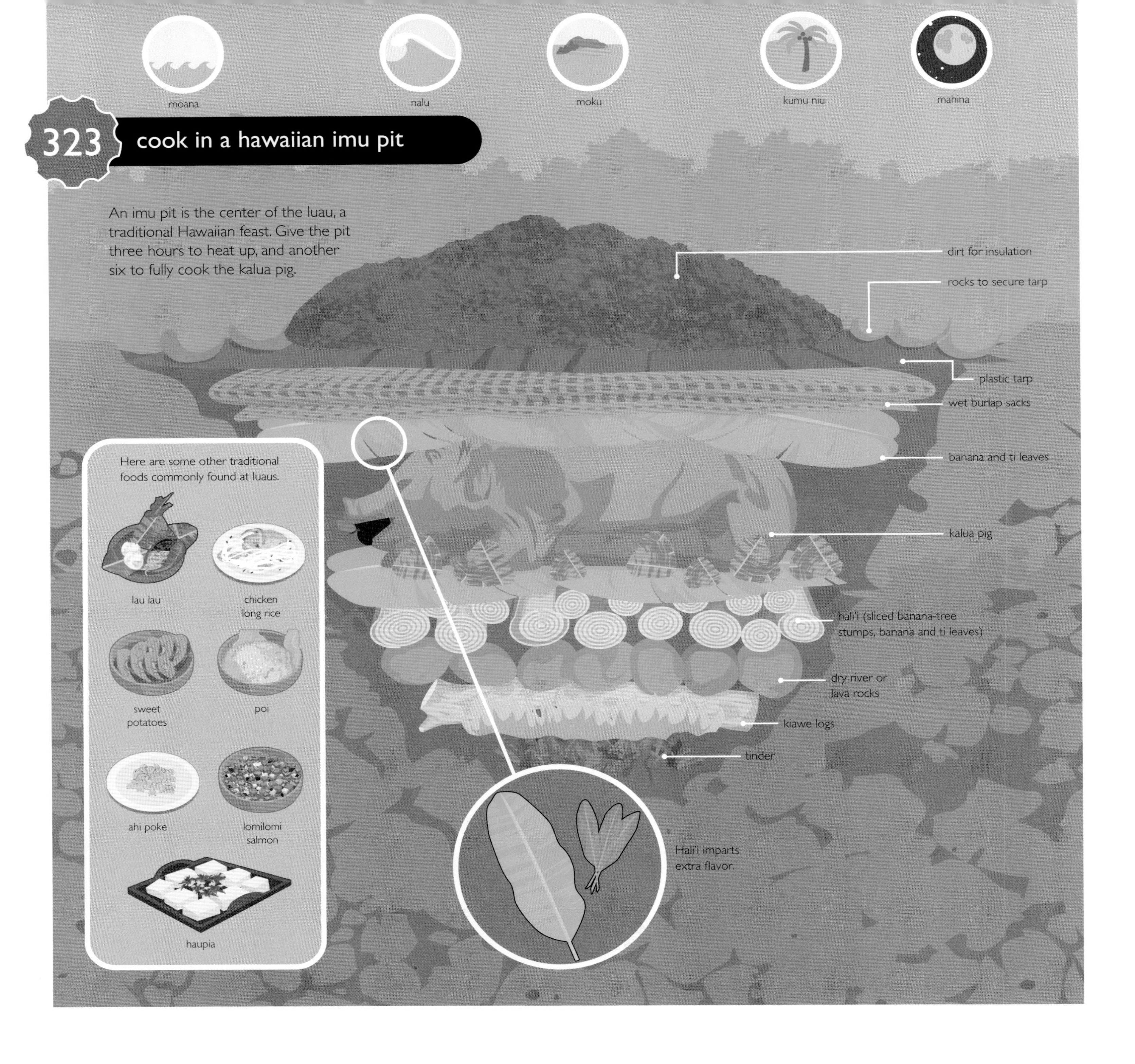

324 cut festive papel picado

Fold tissue papers in half twice.

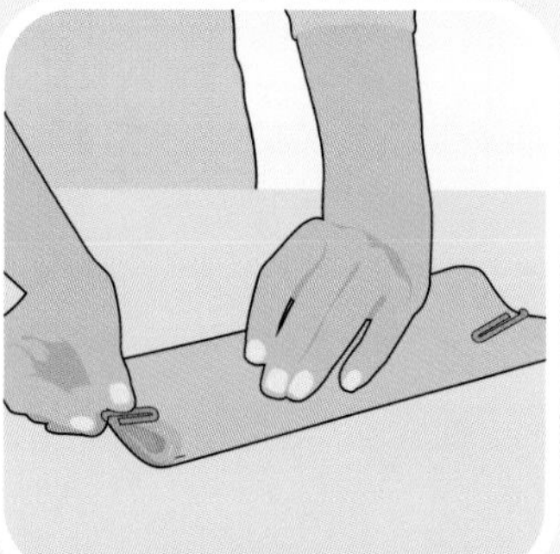

Clip the ends.

Draw a pattern.

Cut out your pattern.

Use a hole punch.

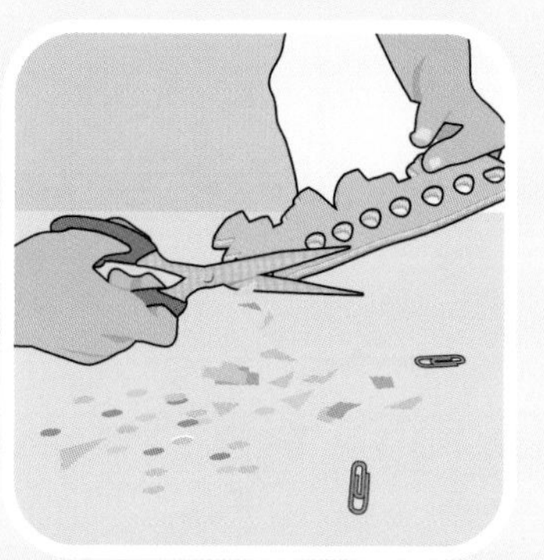

Cut pattern in top, bottom edges.

Unfold and separate.

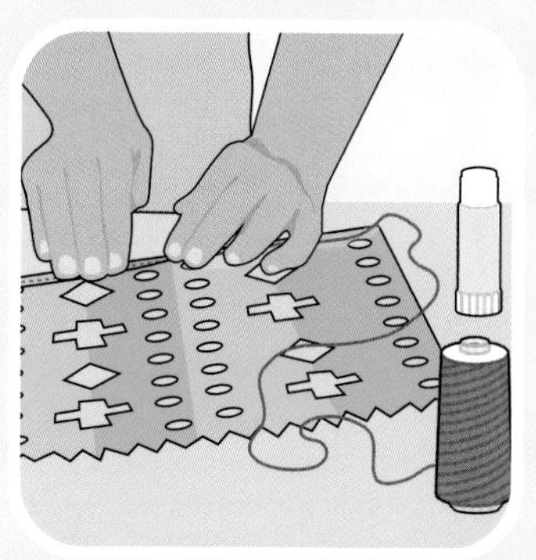

Glue string to top; fold over.

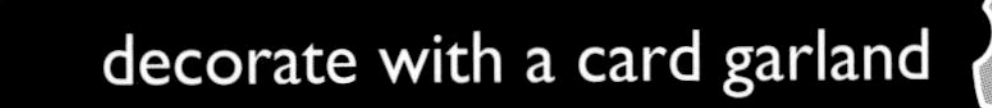

decorate with a card garland 325

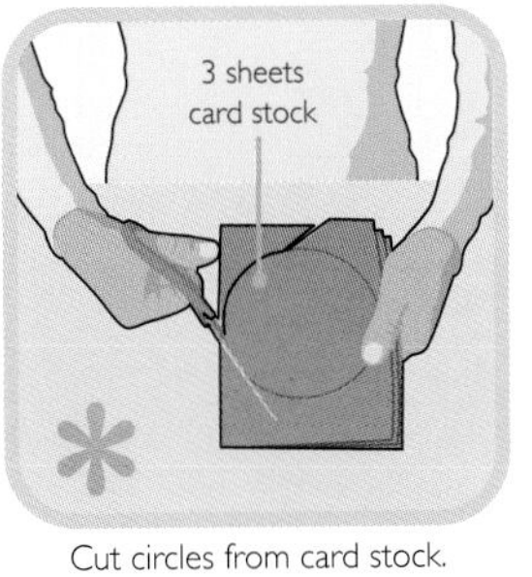

Cut circles from card stock.

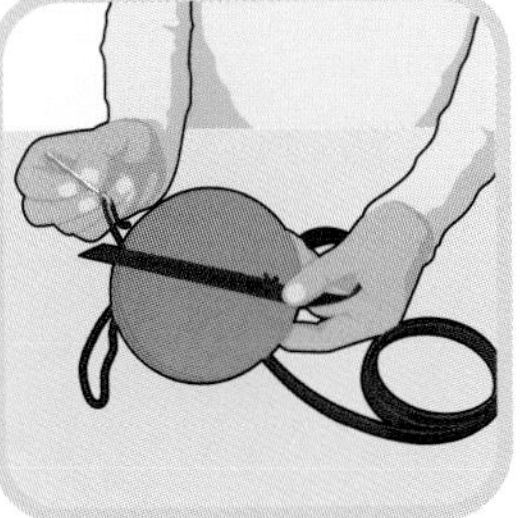

Stitch ribbon to stack.

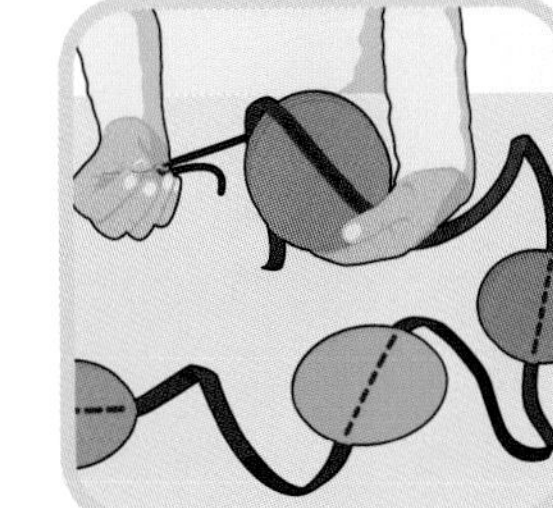

Add more sets of three.

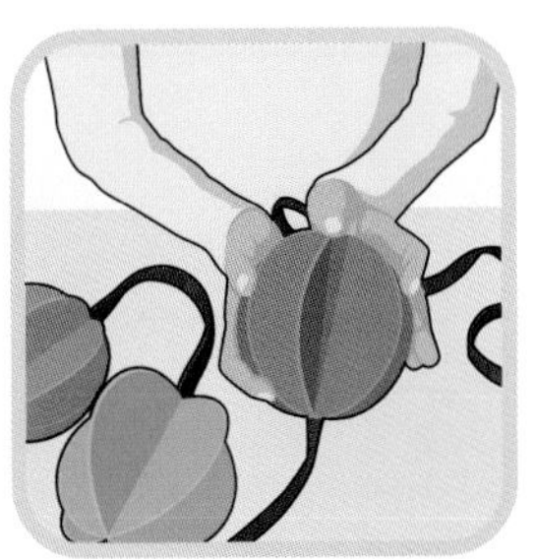

Unfold into spherical shape.

Hang ribbons.

Instead of plain card stock, use holiday greeting cards for a festive recycling project.

fold paper stars 326

Fold paper in half; cut along fold.

Fold in half.

Open.

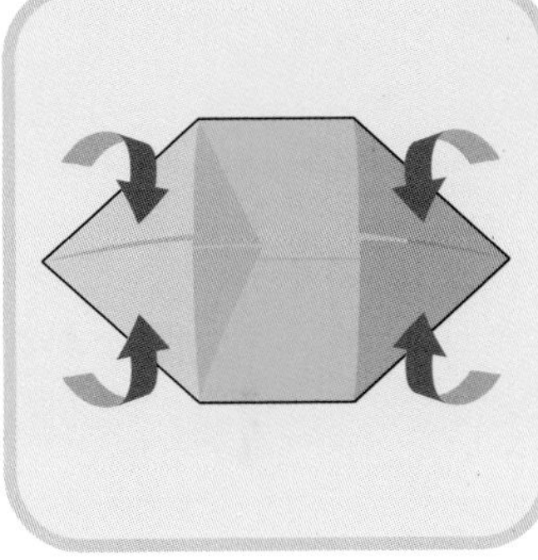

Fold each corner into center.

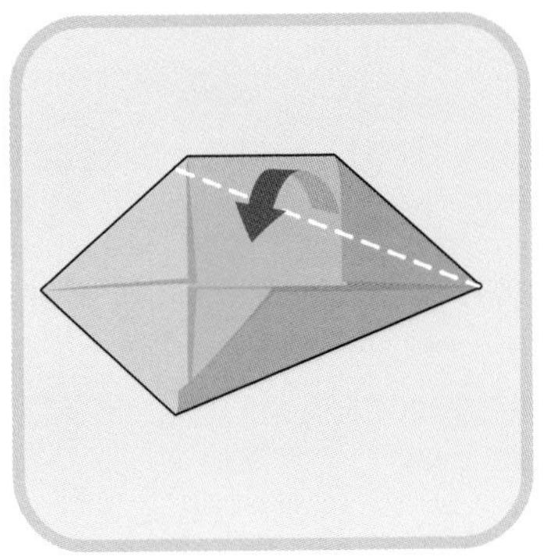

Fold two corners into center.

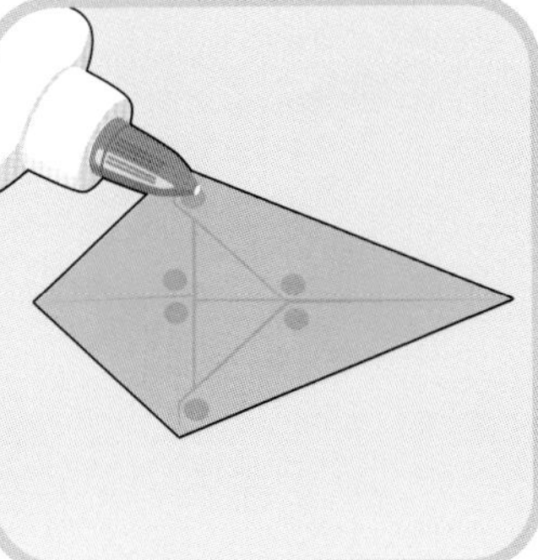

Glue down each corner.

Fold eight shapes; glue into a star.

Display.

327
have a hopi sunrise naming
Give baby her first name and introduce her to nature and society.
After spending her first nineteen days indoors, baby greets the sun and learns her name.
While indoors, baby rests between two perfectly formed ears of corn.
Baby's new name will last until age twenty-one, when she'll receive her adult name.
328
celebrate an orthodox baptism
This ritual welcomes baby into her church community.
Parents and godparents agree to bring baby up within the church.
Baby wears a white christening gown.
The priest anoints the baby with oil and then immerses her three times in holy water.

329
hold a red egg and ginger party
This Chinese celebration marks baby's first month.
Shave baby's head to honor his first full moon.
Serve pickled ginger.
A tiger costume symbolizes protection.
Send guests home with hard-boiled eggs dyed red.
Well-wishers bring gifts of money in red envelopes.

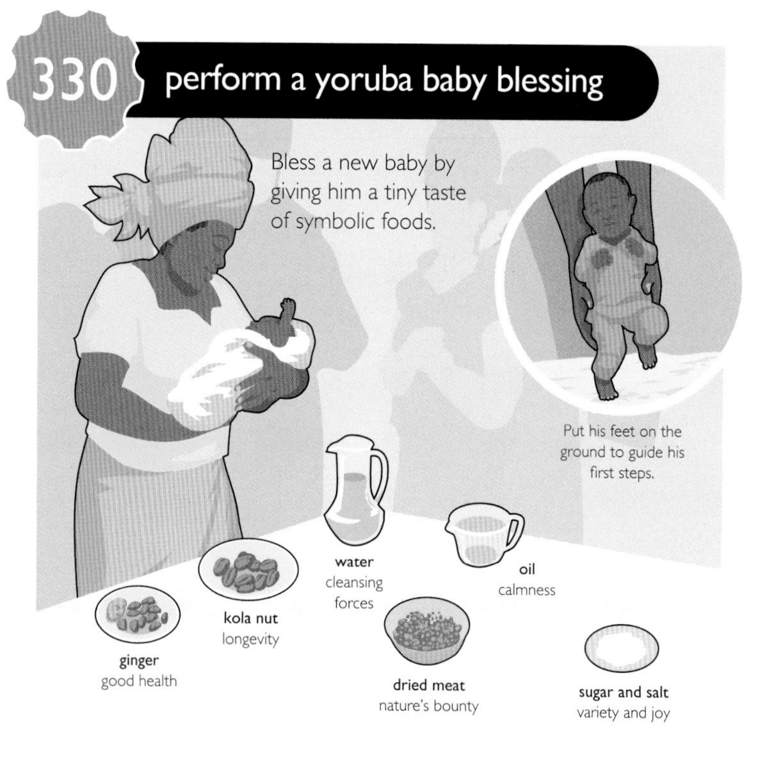
330
perform a yoruba baby blessing
Bless a new baby by giving him a tiny taste of symbolic foods.
Put his feet on the ground to guide his first steps.
water
cleansing forces
oil
calmness
kola nut
longevity
ginger
good health
dried meat
nature's bounty
sugar and salt
variety and joy

hold a hindu naming ritual 331

Apply vermillion to foil bad influences.

Feed honey to promote sweet speech.

Show the baby a rising sun.

Lay on ground to honor Mother Earth.

Reveal the baby's name.

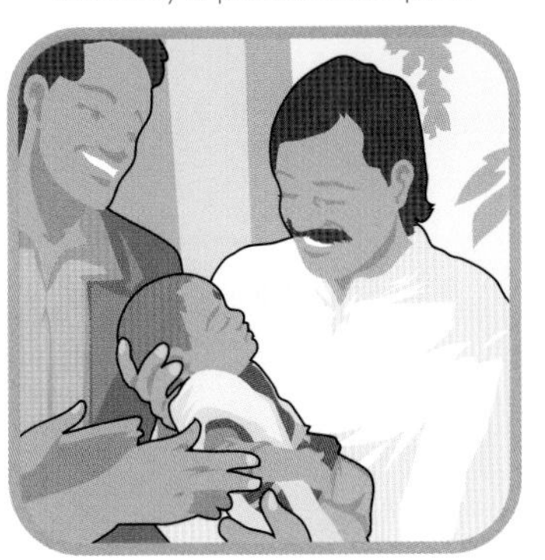

Each relative greets baby in turn.

Introduce baby to nature.

Father tells baby its name.

name baby the egyptian way 332

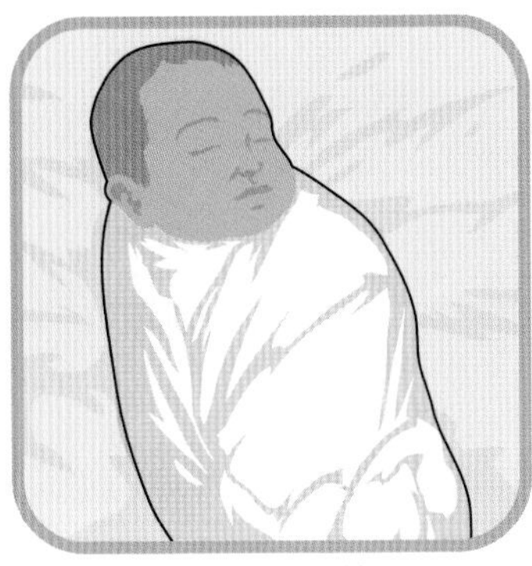

Clothe baby in white.

Pair name choices with candles.

Light candles; wait.

Final lit candle reveals name.

333 chase away the hungarian winter

Citizens of Mohács, Hungary, hold the Busójárás festival to frighten away winter and begin Lent.

Don busó costume.

Ride a corn-husk mobile.

170 get hitched in hungary

Tease bystanders.

Build a bonfire.

Burn a coffin to end winter.

334 dye a river for st. patrick's day

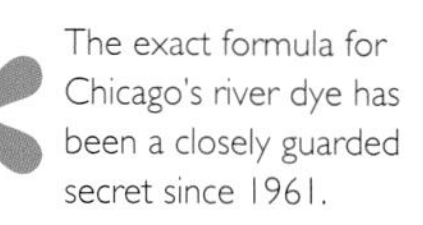

The exact formula for Chicago's river dye has been a closely guarded secret since 1961.

Obtain secret river dye.

Load dye onto a boat.

Sprinkle with flour sifter.

Circle boats to mix dye.

Erin go bragh!

335 leap into the morris dance

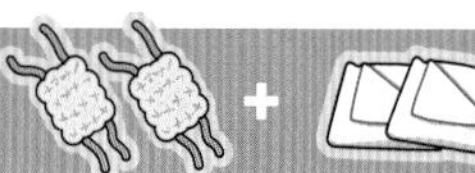

For centuries, villagers in Britain have welcomed the spring with this festive dance.

Hold hankies to sides; lift left leg.

Bring knee up; turn 90 degrees.

Continue spinning, lowering leg.

Leap up; wave hankies above head.

The maypole has been a part of European summertime celebrations for hundreds of years. The colorful dance that circles it occurs on May first or, in some areas, on Midsummer's Day.

Divide dancers into groups.

Take positions, holding ribbons.

Skip in opposite drections.

Go under one dancer's ribbon.

Go over the next dancer's ribbon.

Tie or weight ribbons when complete.

Prepare an even number of ribbons, cutting them to one-and-a-half times the length of the maypole.

indulge

337 pick seasonal produce

Fruits and vegetables are tastiest during the seasons when they naturally flourish. The color bar behind each item on this chart indicates the timing of each food's peak.

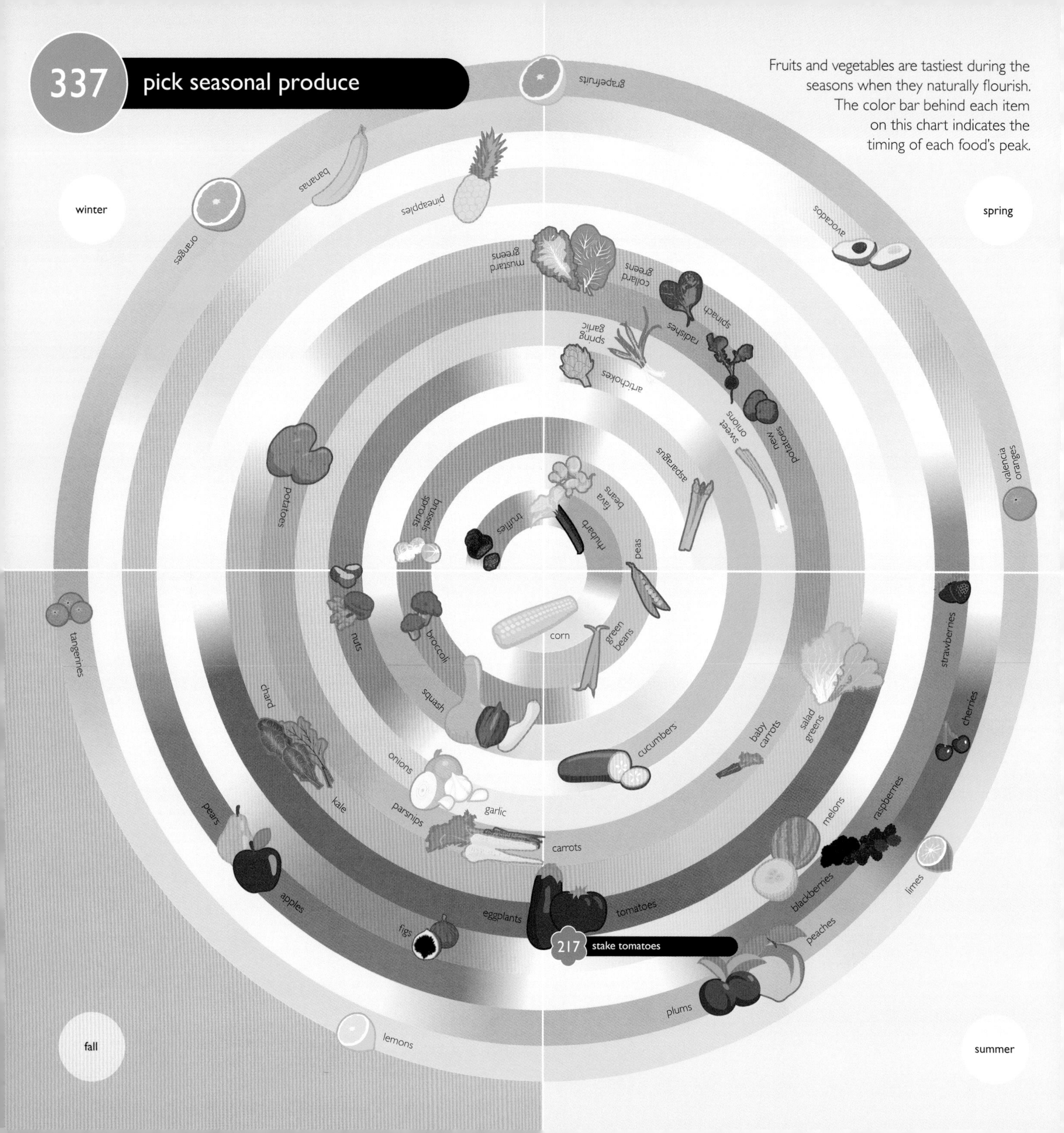

217 stake tomatoes

make a bouquet garni 338

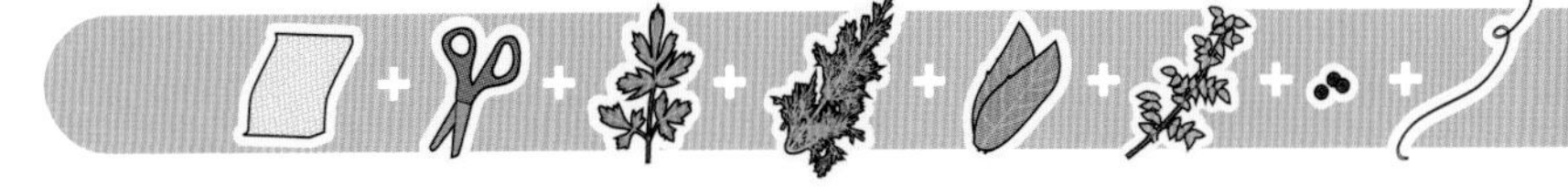

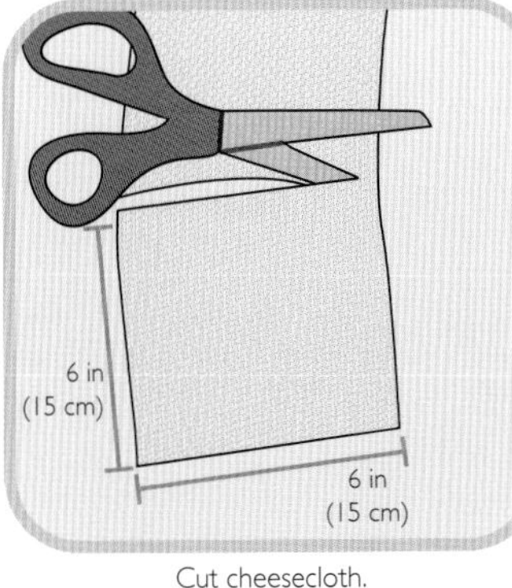

Cut cheesecloth.

Add herbs.

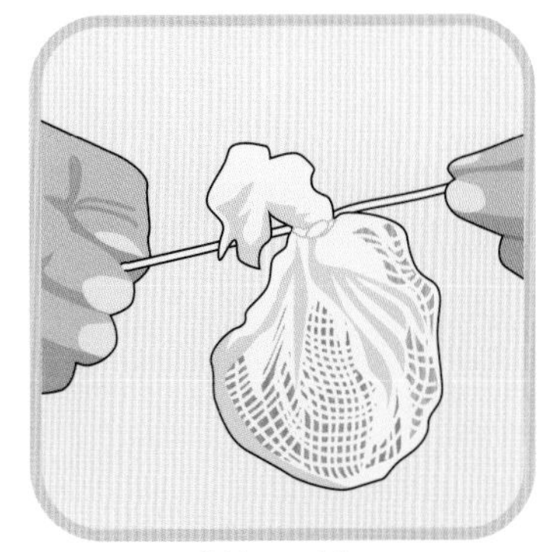
Fold up and tie.

Add to soup at start of cooking.

peel garlic with hot oil 339

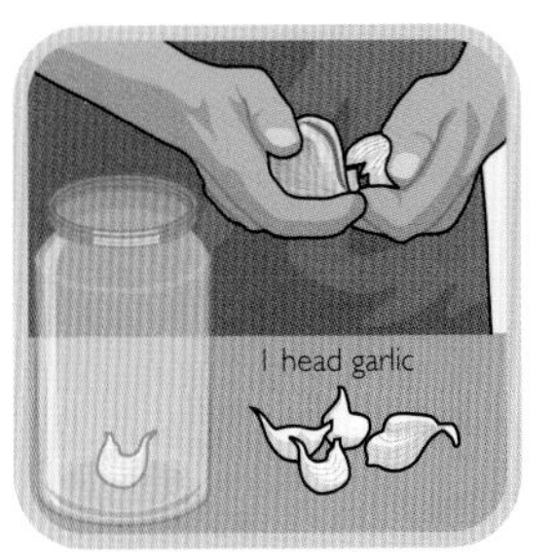

Separate cloves; put in jar.

Heat olive oil.

Pour hot oil over cloves.

Let cool; discard skins.

Strain; reserve infused oil.

prepare compound butter 340

Dice garlic and herbs.

Mix herbs and butter.

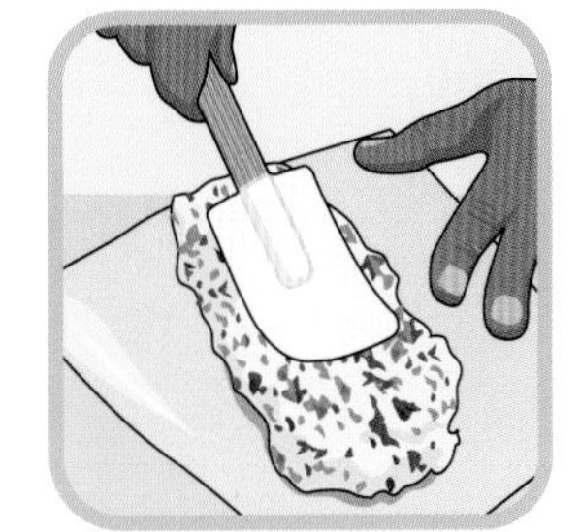
Spread onto parchment paper.

Roll into a cylinder; freeze.

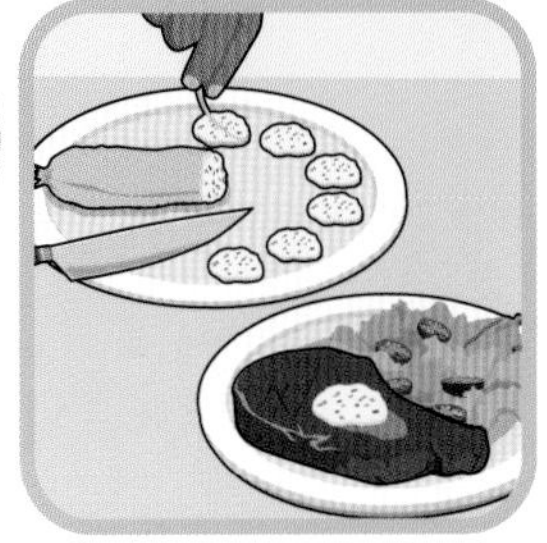
Serve slices on meat or bread.

341 master knife techniques

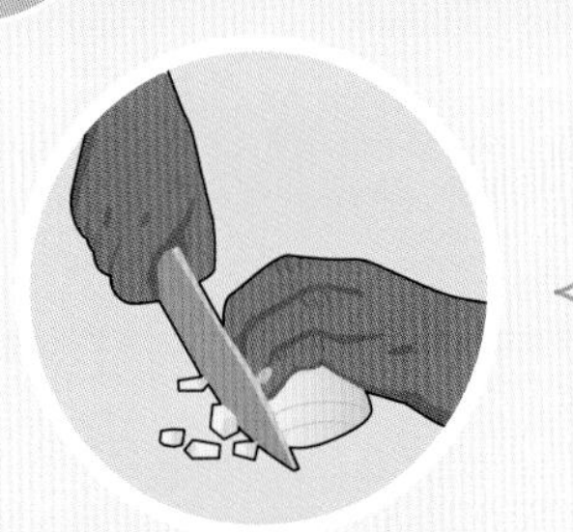

finger curl
To dice, steady the blade against the curled fingers of your free hand.

diced

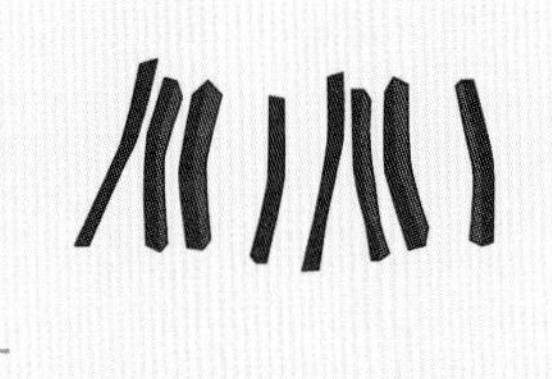

chop and pull
For julienned strips, cut down with tip, then pull back.

julienned

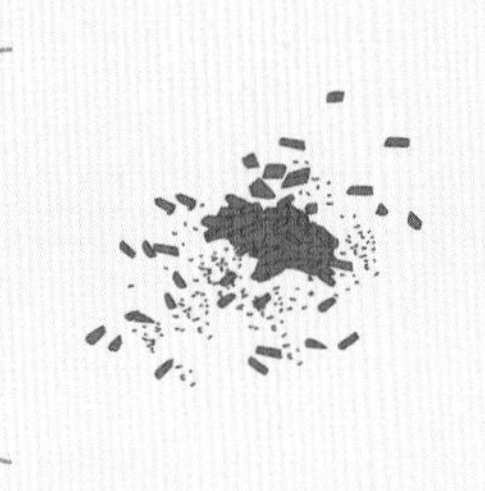

rock
To mince, rock knife back and forth.

minced

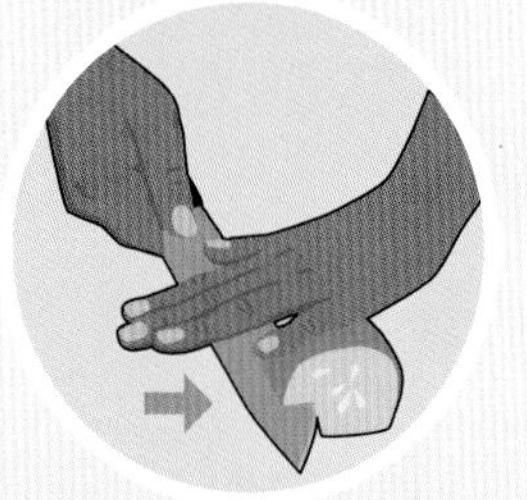

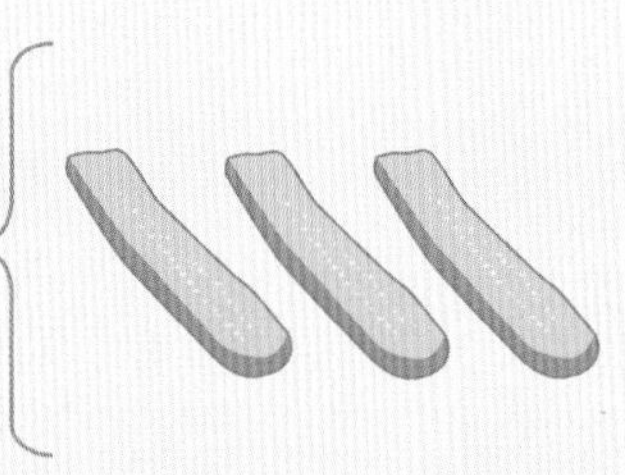

horizontal
For safe slicing, position noncutting hand on top.

sliced

342 dice an onion

343 cut herbs

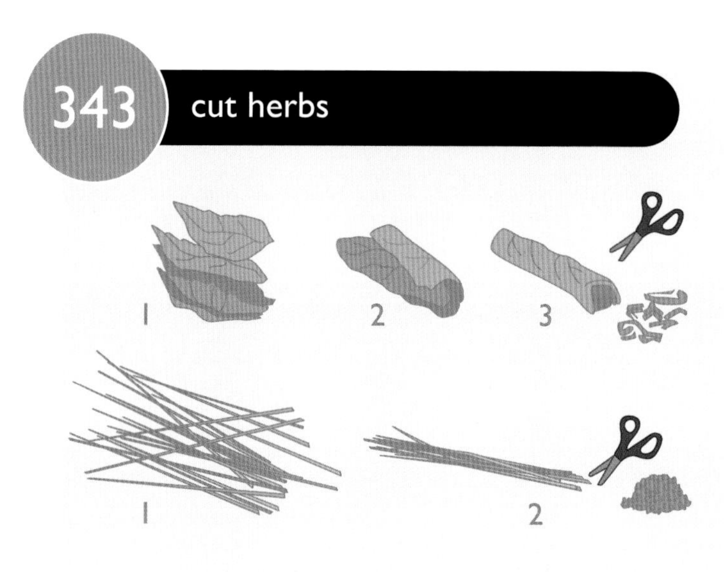

prepare fugu fish 344

Obtain license to prepare fugu.

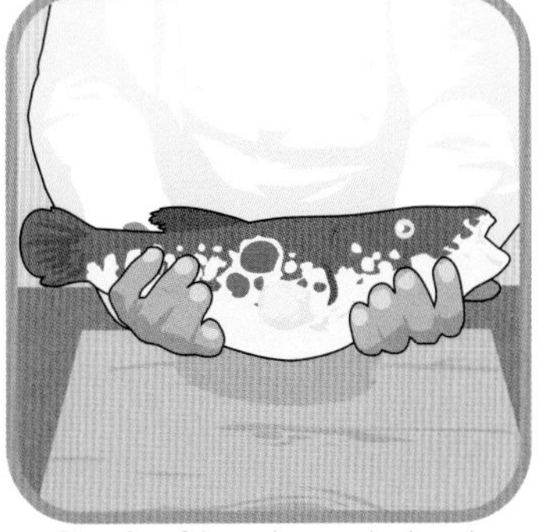

Place fugu fish on clean cutting board.

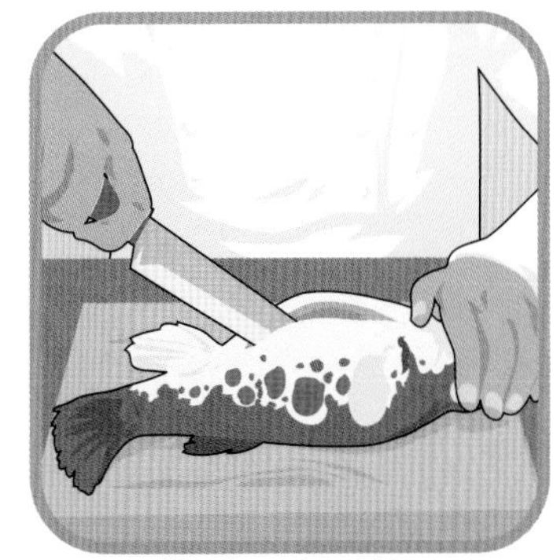

Slice down the middle.

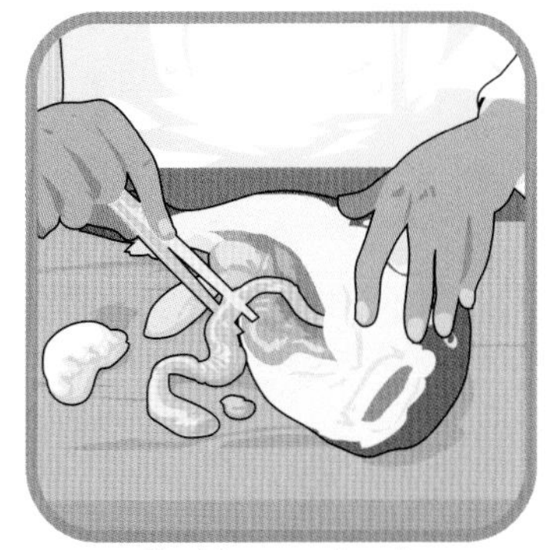

Carefully remove innards.

Cut away scales.

Dispose of refuse properly.

Cut into thin slices.

Serve to brave souls.

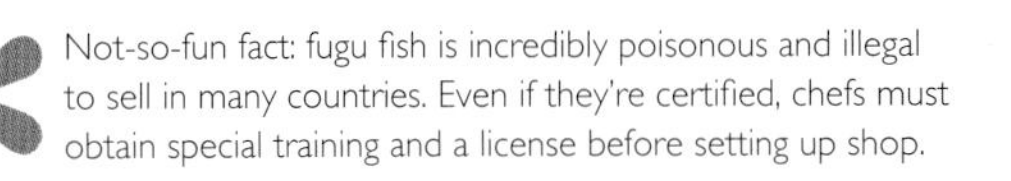

* Not-so-fun fact: fugu fish is incredibly poisonous and illegal to sell in many countries. Even if they're certified, chefs must obtain special training and a license before setting up shop.

create a sashimi rose 345

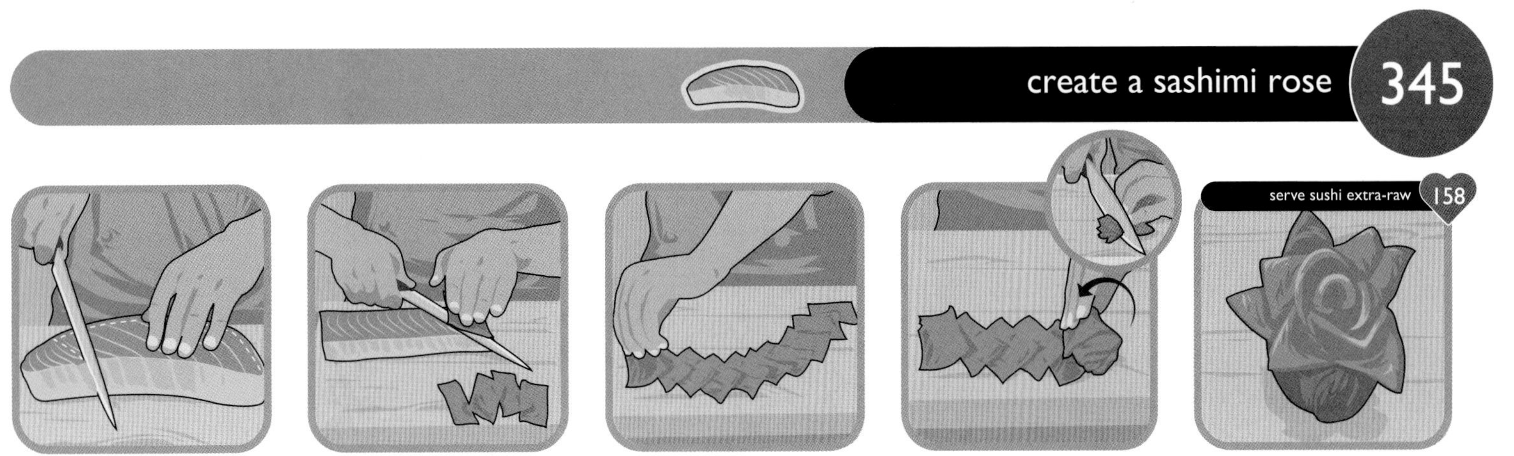

Square off a tuna loin.

Cut into thin slices.

Fan out slices.

Roll into a cone; cut off end.

Serve.

346 chill refrigerator pickles

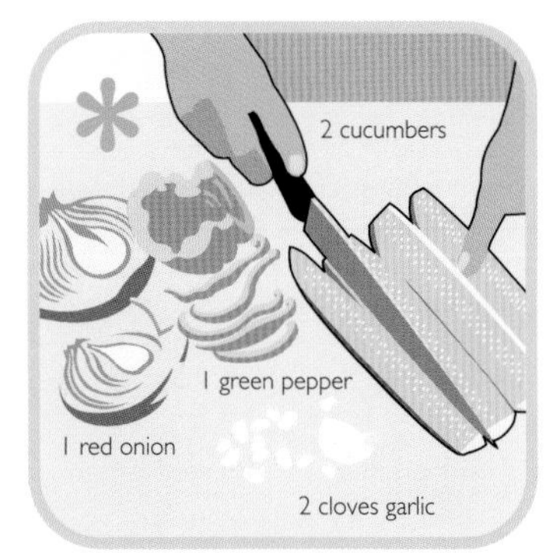

Seed the pepper and cut the veggies.

Place in a clean jar.

Combine over low heat.

Add seasoning. Cook to dissolve sugar.

Remove from heat; let sit.

Cover the veggies with the marinade.

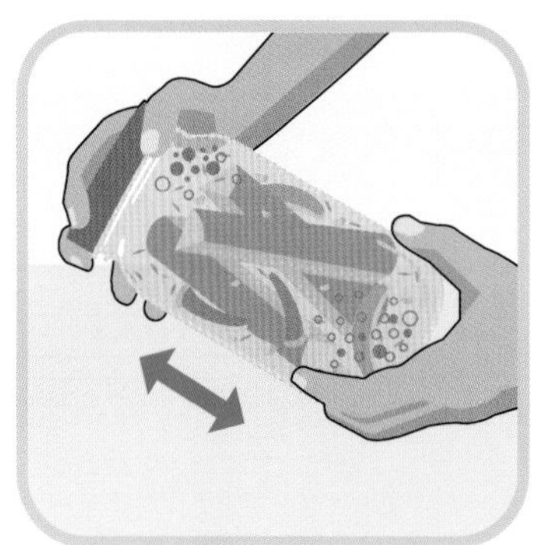

Seal, then shake the jar to mix.

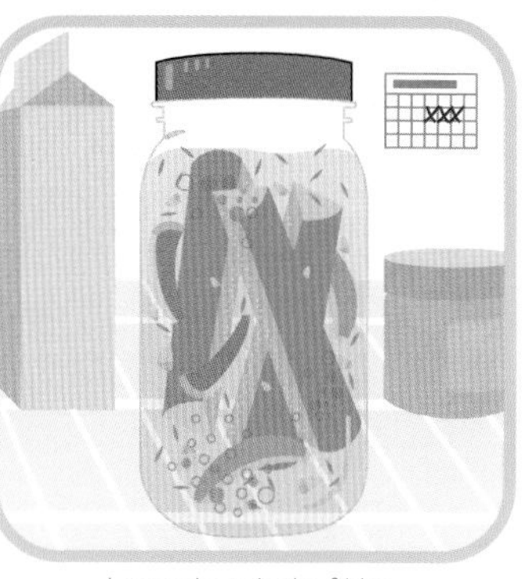

Let marinate in the fridge.

* Pickling isn't just for cucumbers. Try this recipe with other vegetables, or even hard-boiled eggs.

347 ferment hundred-year eggs

Combine equal parts.

Fill pot halfway with soil.

Coat raw eggs; place in pot.

Cover with soil; store.

329 hold a red egg and ginger party

Peel and serve.

make strawberry preserves 348

Wash in warm soapy water.

Put lids, bands in hot water.

Clean and hull berries.

Crush with a fork.

Mix in large pot.

Boil, skimming off foam.

Allow to thicken slightly.

Fill jars nearly to top.

Seal without touching rims.

Submerge in boiling water.

raspberry

jalapeño

tangerine

fig

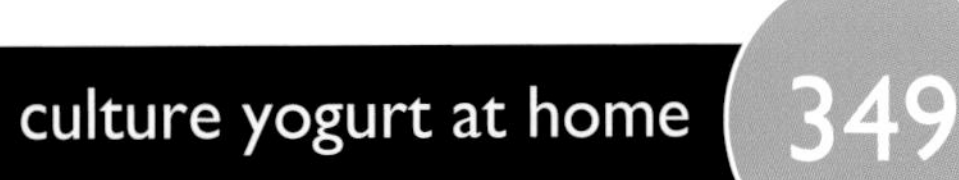

culture yogurt at home 349

Heat milk.

Allow to cool; add yogurt.

Pour into clean jar.

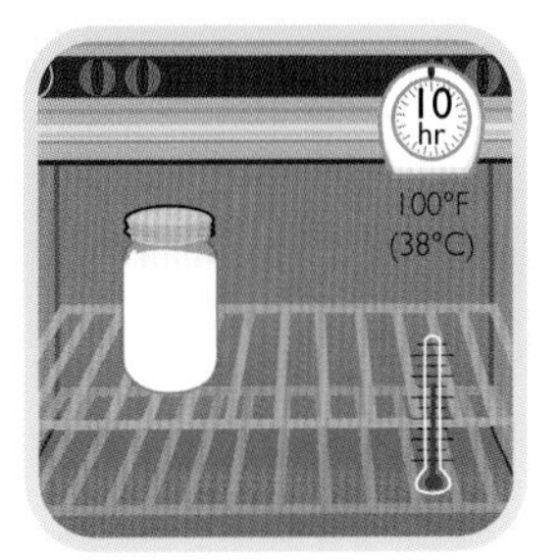

Cook in oven.

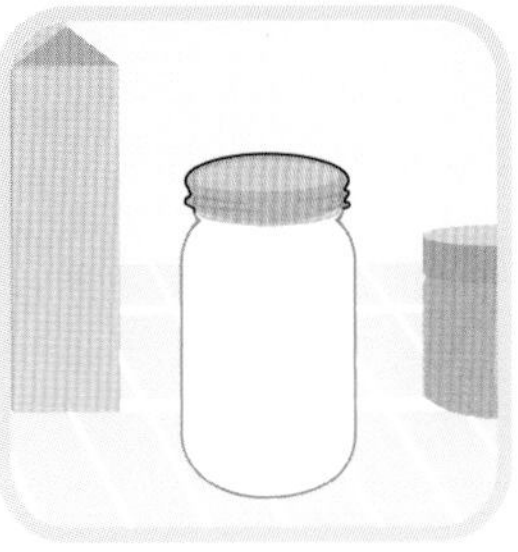
Cool before serving.

350
grill juicy ribs
1
Rinse ribs in cold water; dry.
2
Season generously.
3
1 in (2.5 cm)
water in pan
Put on rack in pan; cover.
2.5
hr
350°F
(180°C)
4
Roast in the oven.
5
Grill and baste until crispy.
6
Slice and serve.
351
roast veggies in foil
1
Place chopped veggies on
foil. Toss in oil; season.
2
Fold up packet.
10
min
3
Place packet in hot coals.

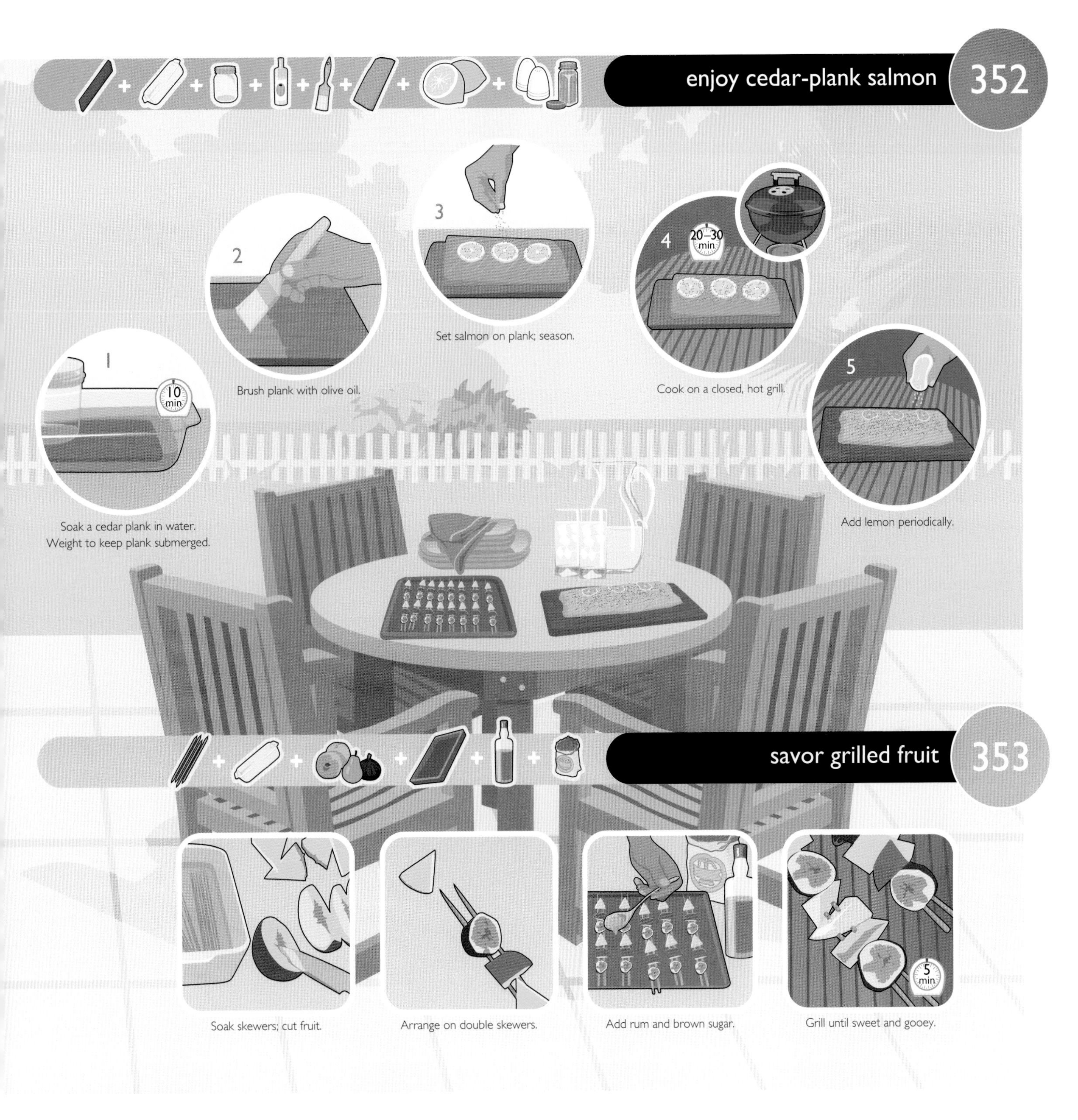
enjoy cedar-plank salmon
352
1
10 min
Soak a cedar plank in water. Weight to keep plank submerged.
2
Brush plank with olive oil.
3
Set salmon on plank; season.
4
20–30 min
Cook on a closed, hot grill.
5
Add lemon periodically.
savor grilled fruit
353
Soak skewers; cut fruit.
Arrange on double skewers.
Add rum and brown sugar.
5 min
Grill until sweet and gooey.

354 bake bread in a clay-pot oven

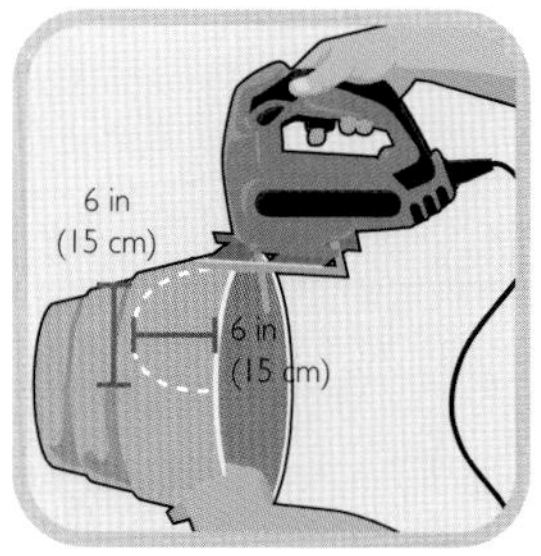

Cut pot with jigsaw.

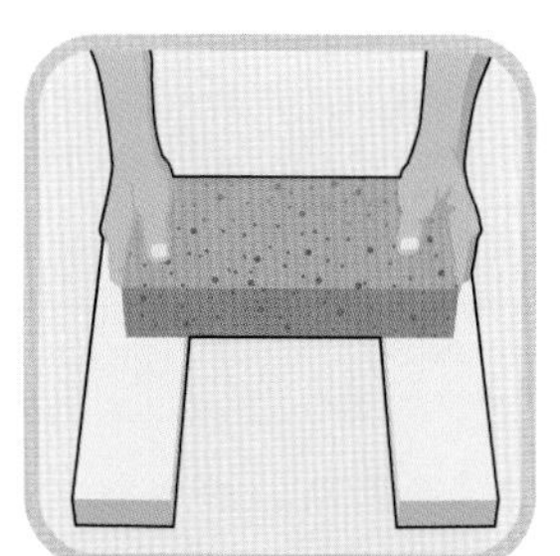
Set paving block on boards.

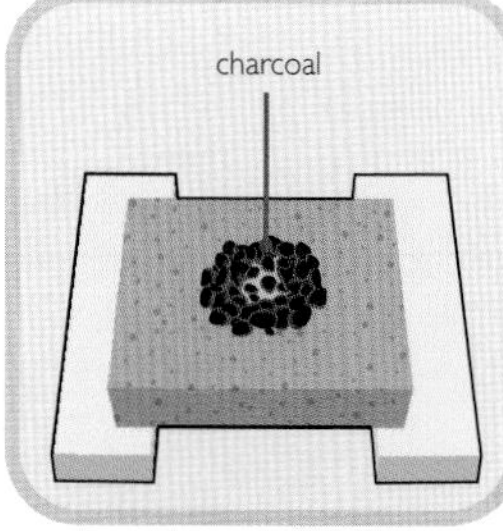

Burn until flames die out.

Place pot; add thermometer.

Add ceramic spacers.

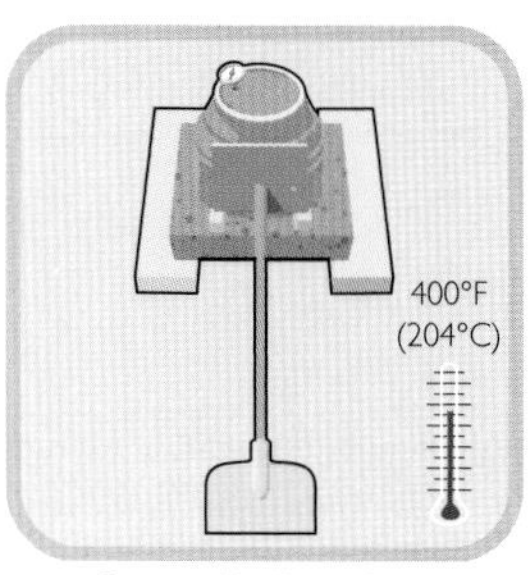

Cover with tile; preheat.

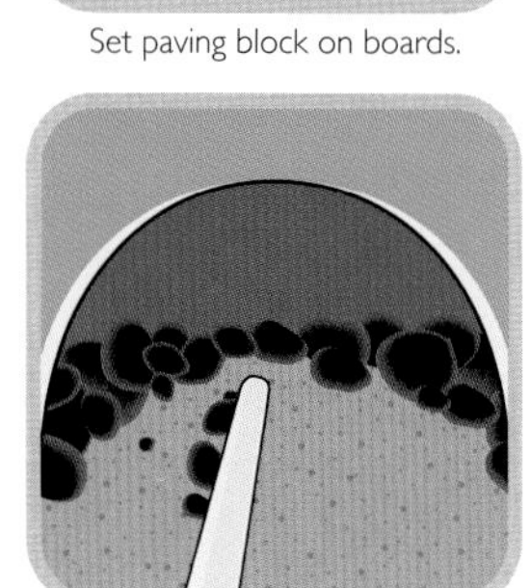
Push coals aside.

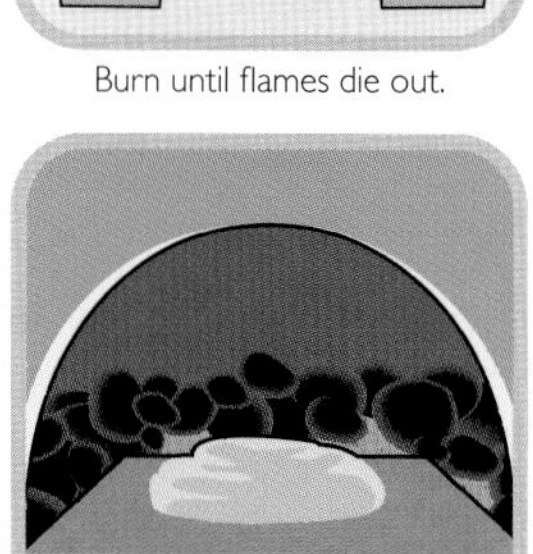
Insert tile and dough.

Replace door; bake.

289 join in with bread and salt

Remove bread.

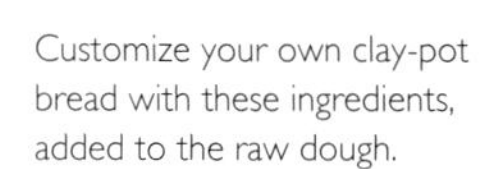

Customize your own clay-pot bread with these ingredients, added to the raw dough.

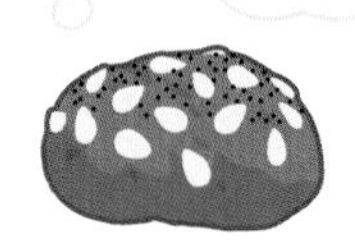
olives, almonds, and poppy seeds

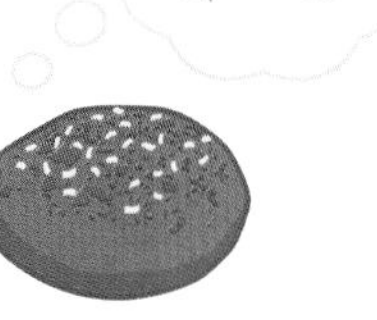
onion and tomato

parmesan and sesame seeds

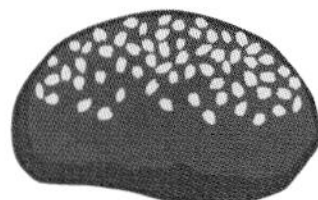

oatmeal and honey

355 cook coconut rice in bamboo

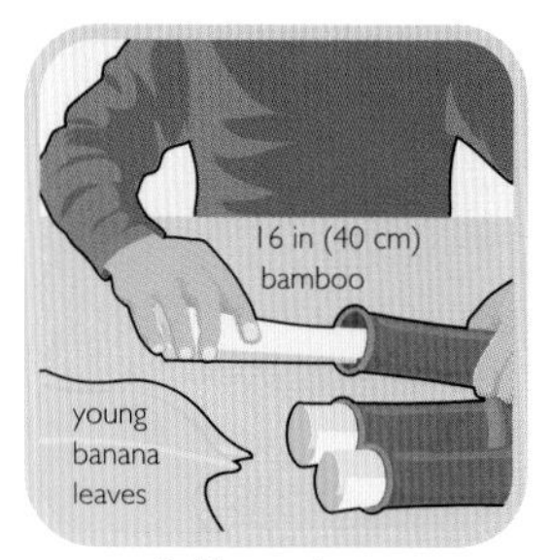

Roll leaves; insert.

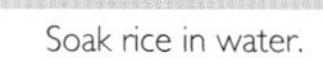
Soak rice in water.

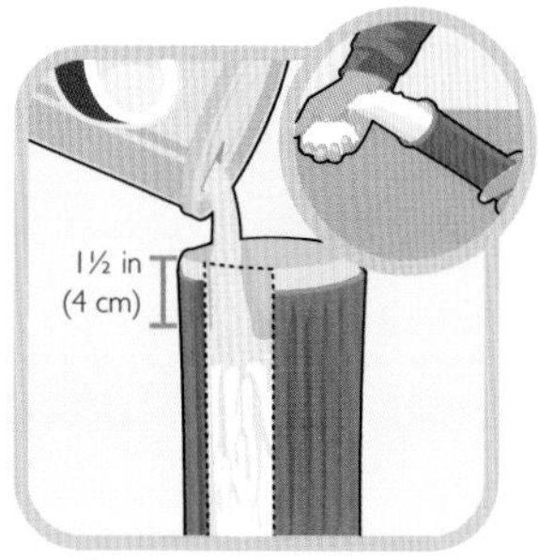

Fill with rice, coconut milk.

Cook, turning occasionally.

Split bamboo; cut leaves.

make a crêpe 356

Combine in blender.

Add; blend until smooth.

Refrigerate batter.

Brush on melted butter.

Hold pan at angle; pour.

Quickly swirl pan to coat.

Cook until batter looks set.

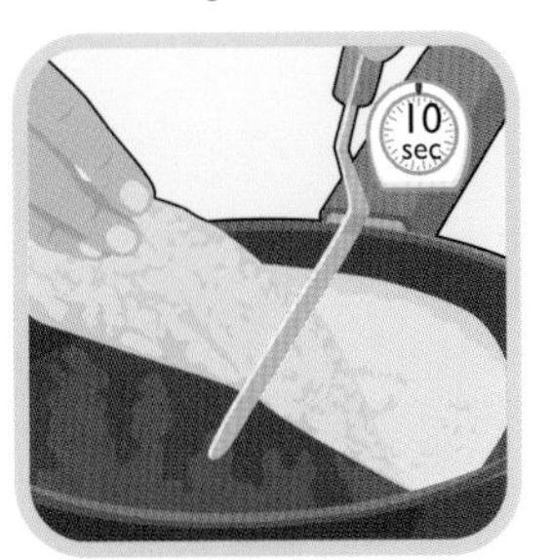

Flip crêpe; cook other side.

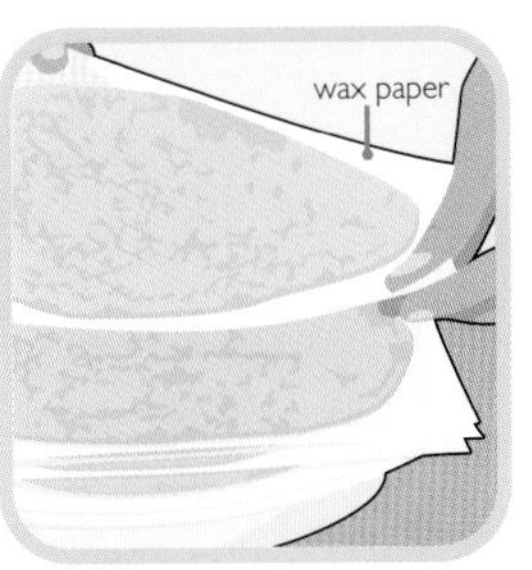

Remove crêpe; stack.

Fill and serve.

A crêpe isn't a crêpe without a fun filling—go sweet or savory with one of these suggestions.

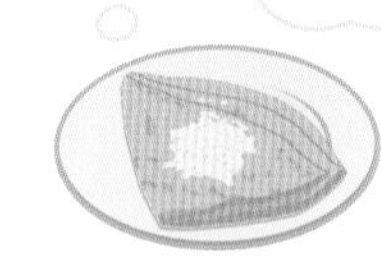
honey and lemon

chocolate and banana

cheddar, roast chicken, and eggs

asparagus, mushrooms, and gruyère

fry doughnuts 357

Cut holes with apple corer.

Boil oil.

Dip doughnut in oil.

Flip to cook other side.

Dip in cinnamon sugar; serve.

358 frost a perfectly smooth cake

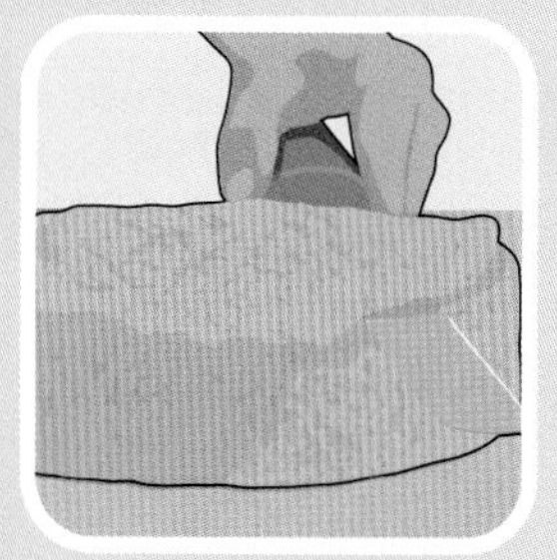
Level a cooled cake.

Set on cardboard on turntable.

Make a thin coat of icing.

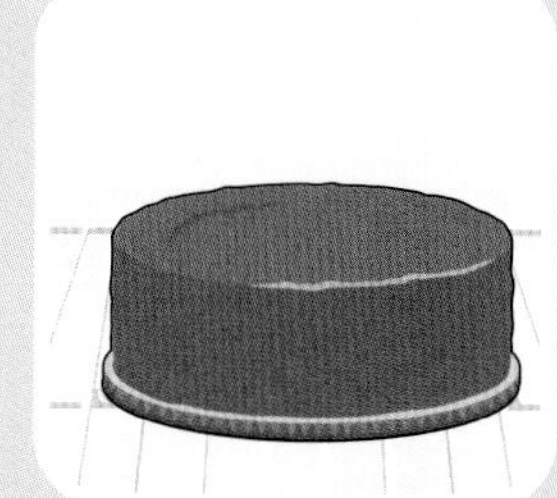
Let icing dry in refrigerator.

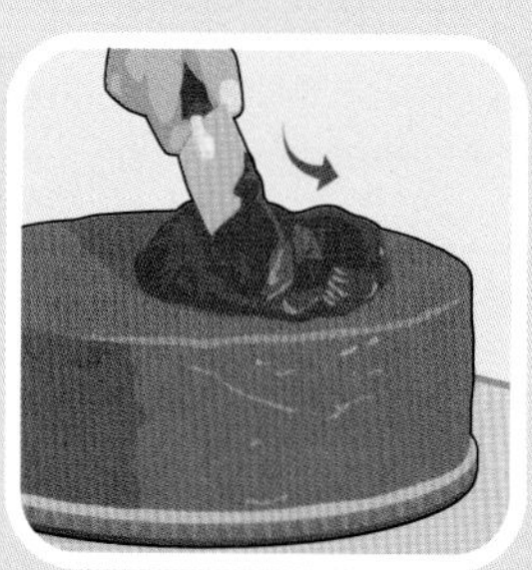
Add more icing; spread out.

Serve your flawlessly frosted cake as is, or consider it a canvas for decorative frosting flourishes. Use a pastry bag filled with icing and a range of specialty tips to get these frosting effects.

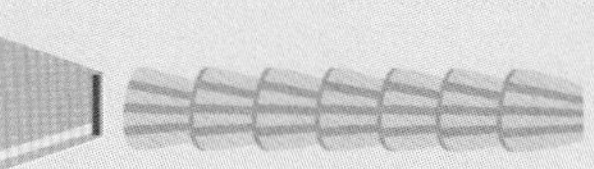

359 make chocolate mint leaves

360 design chocolate lace

Smooth from rim to center.

Cover sides with icing.

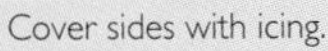

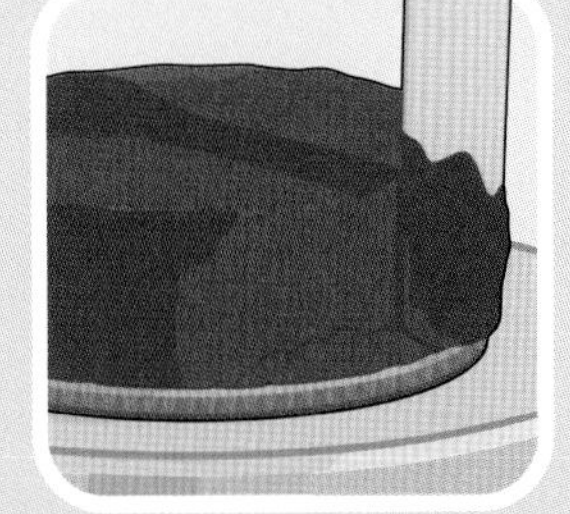
Smooth with knife's edge.

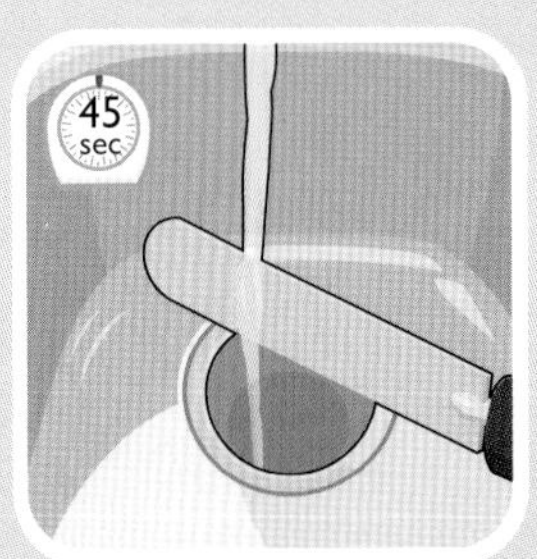

Hold knife under hot water.

Hold edge to cake; rotate.

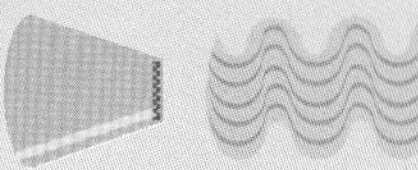

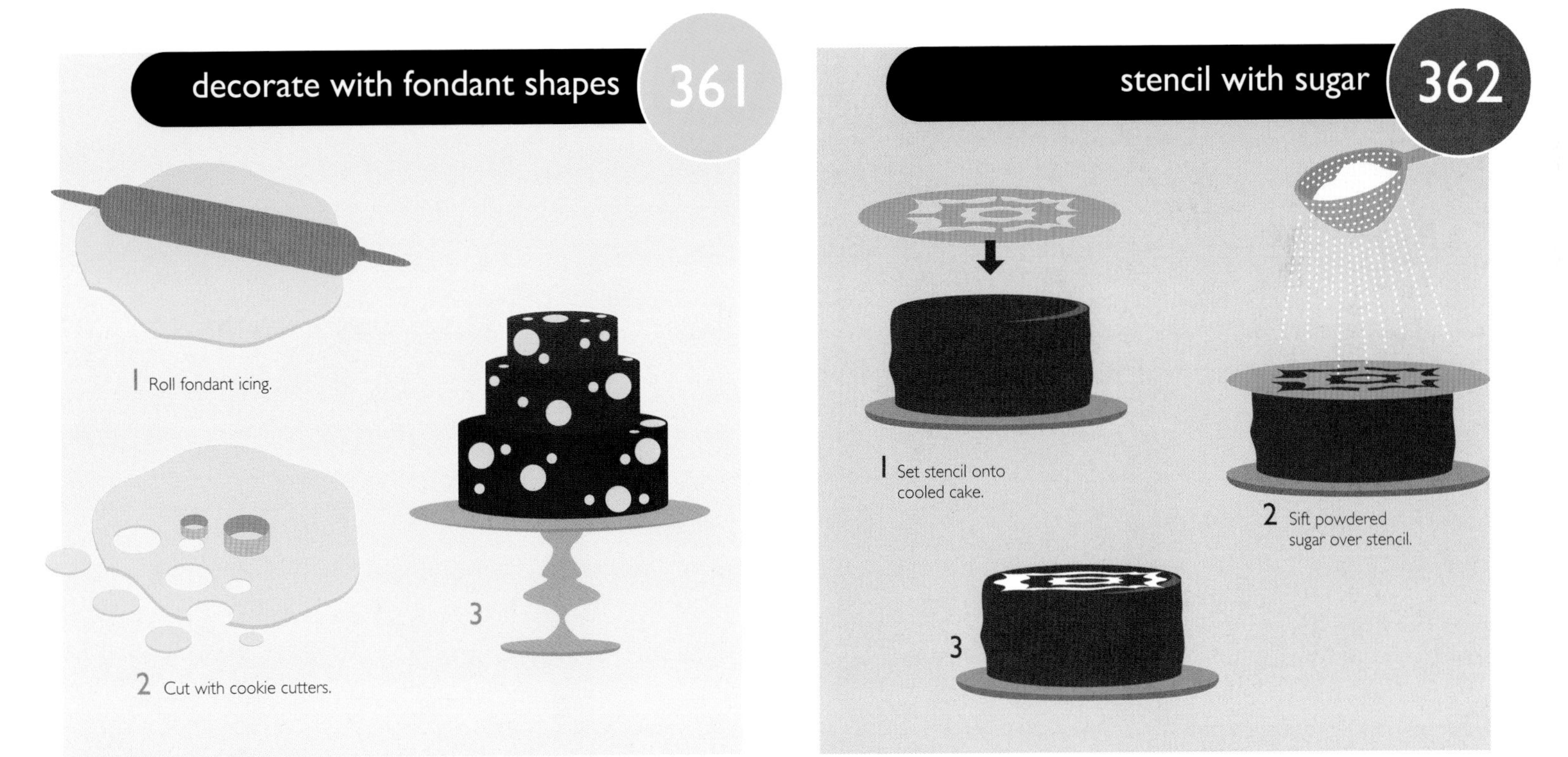

decorate with fondant shapes 361

1 Roll fondant icing.

2 Cut with cookie cutters.

3

stencil with sugar 362

1 Set stencil onto cooled cake.

2 Sift powdered sugar over stencil.

3

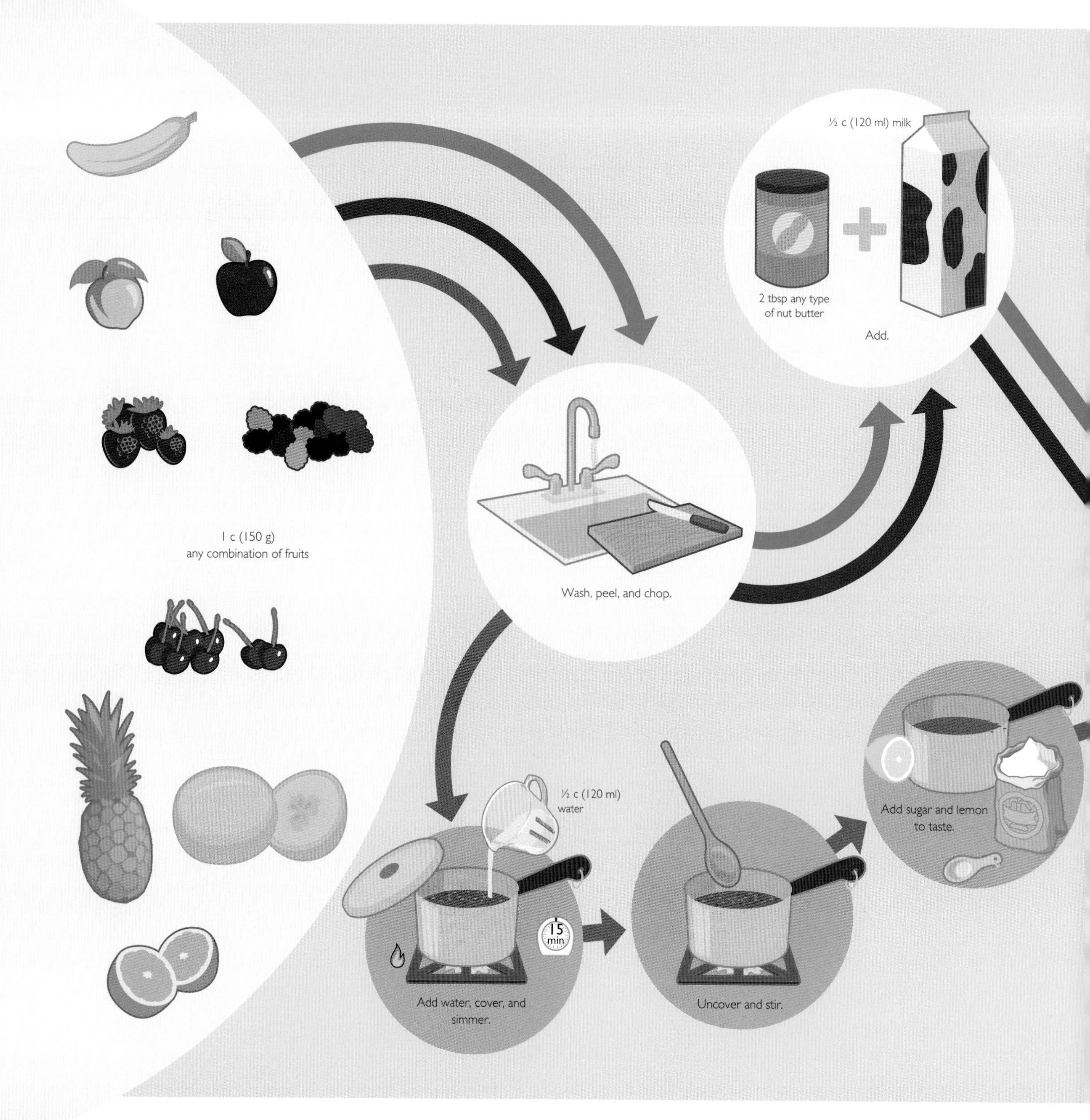

½ c (120 ml) milk
2 tbsp any type of nut butter
Add.
1 c (150 g) any combination of fruits
Wash, peel, and chop.
½ c (120 ml) water
15 min
Add water, cover, and simmer.
Uncover and stir.
Add sugar and lemon to taste.

8 hr

Blend.

364 blend a smoothie

365 make chewy fruit leather

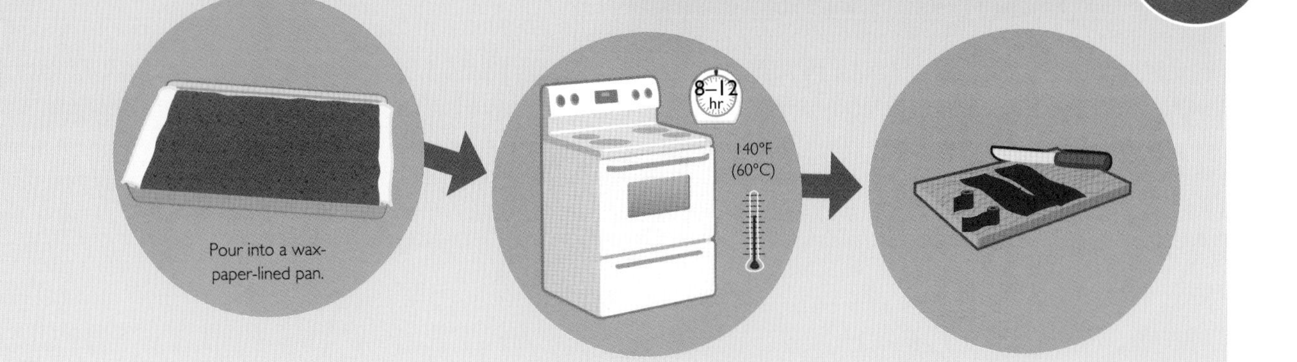

366 concoct dandelion wine

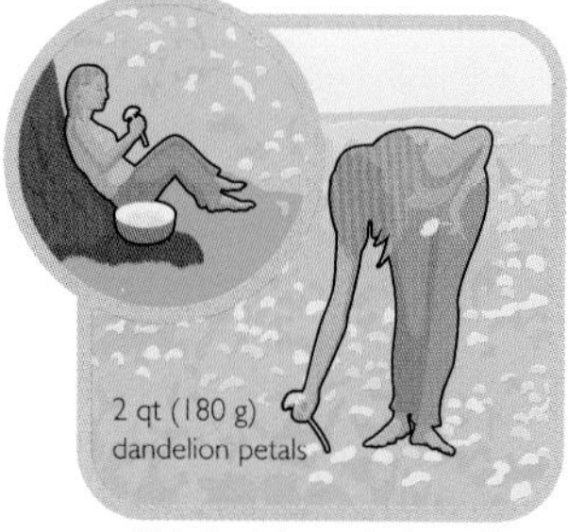

Pick dandelions; pluck petals.

Pour over petals.

Cover with cloth; steep.

Bring to boil.

Add orange peels.

Strain into pitcher.

Stir in sugar.

Stir in yeast, orange juice.

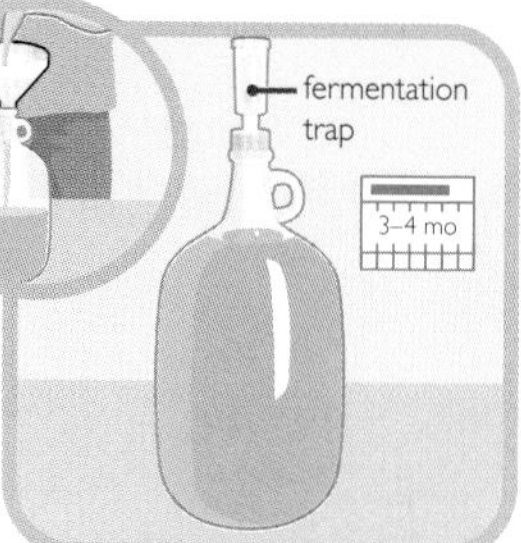

Pour into jug; let ferment.

Rack and bottle; let age.

* To ensure a tasty brew, be careful to remove all the pith from the peels—pith will ruin your wine!

367 brew a restorative tea

make a fruity soda-pop float

Combine over low flame.

Cook until the sugar dissolves.

Strain and let cool.

Add lemon juice to taste.

Fill a third of the way with syrup.

Top off with sparkling water; stir.

Add ice cream.

Garnish and serve.

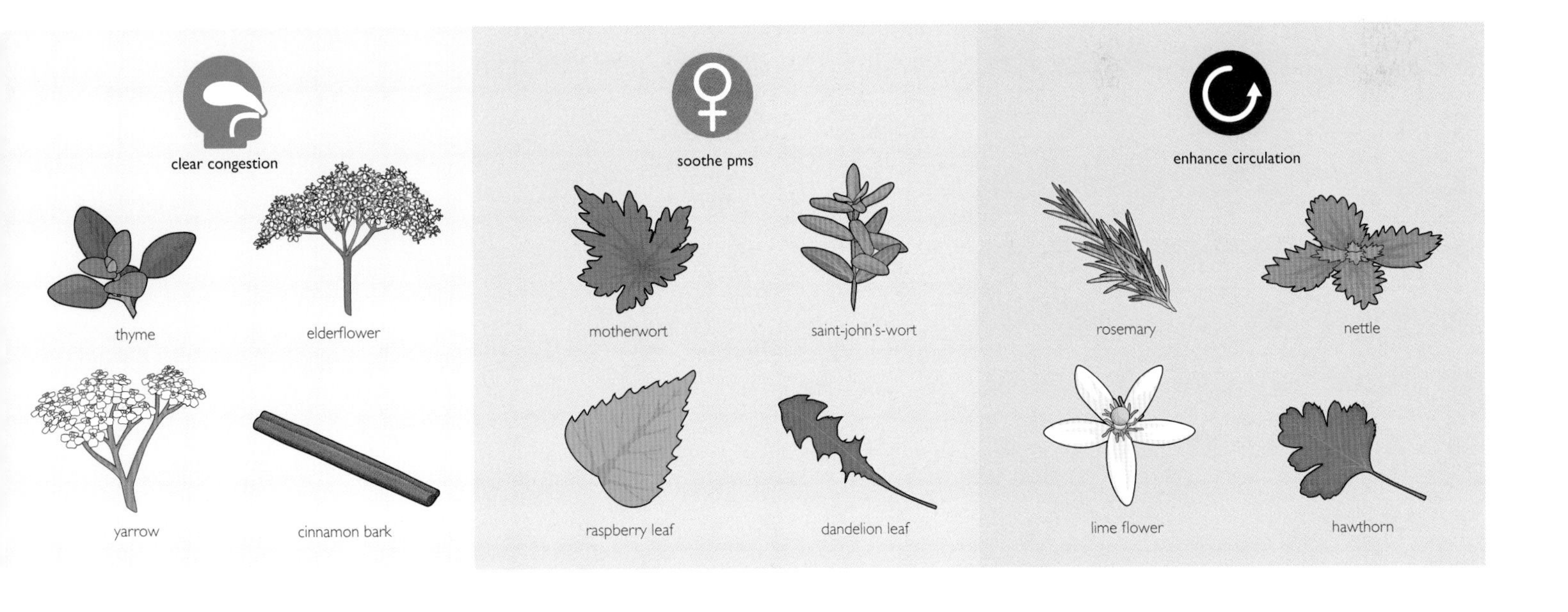

Grate fresh ginger.

Put ginger into a mason jar.

Cover with vodka; seal jar.

Store in a cool, dark place.

Shake once a day.

Filter; discard ginger.

Mix up a gingertini.

Garnish with fresh ginger.

370 garnish with a flaming orange

Hold match near orange peel.

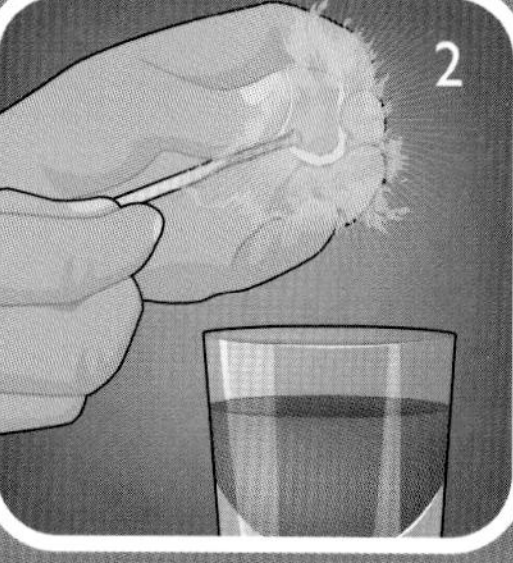

Squeeze, spraying citrus oil.

Rub peel around lip of glass.

Add to drink.

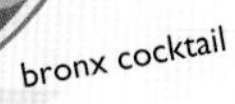

bronx cocktail

2 fl oz (60 ml) gin
½ fl oz (15 ml) dry vermouth
½ fl oz (15 ml) sweet vermouth
1 fl oz (30 ml) fresh orange juice
ice

Shake with ice; garnish with orange peel.

6 pour three drinks at once

serve cocktails with a twist 371

sazerac

1 tsp sugar
1½ fl oz (45 ml) rye whiskey
4 dashes bitters
1 dash absinthe
1 twist orange peel
ice

Crush sugar in a glass with a few drops of water. Add whiskey, bitters, and several ice cubes; stir. Coat a second glass with absinthe and strain in whiskey mixture. Garnish with orange peel.

pimm's cup

ice
2 fl oz (60 ml) Pimm's No. 1
½ c (120 ml) lemon soda
1 orange slice
1 slice lemon
2 cucumber slices
2 tsp chopped mint leaves

Fill a tall highball glass with ice. Add Pimm's and soda; stir in fruit and mint. Garnish with orange peel.

ignite a blue blazer 372

Add lemon twists and sugar.

Pour boiling water into mug.

Quickly add scotch.

Light with long match.

Pour into second mug.

Pour back from greater height.

Repeat with blasé expression.

Pour; extinguish before serving.

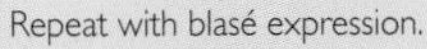

373 brew ethiopian coffee

Roast beans, stirring.

Grind beans by hand.

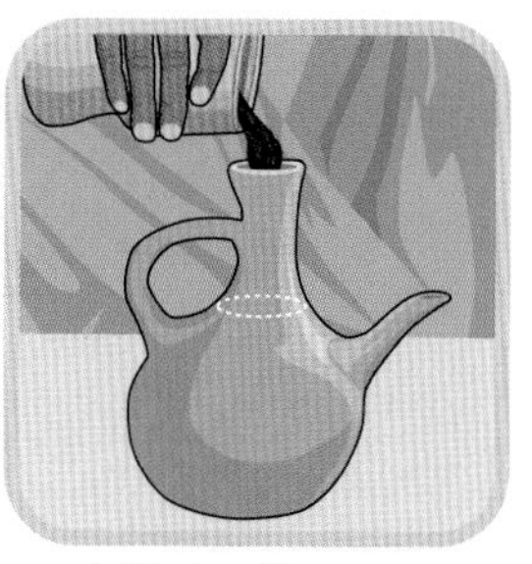

Add to jug of hot water.

Filter several times.

Serve in small cups.

374 serve vietnamese coffee

Cover bottom of glass.

Add coffee to filter.

Assemble filter on cup.

Pour hot water to brew.

Stir before drinking.

375 shake up a greek frappé

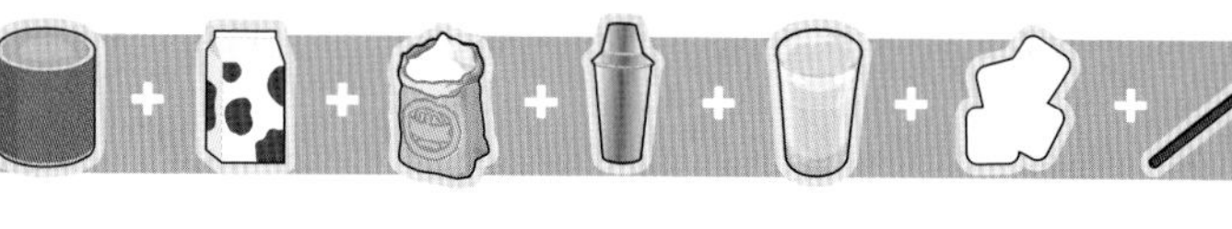

Combine in a shaker.

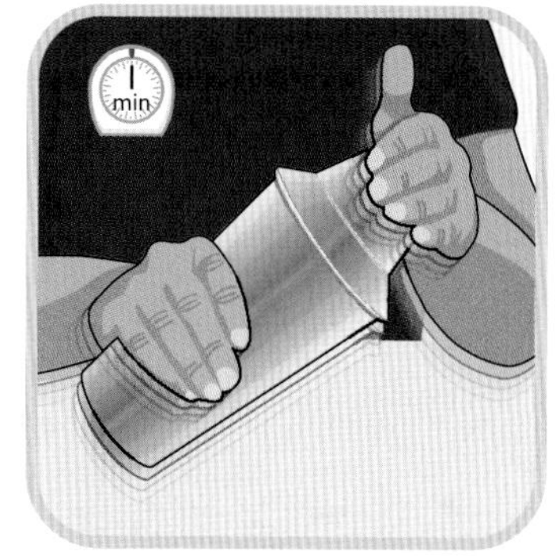

Shake hard, making a foam.

Pour foam over a glass of ice water.

29 make a straw wrapper slither

Drink with a straw.

make a new orleans iced coffee 376

Combine.

Soak overnight.

Strain.

Pour over ice.

Add milk.

froth up a turkish coffee 377

Combine in copper pot.

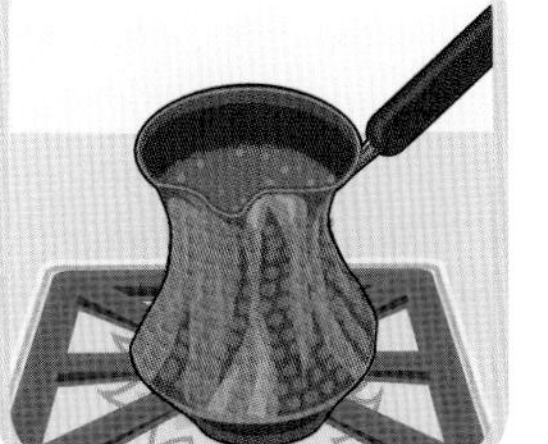
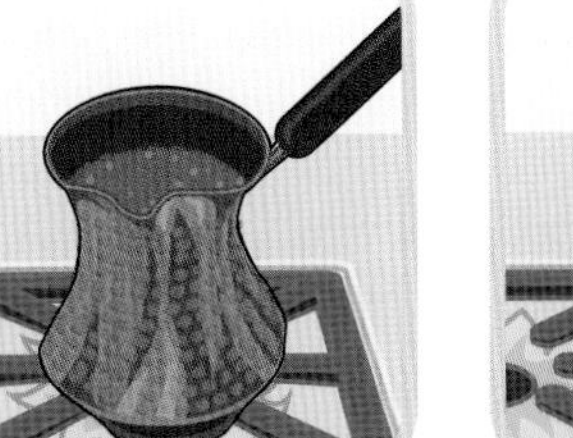
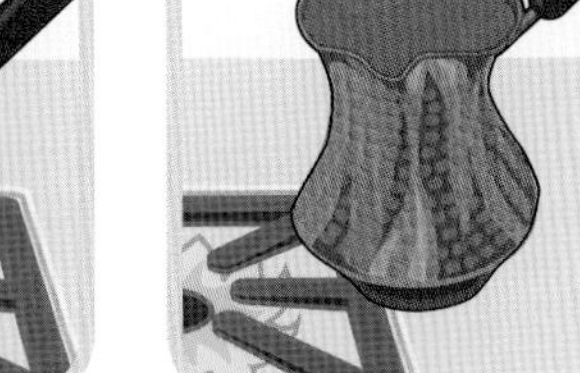
Continue to boil.

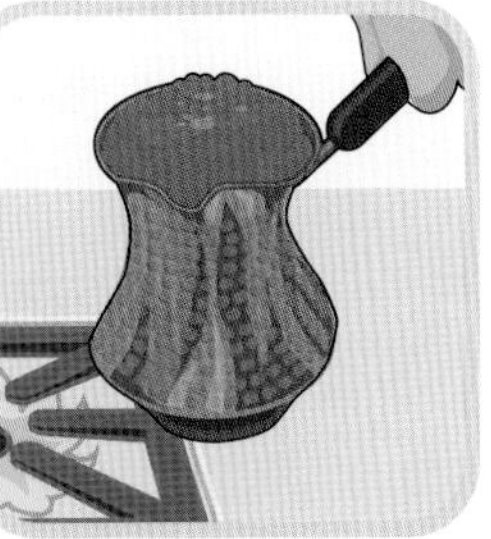
When foam rises, remove.

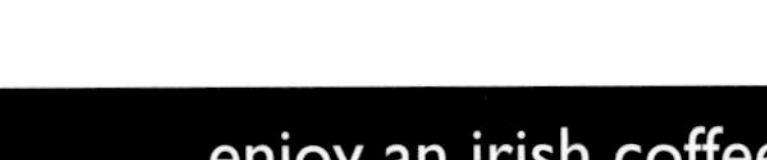
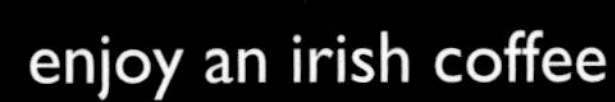
Boil to foam; remove; repeat.

Pour; let sit before drinking.

enjoy an irish coffee 378

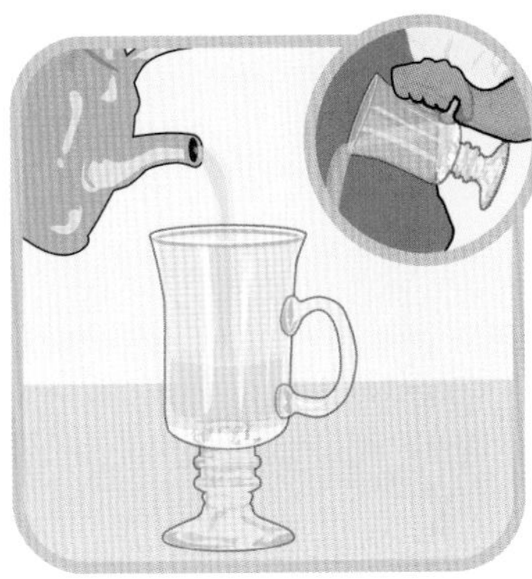
Warm glass mug with boiling water.

Add whiskey and sugar.

Fill nearly to top with coffee.

Carefully float cream on top.

move

379 fight like a cholita

Stun with a high kick.

Run to the ropes.

Climb; turn.

Leap.

Deliver an elbow drop.

380 power slam like a luchador

Rally the crowd.

Stun with a kick to the solar plexus.

Grab opponent by the arm.

Toss him into the ropes.

Grab him as he bounces back.

Use his momentum to swing him up.

Complete the power slam.

Bask in the glory.

spin a capoeira helicóptero 381

Cartwheel onto hands.

Open legs.

Twist legs clockwise.

Bring right leg down.

Land left leg; stand.

Capoeira is an Afro-Brazilian game that combines dance, combat, and music into a lively art form. The dancers perform their stylized, acrobatic sparring encircled by traditional musicians.

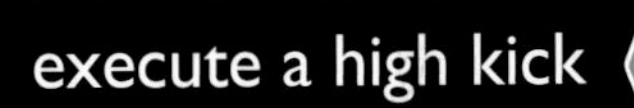

execute a high kick 382

Crouch.

Step onto left foot.

Swing right leg forward.

Leap off left foot.

Kick left foot as right falls.

Land on right foot.

Plant left foot.

Catch balance.

383 do a vertical parry in fencing

Raise foil as foe lunges.

Shift weight backward.

Thrust foe's foil down.

Lunge and score a touch.

Retreat to neutral position.

384 disarm a broadsword opponent

Wait for opponent's lunge.

Touch inside blade; step in.

Bring up sword, twist.

Push down and away.

Pry sword from his grasp.

385 joust like a knight

Hold lance upright in right hand.

Ride hard, lowering lance to level.

Aim for opponent's torso.

Lean backward on impact.

386 jump mount a pony

Grab reins and mane.

Lunge; grab shoulder.

Swing left leg over back.

Drop onto pony's back.

Sit up.

387 execute a jump on horseback

Lean over; grab the mane.

Sit very still, reins lax.

Stretch arms forward.

Swing legs forward; sit up.

Tighten reins.

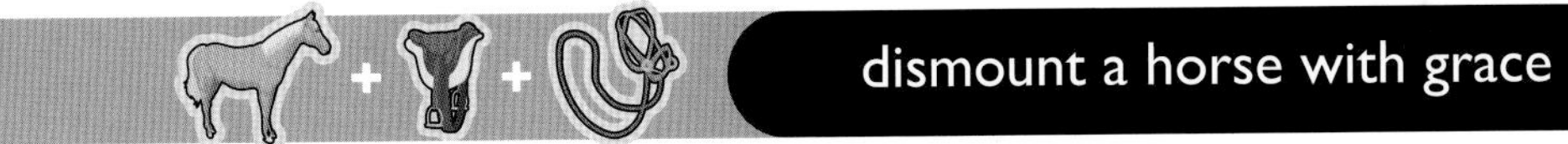

388 dismount a horse with grace

Take feet out of stirrups.

Reins in left hand, swing left leg over.

Bend knees as you land.

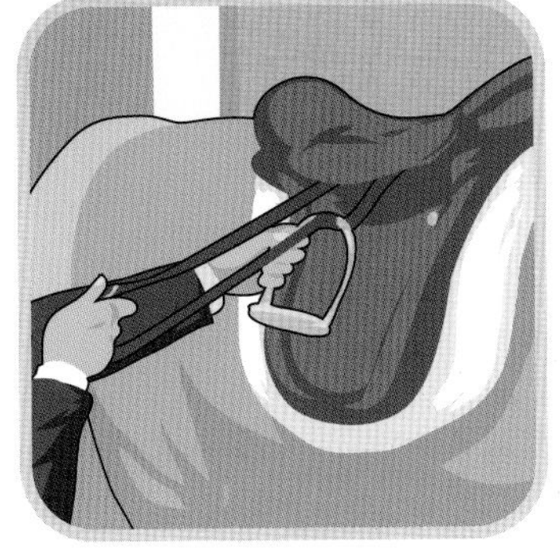

Shorten stirrups.

389 toss horseshoes

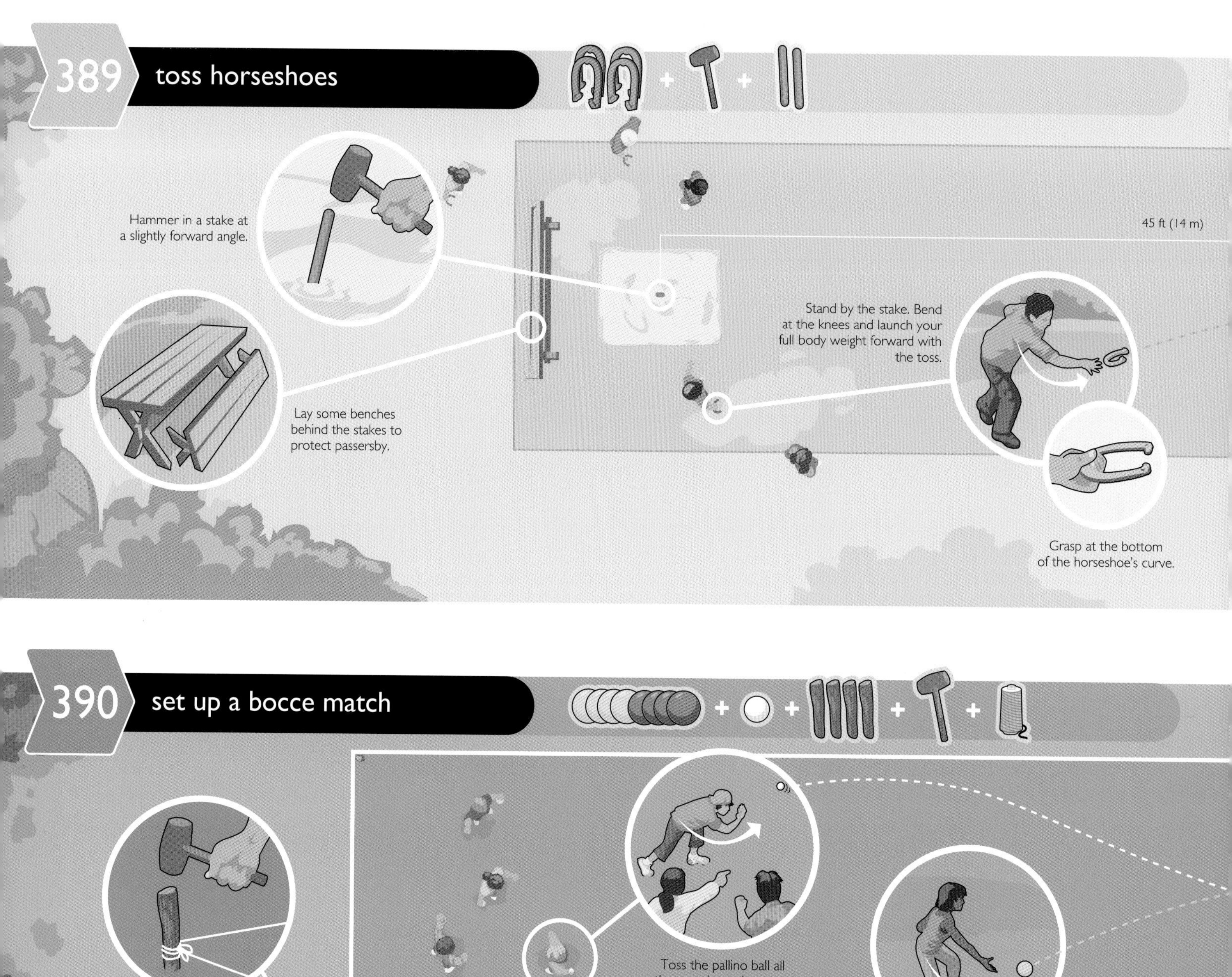

390 set up a bocce match

Mark off the field with stakes and a rope.

Throw bocce balls underhand, aiming for the pallino.

11 do a draw shot in pool

Bump opponent's bocce balls away from the pallino.

1 point

1 point

0 points

Switch sides of the court after each round. The first team to reach thirteen points wins.

What's the score? Check which team's bocce ball is closest to the pallino. Then imagine a circle around the pallino starting at the other team's closest ball. Every ball inside the circle scores one point.

391 ace a cherry drop

Hang from bar; bring toes up.

Hook legs over bar.

Let go, hanging by knees.

Swing body to build momentum.

At height of swing . . .

. . . unbend knees.

Tuck legs toward body.

Land with knees bent.

392 skin the cat on the high bar

Hang; kick back to build momentum.

Swing legs up and under bar.

Let your legs drop down.

Let go; land with knees bent.

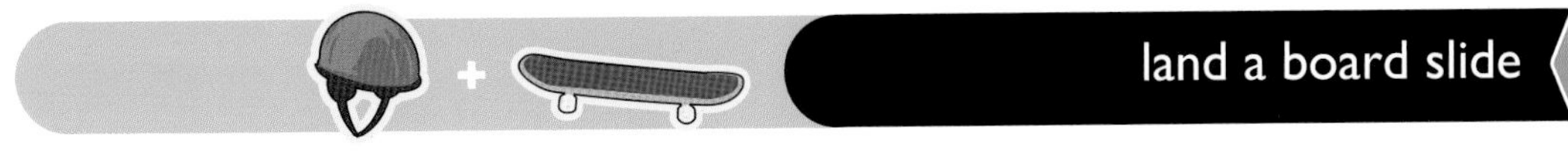

393 land a board slide

Build speed; approach rail.

One-eighty onto rail.

Bring arms out for balance.

Twist forward off rail.

Bend knees at landing.

394 trackstand with no hands

Start with feet parallel.

Turn front wheel slightly.

Shift feet to balance.

Remove hands; sit upright.

Replace hands on bars.

395 skate backward

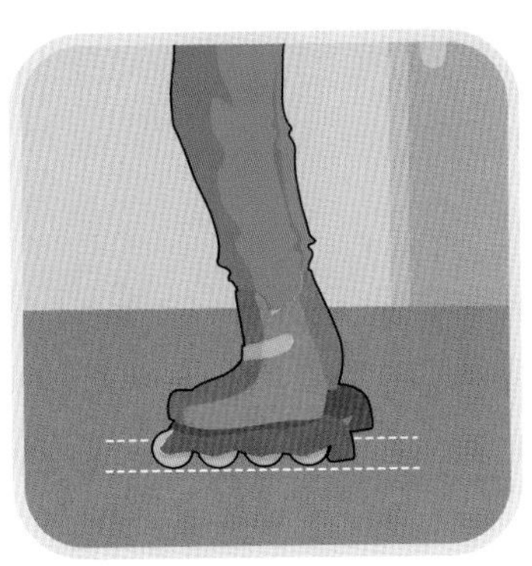

Stand with feet parallel.

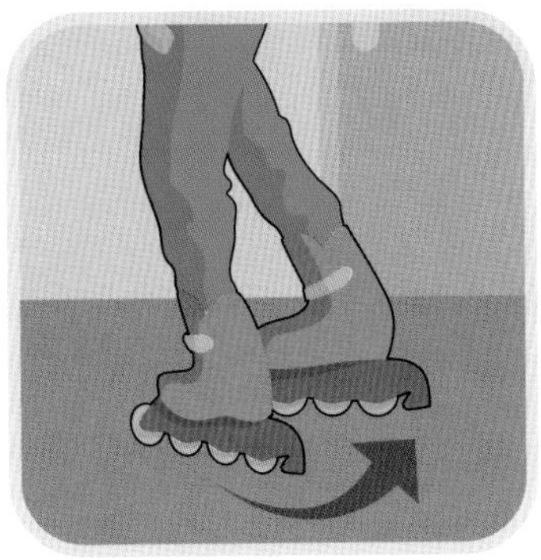

Push one foot out to side.

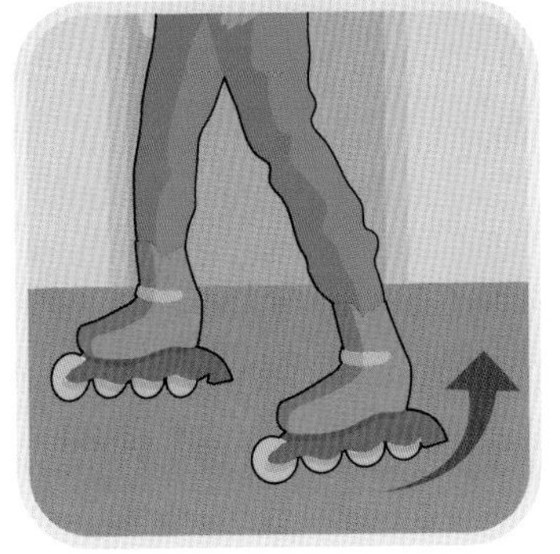

Bring leg back to parallel position.

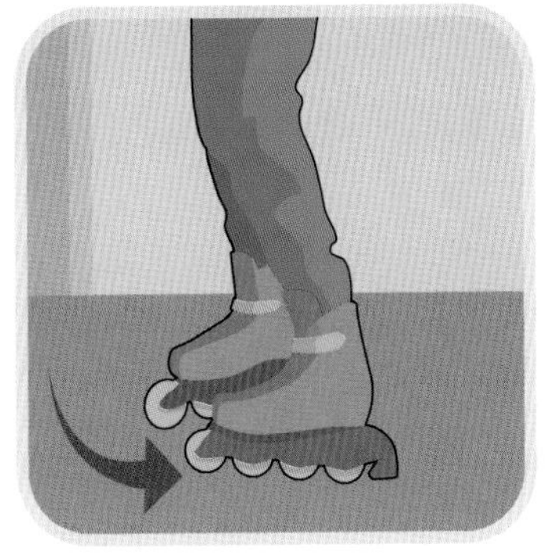

Repeat.

396 bowl an outswinger in cricket

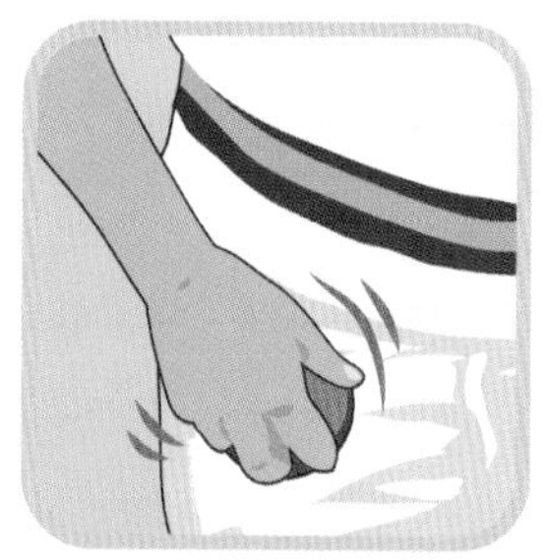

Shine ball on pants.

Hold with seam vertical.

Run toward batter.

Pitch down the middle.

Ball will curve at last second.

397 pitch a curveball

Grip with seam upright.

Conceal behind back.

Step right foot backward.

Wind up.

Pitch; snap wrist toward plate.

398 serve a volleyball

Stand at the end line; aim serve.

Lower ball and draw arm back.

Toss ball; keep your eyes on it.

Strike at high point; follow through.

399 complete an alley-oop

Run; signal for ball.

Ball is thrown above basket.

Leap toward ball.

Catch in midair.

Dunk.

400 fire a slap shot

Place puck between feet.

Raise stick above shoulder.

Strike ice in front of puck.

Slap puck; rotate lead wrist.

Follow through.

401 throw a perfect spiral

Line up shoulders with receiver.

Bring ball behind ear.

Throw, rolling ball off fingers.

Follow through, bringing wrist down.

402 protect your lacrosse stick

Cradle stick vertically.

Assume a defensive stance.

Zigzag to confuse opponents.

Shield stick from opponent.

403 confuse with a soccer step-over

Dribble toward defender.

Line up for a big kick.

Swoop leg around the ball.

Give a soft-heel stop.

Change direction.

404 shoot a screwshot in water polo

Charge at goal, dribbling ball.

Reach under ball midstroke.

Bring ball in line with ear.

Shoot.

Score!

405 form a rugby scrum

In union-rules rugby, a scrum is formed when the forwards of each team link up and face off with their team members behind them. The ball is dropped into the space between the teams, and each tries to get possession by kicking the ball back into their line.

The hooker stands between two props and prepares to link up.

The players link up for stability.

The second row forms.

They crouch behind the front row.

With heads between the front rows' hips, the second row leans in and grips.

Two crouching flankers surround the middle row, and the eightman huddles in back.

406 pole vault like an olympian

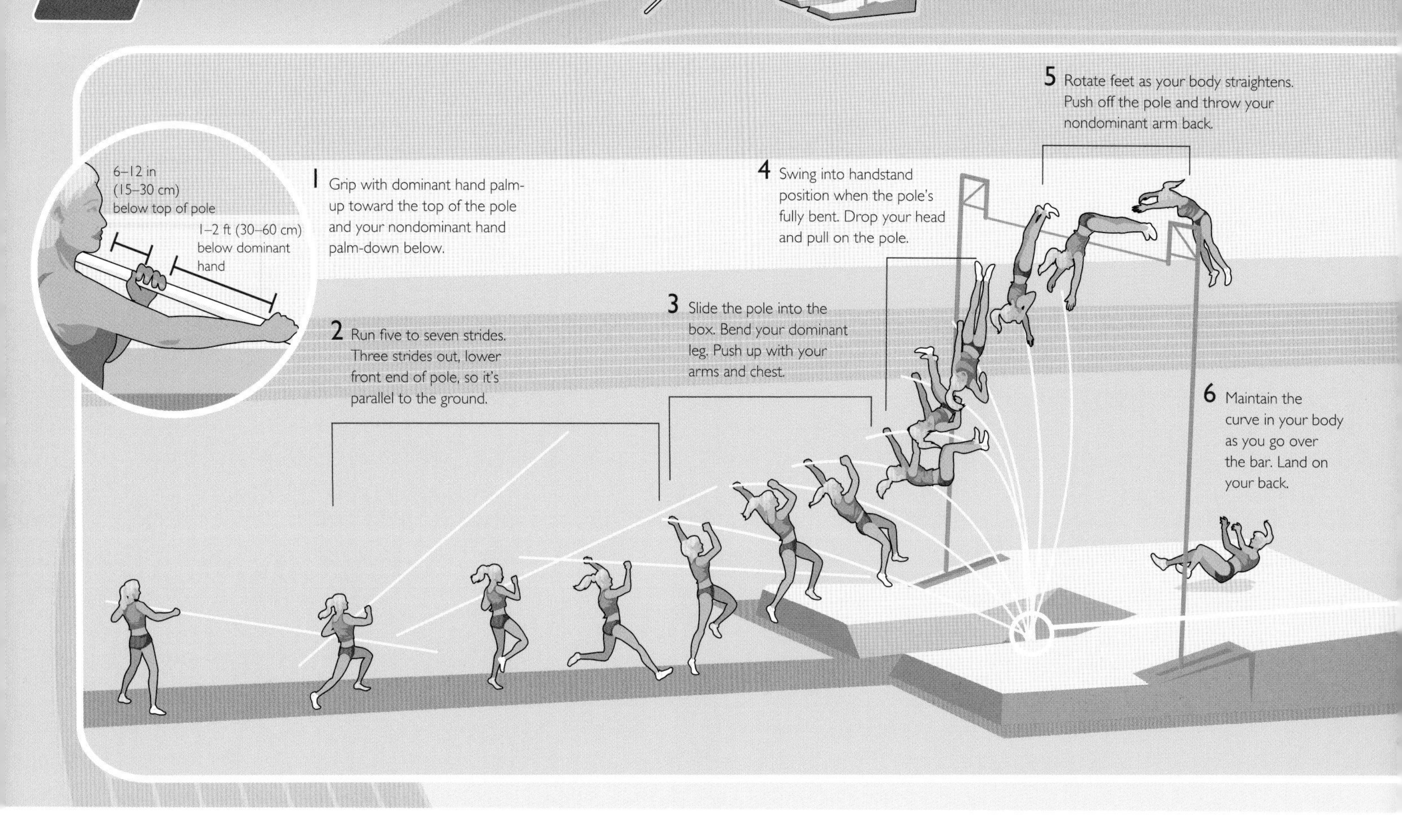

407 excel at shot put

Assume the stance.

Bend to pick up shot put.

Stand; raise overhead.

Place at back of jaw.

Bend center leg.

The Olympians make it look so easy! Be warned that learning to pole vault is a long process, with multiple stages that get you geared up for bigger and bigger jumps. Here are the basic movements involved.

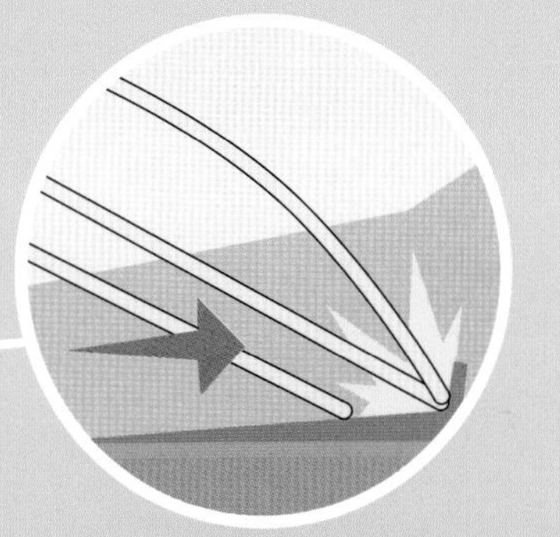

Aim your pole for the middle of the box. It will slide back and bend just right.

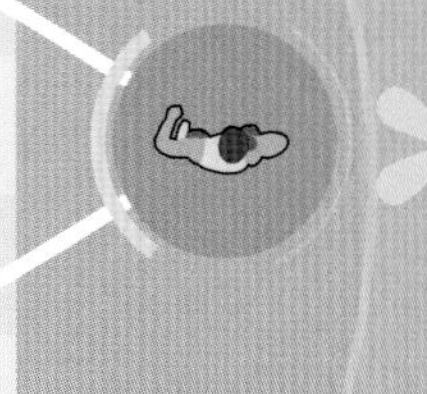

Your starting position in shot put is perpendicular to the field, with your dominant foot in the center and your other foot along the outside ring. After you've released the shot put, rotate back to this position—otherwise, you'll be disqualified!

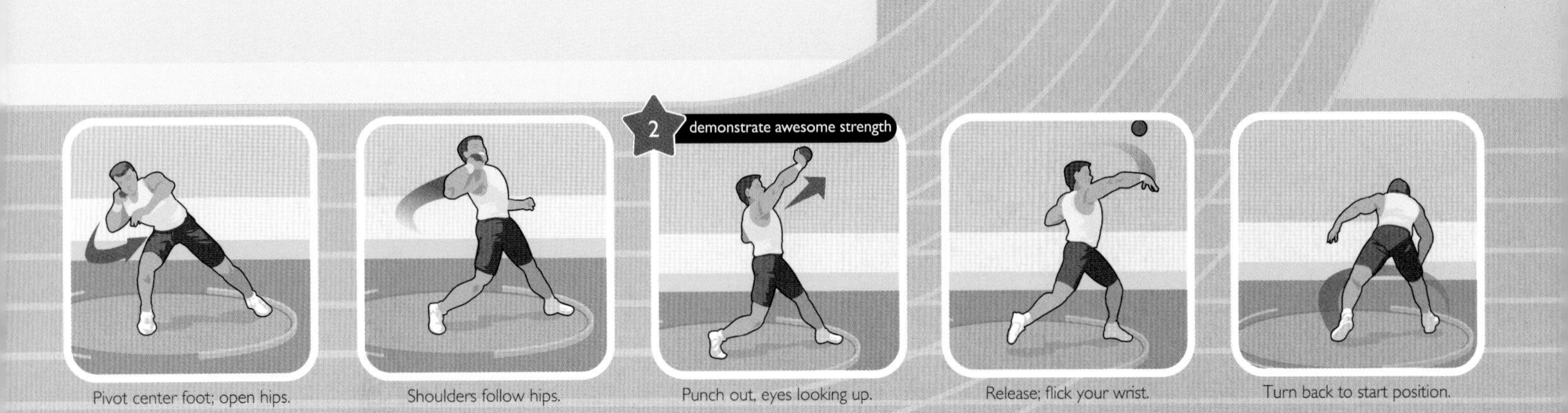

Pivot center foot; open hips.

Shoulders follow hips.

Punch out, eyes looking up.

Release; flick your wrist.

Turn back to start position.

408 tone arms with an exercise band

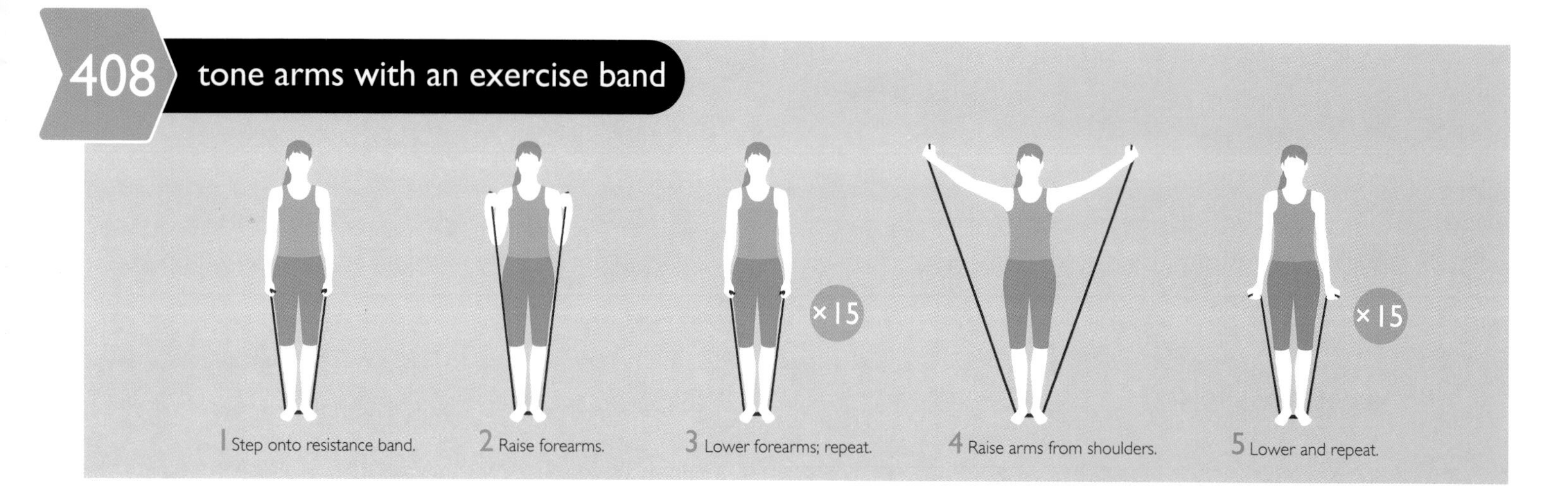

1 Step onto resistance band. 2 Raise forearms. 3 Lower forearms; repeat. 4 Raise arms from shoulders. 5 Lower and repeat.

409 sculpt legs with a kickback

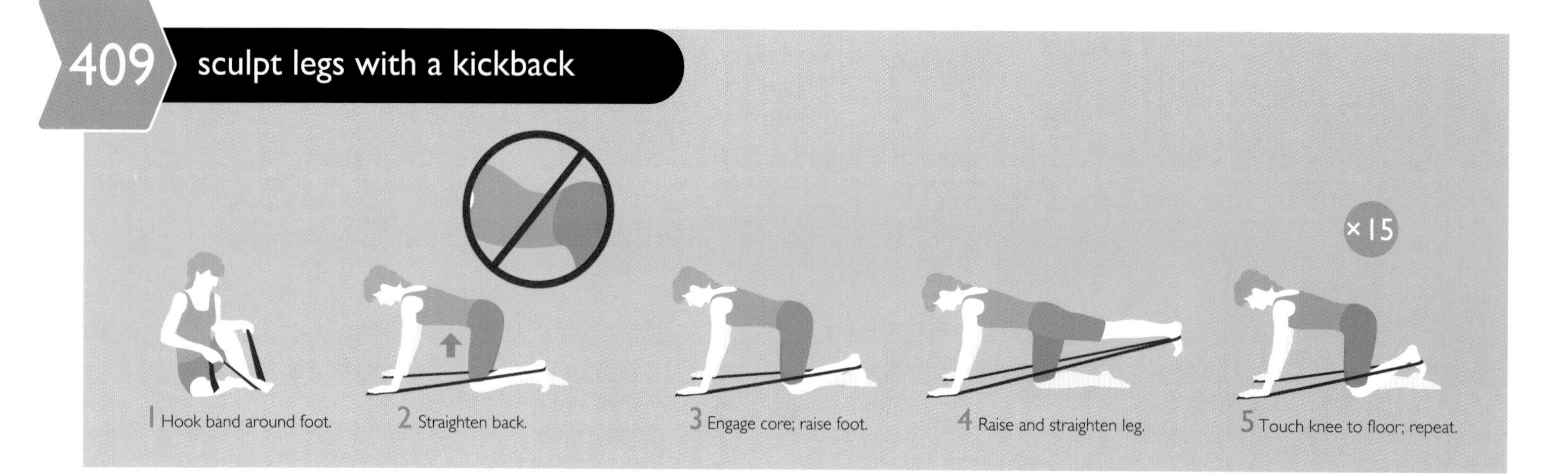

1 Hook band around foot. 2 Straighten back. 3 Engage core; raise foot. 4 Raise and straighten leg. 5 Touch knee to floor; repeat.

410 work abs with a pilates 100

1 Lie with knees bent. 2 Engage core; raise neck and shoulders. 3 Pump arms. Exhale slowly during ten pumps, inhale slowly during ten pumps. Repeat for 100 pumps. 4 Try these alternate leg positions for extra core toning.

Who needs an expensive gym membership? With a little creativity, any environment can become your own personal fitness center!

collect water from a branch 274

angled push-ups
Find a well-anchored object that's about waist-height.

chin ups
No bar handy? Use sturdy tree branches to tone your arms.

triceps dips
Raise and lower yourself using a fixed, raised object.

hamstring stretches
Kick your leg up to stretch before and after exercising.

squats
Hold this position against a tree.

side planks
You can strike this abdominal-toning pose on any level surface.

walking lunges
Pick a path and do lunges down its length.

crunches
Find an elevated surface for ab "bicycle" moves.

412 step into ballet positions

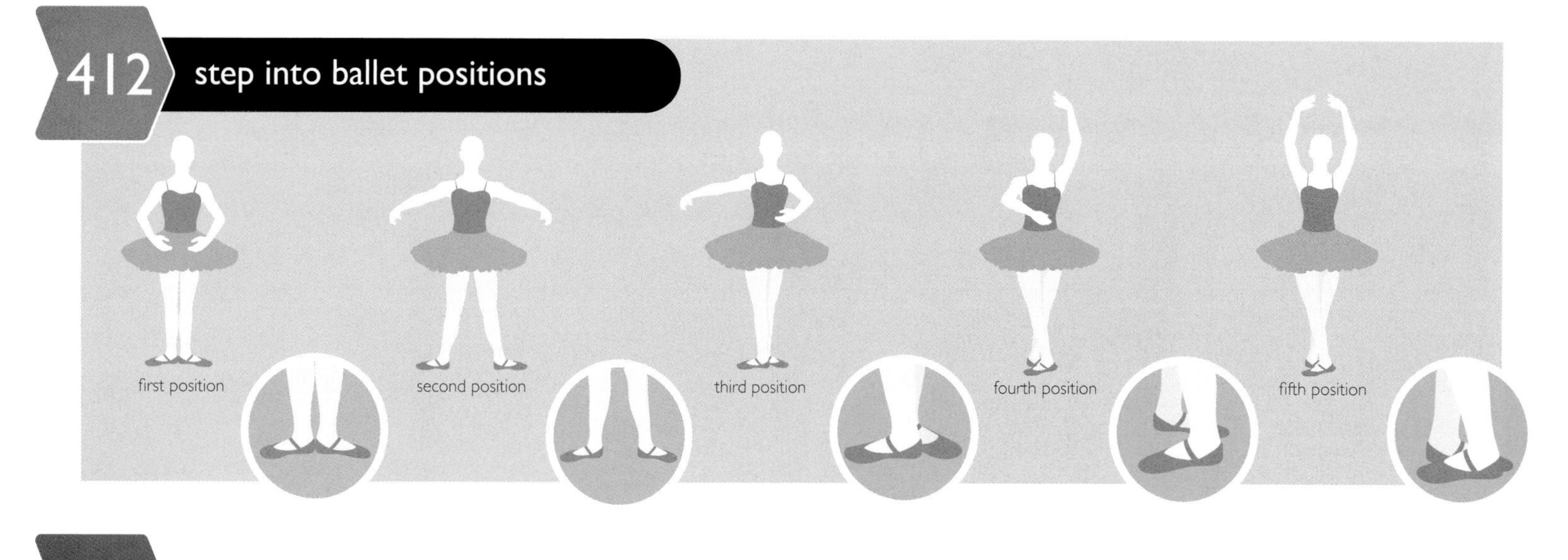

413 twirl a pirouette

1 Start in fifth position.

2 Point to second; raise arms to the side.

3 Point to fourth position; cross arm in front of chest.

4 Bend knees; keep your weight centered.

5 Focus on a point in front of you; this will be your "spot."

6 Lift your right foot to your left knee, rise on your left foot, and close your arms.

7 Whip your head to the right; your body will follow.

8 Whip your head around to find your "spot."

9 Land in fourth position with your front leg bent. Open your arms.

10 Close to fifth position.

do a lindy backflip 414

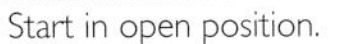

Start in open position.

Bend; lean torso left.

Grab behind her knees.

Straighten up and flip her.

She lands with bent knees.

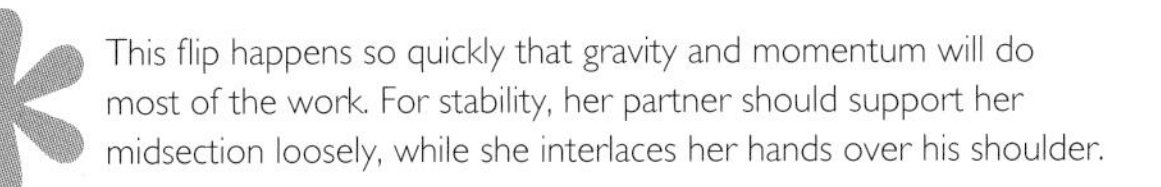

This flip happens so quickly that gravity and momentum will do most of the work. For stability, her partner should support her midsection loosely, while she interlaces her hands over his shoulder.

rock a headspin 415

Prepare to stand on head.

Kick legs up.

Open legs.

Twist hips for momentum.

Snap hips to face front.

Pull hands off ground as you spin.

Slow spin; return hands to floor.

Swing legs clockwise to repeat.

416 leap into a bobsled

Push bobsled in formation.

Pilot hops in first.

Brakemen enter in order.

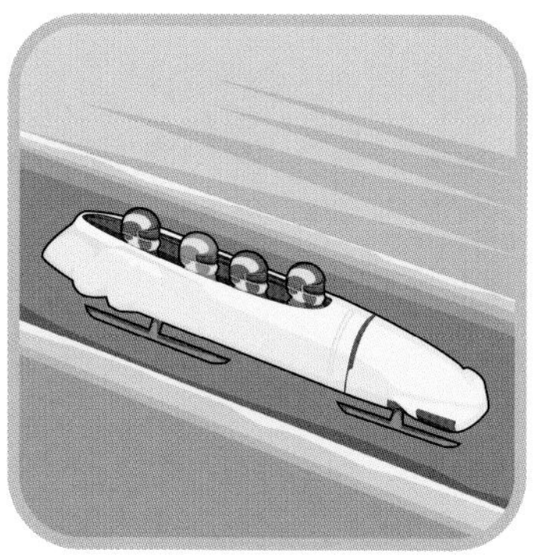
Wheeeeee!

417 get back into skis after a fall

Face skis sideways on slope.

Scrape snow off boot.

Click ski boot into place.

Repeat on other ski.

258 survive an avalanche
Ski on!

418 do a telemark kick turn

Face skis sideways on slope.

Kick up your downhill ski.

Rotate ski 180 degrees.

Swing other ski around.

Plant ski.

419 skate a back crossover

Gather momentum going backward.

Cross right foot in front of left.

Bring left foot out to side.

Repeat, crossing right foot again.

420 land a snowboard jump

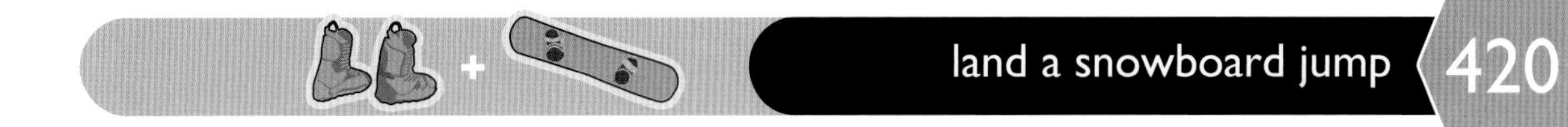

Crouch; shift weight to back.

Lift front foot, board nose.

Level board, raising knees.

Aim nose down to land.

Ride out any wobbles.

421 climb a hill in snowshoes

Adjust poles to hill slope.

Plan to zigzag uphill.

To switch back, plant poles.

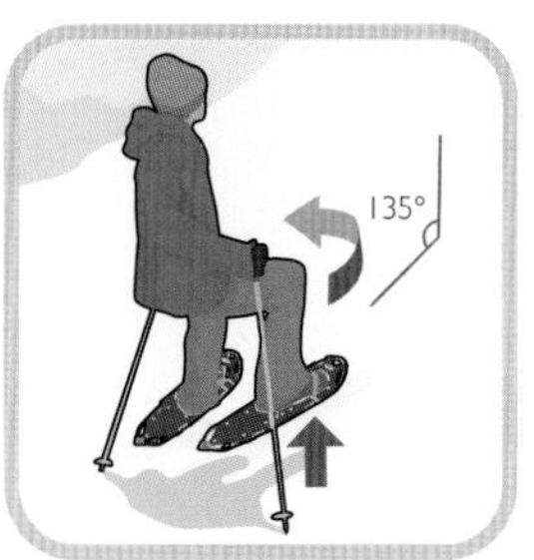

Jump up and twist.

Continue.

422
jump out of an airplane
13,000 ft
(4 km)
1 Leap from the plane.
2 Enjoy your free fall!
2,500 ft
(750 m)
3 Pull out the pilot chute.
2,200 ft
(650 m)
4 The pilot chute catches air . . .
423
fly in formation
tandem
head-down ring
star
solo surfing

53 parachute into a new life
5 . . . pulling the canopy from your pack.
canopy
6 Use the steering lines. To veer left, pull on the left control line.
7 Bend your knees and hit the ground running.

resources

tools

bamboo slice
bubbles
catnip
cucumber
water bowl
glass pitcher
baby bag
plastic tarp
ballast
pole vault bar
blankets
art
bamboo leaves
mug
nature magazine
cash
glass baking pan
liquid starch
khadag (ceremonial scarf)
towel
glass platter
car seat
tea
parsley
dog carrier
cd case
comb
rubbing alcohol
tissue paper
lacrosse helmet
soil
pond form
beads
bamboo
sponge
dead fish
dill
gardening gel
yogurt
biscuit dough
sparkling water
hockey mask
cooler
toy robots
ruler
aloe vera
oven mitt
green pepper
embroidery floss
water bottle
art epoxy
dish cloth
chalk
sheet
teapot
straw
seed packet
art history books
thyme
fringe
shot glass
blow torch
jeans
paintbrush
dart
skydiving suit
acacia branch
flowers
lighter
favorite herbs
hair gel
spray bottle
art knife
snowboard
vegetable shortening
hair clip
beads
rosemary
lingonberry
bougainvillea seedlings
foam ring
bar glasses
gallon of water
calculator
petroleum jelly
etching cream
rubber glove
fabric
ceramic tiles
candy
moss
fuse
mason jars
toothpaste
powdered sugar
pendant lamp
rocks
tempera paint
tiles
brackets
candle cups
compost
wallpaper
propane tank
powder
thin paintbrush
tourist costume
colored pencil
colored gravel
carbon paper

love potion
mechanical pencil
screwdriver
claw clippers
wig
dog treats
underwear
envelope
chairs
surf kayak
beads
parachute
kite string
tape
lance
trench coat
instant coffee
felt
eraser
demerara sugar
protective chaps
feather
bocce balls
soda bottle
scissors
umbrella
electric guitar
heavy books
paintbrush
metal enamel
ribs
paintbrush
hair-dye brush
condom
magazine
bag
steed
rotten fruit
lacrosse ball
skirt
pom pom
shrimp
life jackets
pruning snips
flowers
bucket
bicycle
dog harnesses
jacket
jam
fabric
card stock
pink candle
camp fire
candle
fake flowers
wire nut
snowshoes
biohazard bucket
ceramic jugs (jabeenas)
berries
old clothes
spray bottle
salmon
gold leaf
chisel
red onion
kite
t-shirt squares
jet pack
playing cards
wire cutters
bar stool
confetti
shot put
skydive helmet
pomade
pillow
cream of tartar
balloons
pins
cricket ball
eyeliner pencil
horseshoes
pigs
tuna steak
printing ink
pliers
ribbon
chicory
car
possessions
ascot
decorative hole punch
zipper
river-dye case
basketball
polyurethane
wire strippers
paillettes
credit card and cash
quilling strips
helium tanks
coffin
electrical wire
hair curlers
washcloth
folding chair
varnish
roofing cement
ribbon
painting
strawberries
pants
bucket
plumeria
salami
mineral makeup
hobo bindle
ibuprofen
caulk

grinder	vegetables	carving tool	couch	spatula	logs	charred stick	scotch	old frame	roofing gravel	stake
orange cones	measuring tape	clay	pony	pickling spice	cedar plank	mallet	teddy bear	block of wood	bay leaves	spatula
orange juice	match	ibrik	paintbrush	secret river dye	cardboard	cardboard tubes	brush	cardboard disc	tied stakes	paper bag
pumpkins	whiskey	mechanical bull	kitchen chair	pipe cleaners	log	cat toy	lattice	grout	mushrooms	dresser
orange	hair bleach	bbq sauce	saddle	football	pellet gun	toilet paper tube	bobby pins	sandbags	cork	cinnamon
hat	clay pot	coarse salt	vanilla extract	coffee filter	hammer	table	picture frame	fuel cell	faux bois tool	chicha bowl
fruit	wood block	bookshelves	frosting	pan	sheets of cardboard	stick	boards	ancient stone	fiberglass pole	wood blocks
fixed-gear bike	flower pot	quilt	eyeshadow brush	pole	rum	bridle	rolling pin	paving block	hanging tray	insulation
boats	horse	comics	fabric	chopstick	suitcases	match	sandpaper	tile spacers	clay tool	cork board
shell	teak oil	basket	safety gear	paintbrush	monkey bar	paintbrush	bead loom	basket	scratching post	lumber
camp fire	ceramic tile	baseball glove	cocktail	fugu fish	stakes	paintbrush	rice	incense	picture frame	cocoa butter

hinge
vegetable oil
bbq brush
water-polo ball
buttermilk
purim gifts
elastic cord
busójárás mask
museum map
rubber band
pvc endcap
wooden spoon
flour
dough
canoe
lemon
morris-dance bells
contact paper
beef bone
cloth strips
ping-pong balls
paper towels
plywood
ice cream
cheese cloth
stickers
butter
canopy strap
empty gift box
brayer
small cups
notebook
pvc pipe
honey
coconut milk
craft stick
banana peel
lemon juice
double-stick tape
beer mug
sand
soothing balm
funnel
cotton ball
mortar and pestle
sponge
insulation
tack
liquid silicon
foam pad
antique writing implements
interfacing
paint
plate
talcum powder
garlic
beeswax
coconut oil
dandelions
glue stick
travel documents
baseball
parchment paper
junker car
can of beer
lacrosse stick
skewer
pastry brush
olive oil
grout sealer
lemon peel
helmet
embroidery hoop
sewing machine
mentos®
velcro®
sugar cubes
kiln shelf
yeast
wood glue
drill
corn-husk mobile
skis
planks
brewed tea
garlic
fabric pencil
ice skates
hockey stick
cake
clothespins
box cutter
avalanche beacon
bobsled
wooden spacers
wax paper
fugu license
salt and pepper
paper
sand
cream foundation
ginger root
pencil
beads
tape measure
mongolian breads (ul boov)
wax
ceramic buttons
electric mixer
hair wax
brown sugar
rope
anchor
tin snips
seeds
chainsaw
linoleum block
cloth bandage
light timer
volleyball
spray bottle

pallino
condensed milk
battery
jug
glass bottle
hydrochloric acid
lime
ladle
epee
fork
airplane
paper towel
string
airplane
liquid starch
wine bottles
lip-gloss container
dogs
hoe
gardening shears
guitar string
nails
marker
stump remover
thermometer
liquid dish soap
clear nail polish
shrink plastic
cosmetic scissors
pipe strap
putty knife
glue
duct tape
pen
milk
craft glue
glass of water
martini glass
doily
garbage can
engagement ring
hole punch
paper lantern
hunting knife
white glue
drawing
makeup sponge
glass
wine glass
tea towel
armor
thermometer
staple gun
chalk line
dog food
electric toothbrush
teacups
ice-cream scoop
vodka
small cups
cassette tape
led wires
makeup base
quilt batting
pin
faucet
eggs
soccer ball
plexiglass
fermentation trap
glass pitchers
floss
shovel
coat rack
newspaper
screws
spatula
cigarette
composition book
cling wrap
glass bowls
glass
tea lights
butter knife
pole
plastic
eye hook
paint
basketball hoop
iron
ice
jar
irish-coffee glass
needle
caulk
tape
cell phone
salsa cup
lipstick
googly eyes
fan
plastic bag
jar
bowl
acrylic gloss
mirror
cake knife
craft ball
string
hair spray
peeler
sailboat
white vinegar
healthy snacks
bottles
lei needle
grater
level
knife
whisk
broadsword

apple corer
cocktail shakers
cake turntable
quilling tool
step ladder
coin
cowboy hat
cell phone
lead fishing weight
baking pan
blender

paper clip
waterproof finish
mugs
seasoning
antiquities
tea
haircutting scissors
coffee grinder
life jacket
coffee
period-appropriate tools

needle-nose pliers
wrench
flour sifter
trowel
tongs
sieve
saw
hockey skates
harness
flight suit
licorice

makeup brush
flask
clamps
roofing nails
ski boots
pond pump
vietnamese coffee filter (phin)
snow shovel
keys
jetpack helmet
printer

hair dryer
trowel
leaf blower
hair clips
roller blades
foam insulation
clippers
french press
train
dart board
peppercorns

ski poles
grout float
sugar
garbage can
foam brush
seam ripper
shaker
magnet
combs
eyedropper
binder clip

mesh
stepping stone
safey pins
copy machine
collapsible shovel
screwdriver
file
guitar strap lock
glue gun
old t-shirts
stroller bag

fan
stencil
gold leaf sizing
spoon
scorpion
hose
tongs
drill with holesaw
wire brush
electrical tape
charcoal

paint tray
lacy tablecloth
grout
tweezers
vermillion
flashdrive
double burner
charcoal burner
thread
drill
ink

transistor radio
strainer
mortar
jigsaw
vibrating cell-phone motor
snow probe
hockey puck
elevator
pan with rack
utility knife
disguise

blush
weatherproof black paint
transistor radio
horseshoe stakes
rock
aluminum foil
rocks
skateboard
eyeliner pencil
lingerie
briefcase

index

233

102

285

87

380

404

231

308

149

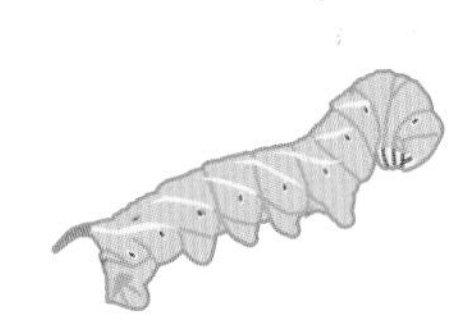
218

d

367

386

103

139

157

e

f

g

60

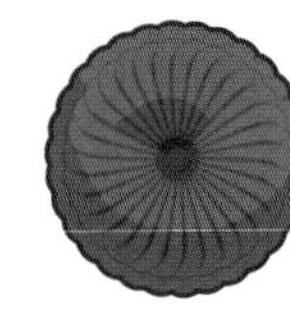

196

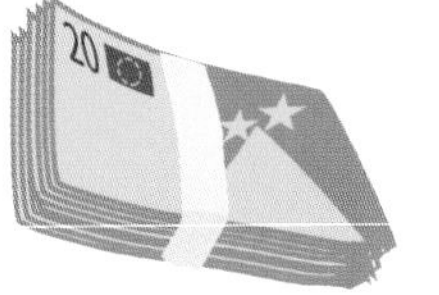

252

9

306

364

239

128

315

171

302

228

356

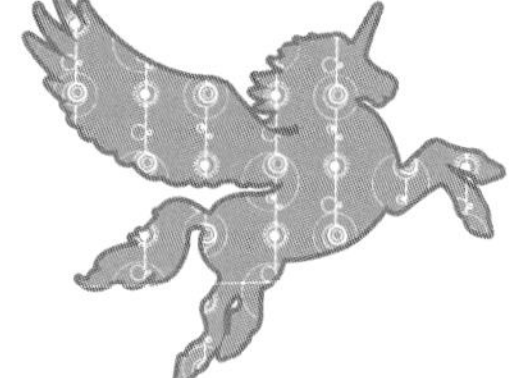
189

68

51

81

49

19

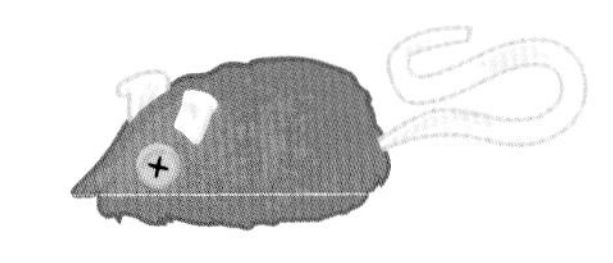
235

q

r

s

53

273

320

411

71

109

72

74

28

149

u

v

w

y

z

333

360

47

136

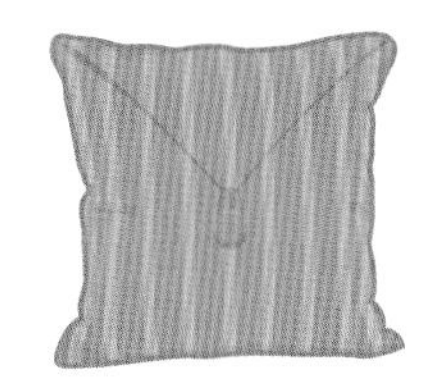

196

show me who

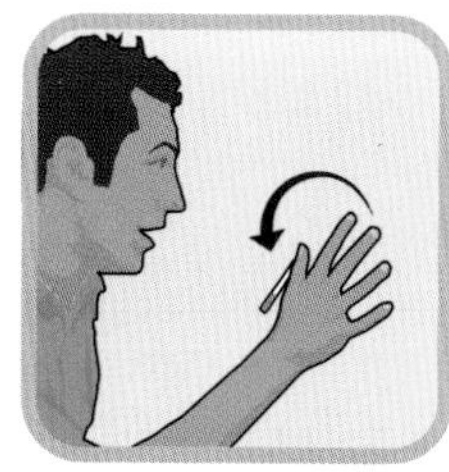
Derek Fagerstrom
Author

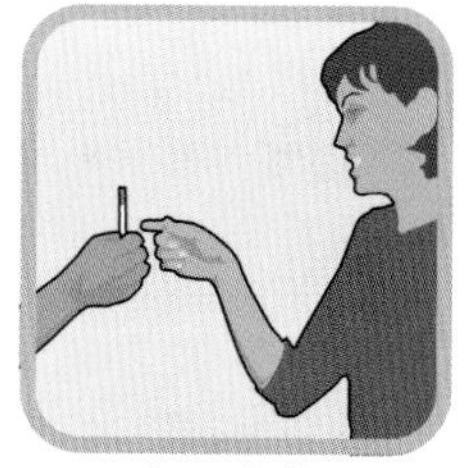
Lauren Smith
Author

Terry Newell
CEO and President

Amy Kaneko
VP; Sales

Mark Perrigo
Director of Finance

Chris Hemesath
Production Director

Roger Shaw
VP and Publisher

Mariah Bear
Executive Editor

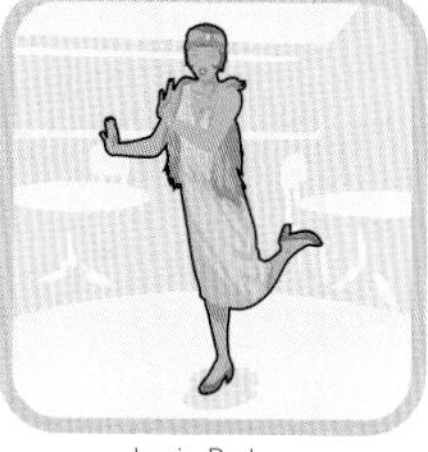
Lucie Parker
Editor

Frances Reade
Project Editor

Emelie Griffin
Editorial Assistant

Charles Mathews
Production Coordinator

Kelly Booth
Associate Creative Director

Marisa Kwek
Art Director

Stephanie Tang
Senior Designer

Delbarr Moradi
Designer

Michelle Duggan
Production Manager

Teri Bell
Color Manager

More Show Me How
Everything We Couldn't Fit in the First Book
Instructions for Life From the Everyday to the Exotic

HarperCollins books may be purchased for educational, business, or sales promotional use. For information, please write: Special Markets Department, HarperCollins*Publishers*, 10 East 53rd Street, New York, NY 10022.

First published in the United States and Canada in 2010 by:
Collins Design
An Imprint of HarperCollins*Publishers*
10 East 53rd Street
New York, NY 10022
Tel: (212) 207-7000
Fax: (212) 207-7654
collinsdesign@harpercollins.com
www.harpercollins.com

Distributed throughout the United States and Canada by:
HarperCollins*Publishers*
10 East 53rd Street
New York, NY 10022
Fax: (212) 207-7654

Library of Congress
Control Number: 2010926436

ISBN: 978-0-06-199879-9
A Weldon Owen Production
415 Jackson Street
San Francisco CA 94111

Printed in Singapore by Tien Wah
10 9 8 7 6 5 4 3 2 1

A **Show Me Now** Book.

Conor Buckley
Illustration Coordinator

Hayden Foell
Illustration Specialist

Raymond Larrett
Illustration Specialist

Jamie Spinello
Illustration Specialist

Ross Sublett
Illustration Specialist

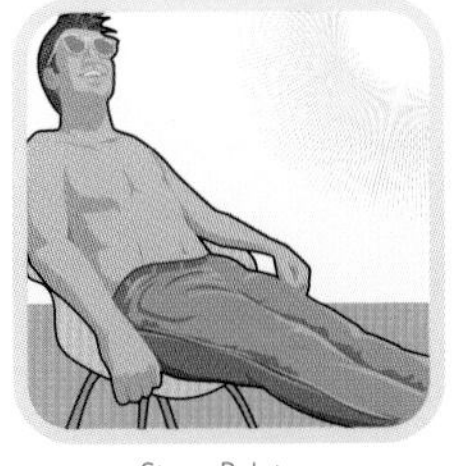

Steve Baletsa
Illustrator

Juan Calle
Illustrator

Sarah Duncan
Illustrator

Britt Hanson
Illustrator

Katrina Johnson
Illustrator

Vic Kulihin
Illustrator

Christine Meighan
Illustrator

Paula Rogers
Illustrator

Bryon Thompson
Illustrator

Lauren Towner
Illustrator

Gabhor Utomo
Illustrator

Tina Cash Walsh
Illustrator

Mary Zins
Illustrator

ILLUSTRATION CREDITS The artwork in this book was a true team effort. We are happy to thank and acknowledge our illustrators.

Front Cover: **Juan Calle (Liberum Donum):** gorilla **Britt Hanson:** airplane, cake decoration **Bryon Thompson:** sari, smoothie, compost tea, hula dancers

Back Cover: **Juan Calle:** jousting, llama spew **Hayden Foell:** blue-blazer drink **Gabhor Utomo:** balloon chair

Key: bg=background, bd=border, fr=frames, ex=extra art

Steve Baletsa: 36–38, 88–90, 95, 99–100, 106, 112, 153, 187, 250–251, 256–257, 275–276, 308, 311, 370, 393, 416–418 **Kelly Booth:** 337 **Juan Calle:** 5–6, 12, 15–18, 22, 24–26, 40–42, 46, 53–54 fr, 75, 154–156, 230–232, 248, 253–254, 258–270, 277–279, 286–288, 289–292 fr, 295–298, 299–302 fr, 306–307, 309–310, 312, 317, 327–336, 344–345, 369, 379–392, 396–401, 405–407, 414–415 **Tina Cash Walsh:** 8, 39, 59, 111, 113, 133, 219–221, 341–343, 419–421 **Sarah Duncan:** 35, 43, 84, 124 **Hayden Foell:** 72–74, 86, 157, 173–174, 178–179, 184, 213–214, 217–218, 238, 252, 285, 338–340, 371–378, 394 **Britt Hanson:** 20 bg, 33, 53–54 bg, 101–102, 114–117 top, 141–144, 149–150, 158–161, 165–172, 180–182, 198–205, 244–245, 271–272, 289–292 bg, 299–302 bg, 359–362 **Jessica Henry:** 13, 97 **Pilar Erika Johnson:** 7 **Joshua Kemble:** 71 **Vic Kulihin:** 20, 23, 70, 77, 175–177, 215–216 **Raymond Larrett:** 34, 66–69, 78–79, 85, 87, 114–117 bottom, 190–191, 235, 237, 246–247, 273, 318, 324, 354 fr, 355 **Christine Meighan:** 44, 57–58, 60, 98, 105, 127, 129, 192–193, 195–197, 315–316, 323, 325–326, 347–348, 354 ex, 356–358, 366, 368, 395 **Hank Osuna:** 346 **Paula Rogers:** 96, 107, 233–234, 240–242, 313–314 **Jamie Spinello:** 81–83, 106, 108, 110, 126, 128, 151–152, 162, 183, 186, 188–189, 194, 293–294, 349, 402–404 **Ross Sublett:** 284 **Bryon Thompson:** 19, 109, 118–121, 134, 163–164, 185, 212, 222–223, 226–229, 236, 243, 249, 305, 322, 363–365, 408–410, 412–413 **Lauren Towner:** 103–104 **Gabhor Utomo:** 2–4, 14, 21, 27–32, 45, 47–49, 50–52 bg, 55–56, 61–65, 88–90 bg, 91–94, 136–140, 145–148, 206–211, 239, 255, 274, 280–283, 303–304, 319–320, 350–353, 411, 422–423 **Mary Zins:** 9–11, 50–51 fr, 76, 80, 122–123, 125, 130–132, 135, 224–225, 321, 367

additional thanks

Research, verification, and fact-checking were performed by a host of experts and passionate practitioners. We are especially grateful to Donita Boles for culinary expertise, Elizabeth Dougherty for parenting input, Renée Myers for diverse cultural pointers, Lou Bustamante for mixological advice, Joseph Pred for medical guidance, and Marc Caswell and Joseph Judd for cycling know-how. Many other experts, in everything from swing dance to personal safety, gave advice and input, to the great improvement of this book.

The majority of the Show Me Team is pictured on the preceding pages. Others who contributed to the production of this book are Meghan Hildebrand, Katharine Moore, Julie Vuong, and Mary Zhang. Our storyboarding dream team was Sam Belletto, Esy Casey, Sarah Lynn Duncan, Chris Hall, Sheila Masson, Paula Rogers, Jonathan Shariat, Brandi Valenza, and Kevin Yuen. Marisa Solís added copyediting expertise, Michael Alexander Eros gave production help, Gail Nelson-Bonebrake proofread every word, and Marianna Monaco rocked out the index. Many thanks to all. And a particularly heartfelt thanks to the dozens and dozens of bloggers who discovered, loved, and praised the first book. This international phenomenon couldn't have happened without you and your enthusiastic support.

show us how

Do you think you have a way to do one of the things in this book better, faster, or smarter? If so, we want to hear about it! Send us an e-mail at info@showmenow.com, and your ideas could be featured in the next edition of this book. Send photos and/or a video, and we may even depict you showing us how.

join the team

Is there something that you think should have been in this book? Something you or your friends know how to do and want to show off? Our Show Me Team is looking for new members to share their expertise with the world. Please send us your best ideas* and, if we use them, you'll be credited as an official member of this exciting group of experts and enthusiasts. Most recent team members are Pat Phillips, Felicite Briggs, Mandy Maung, Jamie Ditaranto, Harold E. Grupe Jr., Suzanne Kennedy, Dorothea Michel, Boris Palchik, Alicia Sanders, and Alina Ungureanu.

* and maybe some of your second-best ones too.